TRICHODERMA
Identification and Agricultural Applications

Gary J. Samuels
U.S. Department of Agriculture, Agriculture Research Service (retired)
Systematic Mycology and Microbiology Laboratory
Beltsville, Maryland

Prakash K. Hebbar
U.S. Department of Agriculture
Animal and Plant Health Inspection Service, Plant Protection and Quarantine
Riverdale, Maryland

APS PRESS
The American Phytopathological Society
St. Paul, Minnesota, U.S.A.

Cover images:
Top: Conidiophore of *Trichoderma asperellum*.
(Courtesy G. J. Samuels)

Top right: Culture of *Trichoderma atroviride* on potato dextrose agar. (Courtesy G. J. Samuels)

Bottom left: Application of *Trichoderma asperellum* to cacao pods. (Courtesy D. Begoude)

Library of Congress Control Number: 2015908956
International Standard Book Number: 978-0-89054-484-6

Printed in the United States of America on acid-free paper.

The American Phytopathological Society
3340 Pilot Knob Road
St. Paul, Minnesota 55121, U.S.A.

Preface

This book grew out of the need expressed by participants of the First West and Central Africa Regional Training Workshop on Biocontrol of Plant Diseases, which was held in 2001 at the Regional Biological Control and Applied Microbiology Laboratory, Institute of Agricultural Research for Development (IRAD) in Yaoundé, Cameroon. The workshop gave special attention to biocontrol methods for *Phytophthora megakarya*, the cause of cacao black pod disease.

The primary aims of this manual are to provide "faces" for the names of *Trichoderma* species and to summarize the methods by which *Trichoderma* cultures are applied in agriculture.

We have included the most common species as well as several seemingly rare but widely distributed and/or unusual species that do not necessarily fit the morphological stereotype of the genus *Trichoderma*. These are the species that appear in isolations made directly from natural or man-made substrates, and cultures of these species are reported commonly in ecological or medical contexts or find their way into industrial applications. We did not include the many, and possibly greater number of, *Trichoderma* species that are known only in culture as anamorphs of the *Trichoderma* teleomorph genus *Hypocrea* since they are not, or rarely at best, observed apart from their *Hypocrea* stromata.

Although *Trichoderma* species are usually thought of in terms of disease control, disease control is only one element in crop management. Today, we understand that *Trichoderma* species and other beneficial microorganisms can contribute to more than disease control as a part of integrated crop management. While *Trichoderma* species have been applied in biocontrol programs for many years, and mycoparasitism has been seen to be the mechanism of control, increasingly, many apparent biocontrol effects are understood to be the result of interaction at the genomic level between the host plant and the *Trichoderma* species. Moreover, quite apart from biological control, *Trichoderma* species are seen to stimulate the growth and establishment of crop plants. The application of *Trichoderma* species to biological control or to promotion of plant health promises improved crop production with at least a reduction in the many negative environmental and health-related side effects encountered in the application of chemicals to the crops. An objective of the current work is to provide an introduction for those who would like to incorporate a *Trichoderma*-based component into a crop management program. We suggest where to look for *Trichoderma* cultures, how to isolate them, how to narrow the field to one or a few cultures through laboratory and controlled field experiments, and how to prepare inoculum, and finally, we give examples for the application of *Trichoderma* cultures. While our primary desire would be to enable individuals or farmer cooperatives to incorporate *Trichoderma* species into their crop management strategies, realistically, this publication is aimed at institutions, universities, government laboratories, and nongovernmental organizations where there are facilities for sterile techniques, storage of strains, and establishment and evaluation of in vitro and in vivo experiments. The ideal would be collaboration between such a facility and farmers, with the objective of providing material at cost to allow farmers to successfully integrate beneficial *Trichoderma* species into their crop management programs.

The descriptions and illustrations presented here are based on our own observations of reliably identified strains, most of which are available from the Centraalbureau voor Schimmelcultures (CBS), Utrecht, The Netherlands. Over approximately 20 years, many people from many countries provided cultures that have enabled us to develop morphological and phylogenetic species concepts. Notable individuals include Walter Gams and Trix Merckx (CBS), Harry Evans (CABI Bioscience, United Kingdom), Giovanni Vanacci (University of Pisa, Italy), Pierre Tondje (IRAD, Cameroon), Le Dinh Don (Long Nam University, Vietnam), John Bissett (Agriculture Canada, Ottawa), Alan Pomella (Brazil), Carmen Suarez-Capiello (Instituto Nacional de Investigaciones Agropecuarias [INIAP], Pichilingue and Universidad Técnica Estatal de Quevedo, Ecuador), and Enrique Giardini Arevalo (Instituto de Cultivos Tropicales [ICT], Tarapoto, Peru). Several individuals shared insights on the application of *Trichoderma* species in agriculture. We note especially Bryan Bailey (USDA ARS), Roy Bateman (International Pesticide Application Research Centre, Imperial College, United Kingdom), Harry Evans, Gary Harman (Cornell University), Robert Lumsden (World Cocoa Foundation, United States), Orlando Petrini (Switzerland), Alba Marina Cotes Prado (CORPOICA, Colombia), Carmen Suarez-Capiello, and Richard Sikora (Institut für Pflanzenkrankheiten, Bonn, Germany). Susan Kaminskyj (University of Saskatchewan, Canada) provided unpublished observations of *Trichoderma* species isolated from plants grown on Canadian oil shales. Mary Palm (USDA APHIS) and Tim Paulitz (USDA ARS, Washington State University) provided us with important literature. Adnan Ismaiel (USDA ARS) has been a constant and excellent collaborator and advisor to us in molecular/phylogenetic aspects of *Trichoderma* taxonomy. Amy Rossman (USDA ARS) has been an unrelenting force throughout the career of one of us (G.J.S.), and without her encouragement and support over many years, this work could not have been completed. We are grateful to Patty Groome for proofreading the work. Mary Palm was a tremendous help and support in producing this work. Finally, we express deep appreciation to the staff at APS PRESS.

Mention of trade names or commercial products in this publication is solely for the purpose of providing specific information and does not imply recommendation or endorsement by the U.S. Department of Agriculture.

Contents

TRICHODERMA

Identification and Agricultural Applications

CHAPTER 1

Introduction

Introduction to the Genus *Trichoderma*

The genus *Trichoderma*, which was first recognized by the Dutch mycologist C. H. Persoon more than 200 years ago, is one of the core mold genera. These core genera, which include the likes of *Aspergillus*, *Cladosporium*, *Fusarium*, and *Penicillium*, are rich in often ubiquitous species that are well-known for their biochemical abilities and biological proclivities. They frequently blend with human activities for good or ill.

Trichoderma species are most commonly thought of as soil fungi, and the literature, beginning early in the twentieth century and continuing to the present, abounds with enumerations of species from many places, niches, and soil types (Blaszczyk et al., 2011; Hoyos-Carvajal et al., 2009a; Kullnig et al., 2000; Mulaw et al., 2010; Nelson et al., 1987; Oudemans and Koning, 1902; Waksman, 1916; Wuczkowski et al., 2003; for others see reviews in Klein and Eveleigh, 1998; Rifai, 1969). In nature, *Trichoderma* species are often found on organic and inorganic matter. They are notorious for their often devastating effects on commercially grown mushrooms (Fujita et al., 1984; Komatsu, 1976; Komoń-Zelazowska et al., 2007; Park et al., 2006; Samuels et al., 2002; Seaby, 1998), but they also occur on basidiomes of polypores, jelly fungi, and other basidiomycete hosts that occur in forests and other natural habitats (Carey and Rogerson, 1976; Hatvani et al., 2007; Jaklitsch, 2011; Overton et al., 2006a,b). *Trichoderma* is certainly more than a genus of soil fungi and mycoparasites. Species have been isolated from a wide diversity of sources, including immersed marine habitats (Paz et al., 2010; Ruiz et al., 2007; Sperry et al., 1998), water-damaged building materials (Lubeck et al., 2000; Polizzi et al., 2011; Thrane et al., 2001), animal feed (Caballero et al., 2007), and other indoor niches (Beezhold et al., 2008). *Trichoderma* was one of the two genera most commonly isolated as endophytes from sapwood of *Theobroma gileri* in Latin America (Evans et al., 2003), and members of the genus are common endophytes of wood and leaves of other plants, including rubber (*Hevea brasiliensis*) in Peru (Chaverri et al., 2011; Gazis and Chaverri, 2010), cacao (*Theobroma cacao*) in South and Central America and Africa (Chaverri and Samuels, 2013), and roots of coffee (*Coffea arabica*) in Ethiopia (Mulaw et al., 2010; Samuels et al., 2012). Their sexual states, formerly described in the ascomycete genus *Hypocrea*, are commonly found on ascomycete and basidiomycete fungi and wood in varying degrees of decay (e.g., Chaverri and Samuels, 2003; Dingley, 1957; Doi, 1972; Jaklitsch, 2009, 2011; Jaklitsch et al., 2006; Komatsu, 1976; Overton et al., 2006a,b; Samuels et al., 2006b).

The evident success of *Trichoderma* species in the environment is evidently due to the development of a number of mechanisms that enhance their ability to survive and proliferate. These mechanisms include an aggressive ability to attack or inhibit other fungi and the production of hydrolytic and other enzymes that degrade complex carbohydrates. These mechanisms have been exploited for human benefit. Volatile organic compounds produced by *Trichoderma* species can inhibit the growth of wood decay fungi (Dennis and Webster, 1970), and enzymes such as cellulases and hemicellulases are used in commercial applications, including the enhancement of the nutritional value of animal feed, the improvement and enhancement of yields of plant oils and juices, and the production of biofuels from plants (Kubicek et al., 2009; Lantz et al., 2010). *Trichoderma* species mobilize the uptake of minerals and plant nutrients that are not normally available to crops in agricultural systems (Harman et al., 2004). At the same time, they offer the possibility of a role in the bioremediation of soils that contain toxic elements and compounds, such as cyanide (Ezzi and Lynch, 2005) and oil hydrocarbons (Argumento-Delira et al., 2012; Hadibarata and Tachibana, 2009). Susan Kaminskyj (*unpublished data*) isolated *Trichoderma* species from the roots of plants that were naturally able to colonize oil sand tailings that had been depleted of mineral nutrients during bitumen extraction; these strains could be reintroduced into tomato seedlings, which were then able to grow better than the control plants on the depleted tailing sands.

The antagonistic or parasitic effect of *Trichoderma* species against plant-pathogenic fungi and, later, nematodes, brought about by glycolytic and chitinolytic enzymes, peptaibiotics, and antifungal antibiotics has been recognized since the 1930s (Chutrakul et al., 2008; Degenkolb et al., 2006a,b, 2008b; Duclohier, 2007; Kraus et al., 2006; Nalimova, 2007; Röhrich et al., 2012; Sharon et al., 2007; Sikora et al., 2008; Weindling, 1932, 1934, 1941; Weindling and Fawcett, 1936; see chapters in Harman and Kubicek, 1998). Now *Trichoderma* species are exploited as biocontrol components of integrated crop management, which also includes good farm management and the promotion of plant health (e.g., Tondje et al., 2007).

Trichoderma species have the ability to invade and become established within healthy plants as opportunistic, avirulent symbionts (Harman et al., 2004) living endophytically within sapwood (Evans et al., 2003; Gazis and Chaverri, 2010; Verma et al., 2007), leaves (Chaverri et al., 2011; Gazis and Chaverri, 2010; Verma et al., 2007), and roots (Harman, 2011; Harman et al., 2004; Sikora et al., 2008; Verma et al., 2007; zum Felde et al., 2005). The interaction between these endosymbionts and their host plants at the molecular level (Bailey et al., 2006, 2008) can improve plant health and, at the same time, induce systemic resistance to attack by pathogens (Hanada et al., 2008; Harman, 2011; Harman et al., 2004; Vinale et al., 2008) and possibly stimulate the production of diffusing compounds by the endophyte-infected plant, stimulating in planta suppressiveness that reduces damage caused by pathogens and pests such as nematodes (Sikora et al., 2008).

Some *Trichoderma* species, especially members of the Longibrachiatum clade (Samuels et al., 1998, 2012), can grow and sporulate at human body temperature (37°F) and higher and can cause opportunistic fungal infections of humans, especially immunocompromised patients, or hypersensitive reactions (Druzhinina et al., 2008; Kantarcioglu et al., 2009; Kredics et al., 2003; Kuhls et al., 1999; Samuels and Ismaiel, 2011; Sandoval-Denis et al., 2014; Tang et al., 2003). The enzymes produced by *Trichoderma* species are strong sensitizers that have lead to hypersensitive reactions that can develop over time (Beezhold et al., 2008; Caballero et al., 2007). Conidia of *Trichoderma* species are produced in mass fermentation as inoculum for industrial processes. While reports of negative effects of this process are infrequent, it is critically important to recognize the real possibility of an effect on technicians and others who work with the material. Care must be taken to provide adequate ventilation while still working in a sterile laboratory environment.

Biogeography and Host Specificity

Those who read the earlier literature could develop the impression that a few *Trichoderma* species are ubiquitous and that there is no sense of geographic or niche-directed speciation. Now that there is a well-developed taxonomy for the genus, we begin to see that some species have a biogeographic bias and, to some extent, preferences for substrate.

Most published lists of *Trichoderma* species are enumerations of cultures isolated from soils or other natural or man-made substrates. From these, it is impossible to infer more than the presence of a species in a habitat, whether the species was growing, occupying space, and competing with other organisms for nutrients or merely represented by an ungerminated propagule. Equally, in most cases, stromata of the *Trichoderma* (*Hypocrea*) species are found on wood that is colonized by other fungi and sometimes on the stromata or mycelium of those other fungi. While the *Trichoderma* stroma may be formed on another ascomycete, the putative mycelium is not seen and the *Trichoderma* species could be deriving nutrients from almost anywhere. The range of substrates for sexual reproduction and perithecial formation may be far narrower than that supporting asexual reproduction, and sexual reproduction in some *Trichoderma* species is clearly host limited. For example, *T. avellaneum* only produces perithecia on mushrooms of *Marasmius subnudus* (Carey and Rogerson, 1976), and *T. pulvinatum* is apparently restricted to two bracket fungi, *Fomitopsis pinicola* and *Piptoporus betulinus* (Jaklitsch, 2011; Overton et al., 2006a), but these fungi have only been found on these specific substrates and their anamorphs have not been found apart from the teleomorphs. *Trichoderma stromaticum* is a parasite of the cacao (*Theobroma cacao*) pathogens *Moniliophthora perniciosa* and possibly *Moniliophthora roreri* in parts of tropical Central and South America (Samuels et al., 2000, [2011 online] 2012); both the anamorph and teleomorph sporulate on cacao tissue infected with *Moniliophthora perniciosa*, and *T. stromaticum* has been isolated as an endophyte from sapwood of cacao. *Trichoderma aggressivum*, the cause of the devastating green mold disease of commercially cultivated *Agaricus* mushrooms (Samuels et al., 2002; Seaby, 1998), is the only host-specific and presumably clonal (i.e., lacking sexual reproduction) *Trichoderma* species known. It appeared in mushroom compost in the mid-1980s and is only known from that habitat; its closest relatives may be endophytes of tropical trees (G. J. Samuels, *unpublished data*). A recent study of the three *Trichoderma* genomes that have been sequenced (*T. atroviride*, *T. reesei*, and *T. virens*) concluded that mycoparasitism was ancestral in the genus *Trichoderma* because mycoparasitism-specific genes were prevalent in *T. atroviride* and *T. virens* but lost in saprobic *T. reesei* (Kubicek et al., 2011; see also Chet et al., 1998). Indeed, the closest relatives of the genus *Trichoderma*, genera such as *Hypomyces* and *Sphaerostilbella*, include only mycoparasites. Chaverri and Samuels (2013) concluded that, in fact, in the genus *Trichoderma*, there have been numerous shifts to and from mycoparasitism and saprobism, but the endophytic habit is generally more derived than the mycoparasitic or saprobic habits, and that, while shifts from mycoparasitism/saprobism to endophytism have been common (and apparently without loss of genes for mycoparasitism), reversions from endophytism have been rare. The significance to humans of this genetic shift is that endophytic strains may represent a little used reservoir of fungi suited to inducing resistance to pathogens and/or stimulating plant health.

Until 1969, there had not been a critical taxonomic revision of the genus *Trichoderma* and most reports in the literature referred to only one species, *T. viride* (Rifai, 1969). The impression from this literature, then, is that this one species has an incredible geographic range and almost limitless potential for biological activity. As taxonomic precision has increased since the mid-1990s, driven largely by molecular phylogenetic analyses, the number of recognized species has exploded, along with an ability to more precisely identify strains that have been isolated in ecological studies. The following discussion and generalizations about distribution and biogeography are based on our experience with *Trichoderma* and our reading of the literature, but we are not aware of any statistically based studies into the biogeography of *Trichoderma* species. While there are many published lists of *Trichoderma* species isolated from diverse substrates, none of the studied followed a statistically based sampling protocol. Thus, conclusions about differential occurrence of species are not completely reliable. Sometimes the lists are accompanied by descriptions of new species (e.g., from Asia; Bissett et al., 2003) or new species are inferred but not described (e.g., from Colombia and adjacent countries; Hoyos-Carvajal et al., 2009a). However, the greatest number of species isolated directly from soil or other substrates are not new; the species diversity appears to be finite. This is in contrast to the high proportion of new *Trichoderma* species that occurred as anamorphs in sexual (*Hypocrea*) lineages found by Chaverri and Samuels (2003) and Jaklitsch (2009, 2011). Thus, the diversity of sexual lineages of *Trichoderma* species appears to be greater than the presumed clonal *Trichoderma* lineages that are isolated directly from soils or other substrates. *Trichoderma viride* and *T. harzianum* are probably the most commonly cited species in the literature; however, most of these reports are based on a combination of misidentifications and new species concepts (Table 1.1). *Trichoderma viride* in the strict sense is now known to be an uncommon species of the Northern Hemisphere (Jaklitsch et al., 2006). The *T. viridescens* complex, recently segregated from *T. viride* (Jaklitsch et al., 2006) and subsequently divided into several phylogenetically defined species (Jaklitsch et al., 2013), also has a northern distribution. *Trichoderma asperellum* and *T. asperelloides*, both segregates from *T. viride*, are cosmopolitan. *Trichoderma harzianum* (Chaverri et al., 2003b; Druzhinina et al., 2010b) has until recently been broadly conceived and phylogenetically diverse; however, Chaverri et al. (2015) have recognized several morphologically indistinguishable species within this complex, as has already been done for *T. viride* (Jaklitsch et al., 2006) and *T. viridescens* (Jaklitsch et al., 2013). The group of strains that clusters around the ex-type strain of *T. harzianum* is northern in origin and, in this strict sense, *T. harzianum*, like *T. viride*, is a rather uncommon northern species. A typical suite of *Trichoderma* species found in north and south temperate zones includes *T. aggressivum*, *T. atrobrunneum* (*T. harzianum* complex; Europe, North America), *T. atroviride*, *T. brevicompactum*, *T. citrinoviride*, *T. gamsii*, *T. guizhouense* (*T. harzianum* complex), *T. hamatum*, *T. koningii* s. str., *T. longibrachiatum*, *T. minutisporum*, *T. polysporum*, *T. saturnisporum*, *T. simmonsii* (*T. harzianum* complex; Europe, North America), *T. tomentosum*, *T. viride*, and *T. viridescens* s. lat. (including *T. olivascens*, *T. paraviridescens*, *T. viridarium*, and *T. virilente*). *Trichoderma aureoviride* and *T. pseudokoningii* are names often seen in the literature but, again, a precise taxonomy tells us that *T. aureoviride* is a species of northern Europe, maybe not occurring apart from its *Hypocrea* teleomorph and observed only in cultures derived from its ascospores. In its broad, morphologically defined sense, *T. harzianum* is cosmopolitan, but *T. harzianum* in the strict sense is a species of the Northern Hemisphere, while several of the new segregate species are apparently exclusively tropical in distribution (Chaverri et al., 2015). *Trichoderma pseudokoningii* is principally found in Australia and New Zealand in cultures derived from ascospores, although it

Table 1.1. Reidentified *Trichoderma* Strains Cited in the Literature

Strain	Name in the Literature	Correct Name	Topic/Product	Literature Cited
PR26-12-6	*T. viride*	*T. rossicum*[a]	Antagonism to *Rhizoctonia solani*	Grosch et al. (2006)
ICC 080	*T. viride* ICC 080	*T. gamsii*[b]	Bioten WP; Tenet WP; Remedier WP	Puglese et al. (2011)
T-25	*T. viride* T-25	*T. asperellum*[b,c]	Tusal, a fungicide on tomato, pepper, cucumber, and courgette	
TV1	*T. viride* TV1	*T. asperellum*[b,c]	Xedavir	Puglese et al. (2011)
ICC 012	*T. harzianum*	*T. asperellum*[c]	Bioten WP; Tenet WP	
T 203 = ATCC 36042 = CBS 396.92	*T. viride*, *T. harzianum*, *T. asperellum*	*T. asperelloides*[d]	Mobilize micronutrients to enhance growth in cucumber	Yedidia et al. (2001)
			Ultrastructural changes during mycoparasitism	Benhamou and Chet (1993)
			Production of a hydrophobin gene involved in root colonization	Viterbo and Chet (2006)
			Parasitize egg masses of root-knot nematode	Sharon et al. (2007)
			Induced systemic resistance to *Pseudomonas* spp. in cucumber	Yedidia et al. (2003)
SKT-1 = FERM P-16510	*T. asperellum*	*T. asperelloides*[d]	Ecohope	Nagayama et al. (2007), Watanabe et al. (2005)
T-22	*T. harzianum*	*T. afroharzianum*[e]	Biological control, plant health; Plant Shield, Root Shield, Trianum	Various
ATCC 74058 = P1	*T. harzianum*	*T. atroviride*[f]	Chitinase enzymology Gene cloning (*ech*42, *nag*1) Mechanism of biological control	Harman et al. (1993), Hayes et al. (1994), Lorito et al. (1993, 1996a,b), Peterbauer et al. (1996)
IMI 206040	*T. harzianum*	*T. atroviride*[f]	Gene cloning (*ech*42)	Carsiolo et al. (1994)
			Light regulation	Berrocal-Tito et al. (1999)
			Mechanism of biological control	Carsiolo et al. (1999)
ATCC 30642 = CBS 396.92	*T. harzianum*	*T. atroviride*[f]	Enzyme-antibiotic synergism	Schirmböck et al. (1994)
GH 11 = G.J.S. 05-92	*T. asperellum*	*T. asperelloides*[d]	Parasitize egg masses of root-knot nematode	Sharon et al. (2007)
T 44 = G.J.S. 05-94	*T. asperellum*	*T. asperelloides*[d]	Parasitize egg masses of root-knot nematode	Sharon et al. (2007)
Tp0390 = IMI 206039	*T. polysporum*	*T. parapiluliferum*[g]	Binab	
Ta040 = IMI 206040	*T. harzianum*	*T. atroviride*[g]	Binab	

[a] Samuels et al., [2011 online] 2012.

[b] European Commission, Health & Consumers Directorate-General. Directorate E – Safety of the food chain. Unit E.3 – Chemicals, contaminants, pesticides. *Trichoderma gamsii* ICC080, *Trichoderma asperellum* T11 and TV1, formerly *Trichoderma viride* strain ICC080, strain T-25 and strain TV1. SANCO/1868/08 – rev. 3. 14 May 2008.

[c] European Food Safety Authority. 2013. Conclusion on the peer review of the pesticide risk assessment of the active substance *Trichoderma asperellum* strains ICC012, T25 and TV1. EFSA J. 11(1):3036. doi: 10.2903/j.efsa.2013.3036.

[d] Samuels et al., 2010.

[e] Chaverri et al., 2015.

[f] Kullnig et al., 2001.

[g] Feng et al., 2011.

has rarely been confirmed from a few locations outside of Australasia. Many *Trichoderma* species are known only from one or a few strains; thus, it is difficult to infer their distribution. For example, *T. andinense* is only known from a single isolation made at an elevation of about 3,000 m in the Venezuelan Andes, and the three known strains of *T. konilangbra* were isolated from soil collected at elevations of 1,700–3,400 m in the Rwenzori Mountains of Uganda in East Africa. *Trichoderma stromaticum* occurs naturally only in South America, where no close relatives for it have been found. The closest relatives of this species are found in West Africa and the Indian Ocean area. Its occurrence in South America as a host-restricted parasite of a mushroom that parasitizes cacao, which is an American tree, is a mystery. On the other hand, few species have been isolated from widely different geographic zones. For example, *T. capillare* is apparently rare but widely distributed and known from two strains, one collected in Europe and the other in Vietnam (Samuels et al., 2012). Species commonly, or only, found in isolations from tropical zones include *T. amazonicum*, *T. asperelloides*, *T. asperellum*, *T. endophyticum* (*T. harzianum* complex), *T. ghanense*, *T. hamatum*, *T. lentiforme* (*T. harzianum* complex), *T. longibrachiatum*, *T. orientale*, *T. parareesei*, *T. reesei*, and *T. stromaticum*. Most of the species in the Longibrachiatum clade are geographically limited. The only cosmopolitan members of this clade are *T. longibrachiatum* and *T. citrinoviride*, but the former is more common in tropical countries and the latter is more common in north temperate countries (Samuels et al., 2012). While the Longibrachiatum clade is primarily tropical, the greatest species diversity appears to occur in the Paleotropics (East Africa and South and Southeast Asia) and Australasia (Samuels et al., 2012), although this may be an artifact of collecting. While statistical analysis of distribution is needed to confirm these observations, the existing evidence points to geographic bias in the distribution of *Trichoderma* species.

CHAPTER 2

Developing *Trichoderma*-Based Products for Application in Agriculture

Interactions Among *Trichoderma* Species, Plants, and Their Pathogens: A Primer

Healthy soils are critical to sustainable agricultural production. Healthy soil is dynamic; it is a balance between its organic and inorganic components. Soil organisms affect soil morphology. They recycle and detoxify, they mobilize nutrients, they antagonize or kill pathogens, and much more. Microbes naturally occurring in soils maintain fertility and, at the same time, contribute to the health and productivity of crops. There is ample documentation for the suppression of disease and the enhancement of plant growth by naturally occurring beneficial soil microorganisms, such as species of *Trichoderma* (Cook and Baker, 1983; Dennis and Webster, 1970; Harman and Kubicek, 1998; Harman et al., 2004; Kubicek and Harman, 1998; Papavizas, 1981, 1985; Whipps, 2001). While agrichemicals are effective and necessary in many instances, they can have a negative effect on the environment, including the soil microbiota, and their runoff can require mitigation. In contrast, beneficial microbes such as *Trichoderma* species can multiply, persist, and proliferate in the soil without apparent negative effect, and runoff is not an issue. Microbial amendments are seen as important elements in achieving sustainable agriculture through promoting plant health and/or controlling pests and diseases as part of an integrated program of crop management. At the same time, microbial amendments offer a safer, more ecologically friendly alternative for those interested in pursuing organic agriculture.

The genus *Trichoderma* includes beneficial microbes, and mycoparasitism and antibiosis are well-known beneficial effects of species in the genus. Weindling's mid-1930s publications defined biological control of plant pathogens in terms of mycoparasitism and antagonism through the use of strains of *Trichoderma* species (Weindling, 1932, 1934, 1941; Weindling and Fawcett, 1936). Since then, numerous publications emanating from laboratories all around the world have described the isolation of *Trichoderma* species from agricultural and other soils and their application to control, mainly, diseases of vegetable crops but also diseases of fruit, grain, and timber species caused by *Rhizoctonia solani* and *Pythium*, *Phytophthora*, *Botrytis*, and *Sclerotinia* species; root-knot nematodes; and other parasites and pests through attack or antagonism. Early efforts in the biological control of plant pests and diseases were reviewed by Papavizas (1985) and in several chapters in Harman and Kubicek (1998) and Papavizas (1981, 1985), and mechanisms of mycoparasitism and antibiosis were the subject of several recent reviews (Atanasova et al., 2013; Brotman et al., 2010; Druzhinina et al., 2011; Harman, 2000; Howell, 2003; Jeger et al., 2009; Pal and Gardener, 2006; Sikora et al., 2008; Vinale et al., 2008; Whipps, 2001; Zinati, 2005). With few exceptions, *Trichoderma* species are neither pathogenic nor toxic to humans, animals, or plants. Given this benign character, and coupled with their antagonistic abilities, ease of manipulation, and ubiquity, extensive international research has lead to the development of many commercially marketed *Trichoderma*-based agricultural products used in the control of plant diseases and/or promotion of plant health. Many of these are listed in Table 2.1 (see also Fravel et al., 1999; Funck-Jensen and Lumsden, 1999; Kaewchai et al., 2009; Woo et al., 2014).

Table 2.1. *Trichoderma*-Based Agricultural Products

Product	Reported Microbe/Strain	Reported Function	Crop	Manufacturer/ Country
AkTRIvator	*T. harzianum* 002/003	Promote growth; suppress soilborne pathogens; stimulate root development; cited for use in hydroponic systems	Crops not specified	Canna, The Netherlands/U.K.
ArborGuard	*Trichoderma* sp.	Promote growth; suppress *Armillaria* spp.	Forestry seedlings	Bio-Protection Research Centre, Lincoln University, New Zealand
Binab line of products, including Binab T and Binab Golf	*T. atroviride* ATCC 20476 and *T. parapiluliferum* IMI 206039	Tree wound pathogens	Ornamental shade and forest trees	Binab Bio-Innovation AB, Sweden
Biobus 1.0 WP	*T. viride*	General topical mycoparasite	Many crops, including citrus, coffee, potato, mango, tomato, cabbage	Nam Bac Co., Ltd., Vietnam
Bio-Cure-F	*T. viride*	General mycoparasite applied topically, as seed dressing and soil drench	General vegetable crops	T. Stanes & Co., Ltd., India
Bioderma	*T. viride*	Wide-spectrum mycoparasite against root and leaf pathogens; stimulate seed germination	Cotton, cereals, pulses, vegetables, oilseeds, fruit plants, floriculture	Biotech International Ltd., India
Bioderma-H	*T. harzianum*	Wide-spectrum mycoparasite against root and leaf pathogens; stimulate seed germination	Cotton, cereals, pulses, vegetables, oilseeds, fruit plants, floriculture	Biotech International Ltd., India
BioHealth TH WSG	*Trichoderma* sp. T-50	Soilborne pathogens; promote plant growth; seed and foliar treatments	Wide range of crops, including vegetables, ornamentals, turfgrass, cereals	Humintech GmbH, Germany
Bio-Humaxin Sen Vàng 6SC	*Trichoderma* spp.	Applied as seed treatment and in potting mixtures	Crops not specified	Công Ty TNHH An Hưng Tường, Vietnam
Biospark Trichoderma	*T. parceramosum* (= *T. ghanense*), *T. pseudokoningii*, and ultraviolet-irradiated strain of *T. harzianum* (Cumagun, 2012)	Promote growth; biofungicide; activator for rapid composting; control damping-off caused by *Phytophthora* spp., *Pythium* spp., *Sclerotium* spp., and *Rhizoctonia solani*; control mango gummosis; control durian dieback caused by *Phytophthora* spp.	Rice, corn, mango, durian, vegetable crops, organic composting	Biospark Corp., Philippines
Bio-Tam (U.S.A.) (also Bioten, Radix, Remedier, Tenet)	*T. asperellum* ICC 012 + *T. gamsii* ICC 080	Control a broad spectrum of soilborne pathogens, including Basidiomycetes, Ascomycetes, and stramenopiles	Various	AgraQuest, U.S.A.
Bioten (Spain) (also Bio-Tam, Radix, Remedier, Tenet)	*T. asperellum* ICC 012 + *T. gamsii* ICC 080	Control a broad spectrum of soilborne pathogens, including Basidiomycetes, Ascomycetes, stramenopiles	Various	Isagro S.p.A., Italy
Eco-SOM Bio-Cure F Trieco	*T. viride*	Promote root growth, suppress the soilborne pathogens *Botrytis* spp., *Pythium* spp., *Sclerotium* spp., *Fusarium solani*, *Sclerotinia homoeocarpa*	Various	Ecosense Lab. India Pvt. Ltd., India

(continued on next page)

Table 2.1. *Trichoderma*-Based Agricultural Products *(continued from previous page)*

Product	Reported Microbe/Strain	Reported Function	Crop	Manufacturer/ Country
Fertisain	*T. harzianum*	Enzymatic extracts stimulate photosynthesis; stimulate resistance against foliar pathogens; increase leaf surface	Vegetables, grape	Biophytech, France
Fulhumaxin 5	*Trichoderma* spp.	Fungicide; control black spot, rust, powdery mildew	Various	Công Ty TNHH An Hưng Tường, Vietnam
Lettucemate Flake and Drench WP	*T. hamatum*	Plant health; seedling mix and seedling drench	Lettuce	Agrimm Technologies, New Zealand
Naturall	*Trichoderma* sp. (+ bacteria)	Plant health	Various	Advanced Biological Marketing, U.S.A.
NLU-Tri	*T. virens*	Plant health; antagonist	Various	Ho Chi Minh University of Agriculture and Forestry, Vietnam
Plantmate Drench, Plantmate Granular	*T. harzianum*	Root zone drench; root zone starter granule; enhance root development	Various	Agrimm Technologies, New Zealand
Plantmate Foliar	*T. atroviride* + *T. harzianum*	Foliar spray; plant health; micronutrient uptake	Various; cut flowers	Agrimm Technologies, New Zealand
Plantsain	*T. harzianum*	Aromatic extract to clear vascular tissue of pathogens; solubilize iron, copper, manganese	Vegetable crops	Biophytech, France
PlantShield (also RootShield, Trianum-P, Trianum-G)	*T. harzianum* T-22 (reidentified as *T. afroharzianum*; Chaverri et al., 2015)	Biofertilizer; drought resistance	Various	BioWorks, Inc., U.S.A.
Promot WP	*T. harzianum* + *T. koningii* (reidentified as *T. guizhouense* and *T. simmonsii*, *T. koningii* not found in the product; Chaverri et al., 2015)	Promote growth; suppress soilborne pathogens	Annual and perennial crops	SW-Düngesysteme GmbH, Germany
Pusa 5 SD	*T. virens*	Seed coating to protect against *Rhizoctonia solani* (Dubey and Singh, 2013)	Mungbean, chickpea	Indian Council of Agricultural Research, India
Radix (Italy) (also Bio-Tam, Bioten, Remedier, Tenet)	*T. asperellum* ICC 012 + *T. gamsii* ICC 080	Control a broad spectrum of soilborne pathogens, including Basidiomycetes, Ascomycetes, stramenopiles	Various	Isagro S.p.A., Italy
Remedier (Italy, Morocco, Slovenia, Turkey) (also Bio-Tam, Bioten, Radix, Tenet)	*T. asperellum* ICC 012 + *T. gamsii* ICC 080	Control a broad spectrum of soilborne pathogens, including Basidiomycetes, Ascomycetes, stramenopiles	Various	Isagro S.p.A., Italy
Rhizoderma	*T. harzianum* Th2	Seed treatment to control *Bipolaris sorokiniana*, *Drechslera tritici-repentis*, *Fusarium graminearum*	Wheat, barley	Rizobacter Argentina S.A., Argentina
Rootmate	*T. virens* G41	Promote plant health; control Pythium stem and root rot; drench on seedlings	Ornamentals, forest trees	BioWorks, Inc., U.S.A.

(continued on next page)

Table 2.1. *Trichoderma*-Based Agricultural Products *(continued from previous page)*

Product	Reported Microbe/Strain	Reported Function	Crop	Manufacturer/ Country
Root-Pro	*T. harzianum* T-35 + *T. harzianum* T-315	Control *Pythium* spp., *Sclerotium rolfsii*, *Rhizoctonia solani*, *Fusarium* spp. at the nursery stage	Various	Agriance, Israel
RootShield (also PlantShield, Trianum-P, Trianum-G)	*T. harzianum* T-22 (reidentified as *T. afroharzianum*; Chaverri et al., 2015)	Biofertilizer; drought resistance; suppress soilborne *Fusarium* spp., *Pythium* spp., *Rhizoctonia* spp.	Greenhouse crops, outdoor nursery plants, greenhouse transplants, ginseng	BioWorks, Inc., U.S.A.
SabrEx PB and SabrEx HC Root Inoculant	*Trichoderma* sp.	Stimulate growth; planter box treatment and seed treatment	Corn, cotton, wheat, other cereals	Advanced Biological Marketing, U.S.A.
Sentinel	*T. atroviride*	Bioprotectant; foliar spray against *Botrytis* spp.	Various	Agrimm Technologies, New Zealand
Sim Derma	*T. harzianum* Kuen 1585	Seed coating; promote root growth; prophylactic against soilborne pathogens	Wheat, onion	Simbiyotek, Turkey
SoilGard	*T. virens* Gl-21	Soilborne pathogens, including *Fusarium* spp., *Pythium* spp., *Rhizoctonia* spp.	Melons, leafy greens, greenhouse ornamentals	Certis, U.S.A.
Solsain S	*T. harzianum*	Promote root development; stimulate resistance to soilborne pathogens; soil application to promote root development; solubilize phosphorus; promote resistance to *Pythium* spp., *Rhizoctonia* spp., *Sclerotinia* spp.	Fruits, vegetables, grape	Biophytek, France
Supresivit	*T. harzianum* (the result of fusion of two "natural" strains; Brožová, 2004)	Stimulate plant growth	Strawberry, tomato	BioPlant, Denmark
T34	*T. asperellum* T-34	Root and foliar applications; antagonist and stimulate growth and resistance; control *Botrytis* spp., *Fusarium* spp., *Rhizoctonia* spp.	Various	Biocontrol Technologies S.L., Spain
Tenet (New Zealand)	*T. atroviride*	Control onion white rot (*Stromatinia cepivora*), Fusarium basal rot	*Allium* spp.	Agrimm Technologies, New Zealand
Tenet (U.S.A.) (also Bio-Tam, Bioten, Radix, Remedier)	*T. asperellum* ICC 012 + *T. gamsii* ICC 080	Control a broad variety of soilborne pathogens, including Ascomycetes, Basidiomycetes	Various	Isagro, U.S.A.; Isagro, Italy
Tory	*T. harzianum* YC 459 (Lee et al., 2006)	Antagonist; foliar application against Botrytis gray mold	Cucumber, tomato	JGreen, Inc., Korea
Trianum-P, Trianum-G (also PlantShield, RootShield)	*T. harzianum* T-22 (reidentified as *T. afroharzianum*; Chaverri et al., 2015)	Biofertilizer; drought resistance	Various	Koppert B.V., The Netherlands
TRi B_1	*Trichoderma* sp.	Soilborne diseases	Crops not specified	Institute of Plant Protection, Vietnam
Trichobiol	*Trichoderma* sp.	Promote plant growth; antagonistic to *Alternaria* spp., *Armillaria* spp., *Botrytis* spp., *Fusarium* spp., *Phytophthora* spp., *Pythium* spp., *Rhizoctonia* spp., *Rosellinia* spp., *Sclerotium* spp.	Various	Bioecológicos Agrícolas, Colombia

(continued on next page)

Table 2.1. *Trichoderma*-Based Agricultural Products *(continued from previous page)*

Product	Reported Microbe/Strain	Reported Function	Crop	Manufacturer/ Country
Trichodex	*T. harzianum* T-39	*Botrytis cinerea*	Various	Makhteshim Agan of North America, Inc., U.S.A.
Trico-ĐHCT (Trico-CTU rice, Trico-CTU Phytoph)	*Trichoderma* spp. (*T. virens*, *T. asperellum*, *T. atroviride*, *T. hamatum*)	Foliar application to control gold leaf rot of citrus; soil application to control soilborne pathogens, including *Fusarium solani*	Citrus, melons, rice	AGPPS, Vietnam; Can Tho University, Vietnam
Trichodry	*T. atroviride*	Stimulate plant health; root zone starter flake in growing media	Used in all bark- and peat-based growing media	Agrimm Technologies, New Zealand
Trichoflow	*T. atroviride*	Stimulate plant health; root zone starter WP	General use, including nursery, greenhouse, turf, orchards, vineyards	Agrimm Technologies, New Zealand
Trichopel	*T. atroviride*	Stimulate plant health; root zone starter granule	General use, including nursery, bulbs, tubers, vines, trees	Agrimm Technologies, New Zealand
Trichosan	*T. harzianum* (reidentified as *T. simmonsii*; Chaverri et al., 2015)	Induce resistance; promote growth	Various	Sautter & Stepper, Germany
Trichospray	*T. atroviride*	Stimulate plant health; foliar booster WP	Foliage, flowers	Agrimm Technologies, New Zealand
Tusal	*T. harzianum* + *T. viride*	Mycoparasite against soilborne pathogens, including *Colletotrichum* spp.	Cucurbits, strawberry, *Capsicum* spp., sugar beet seed, tomato	Certis, Spain
Unite WP	*T. atroviride*	Bioprotectant against damping-off and root rot diseases, including *Cylindrocarpon* spp., *Phytophthora* spp., *Pythium* spp., *Rhizoctonia* spp.; root zone drench	General use	Agrimm Technologies, New Zealand
VI-DK	*Trichoderma* sp.	Mycoparasite of soilborne pathogens	Tomato, beans, potato, tobacco, durian, melon, *Capsicum* spp.	VIPESCO, Vietnam
Vinevax	*T. atroviride*	Bioprotectant against *Armillaria* spp. and dieback pathogens; wound dressing and injection	Kiwi, grape, orchard crops	Agrimm Technologies, New Zealand
Vitalin	*T. harzianum* (reidentified as *T. guizhouense* + *T. simmonsii*; Chaverri et al., 2015)	Strengthen plant; suppress soilborne pathogens	General use	Vitalin Pflanzengesundheit GmbH, Germany
Vitalin MX5	*T. harzianum*, *T. hamatum*, *T. koningii*, *T. reesei*, *T. virens*	Strengthen plant; suppress soilborne pathogens and *Botrytis* spp.	General use	Vitalin Pflanzengesundheit GmbH, Germany

For at least 150 years, beneficial soil microorganisms, including bacteria and fungi such as *Trichoderma* species, have been known to promote plant growth. In fact, a greater percentage of microbes in the rhizosphere are capable of benefitting plant growth than are antagonistic or parasitic to plant pathogens (Berg, 2009; Whipps, 2001). Although there is still much to be learned about how *Trichoderma* species affect plant health, close communication between the proteome and metabolome of the *Trichoderma* species and the host plant has been confirmed. Several reviews summarize what is known about these molecular-level interactions (e.g., Druzhinina et al., 2012a; Harman and Kubicek, 1998; Harman et al., 2004; Hermosa et al., 2012; Lorito et al., 2010; Shoresh et al., 2010; Vinale et al., 2014). Today's new view is that *Trichoderma* species are plant symbionts, able to asymptomatically colonize leaves, roots, and the sapwood of stems. Although we do not know the role played by endophytic *Trichoderma* species occurring in plant tissues other than those occurring in the roots, we do know that those colonizing intercellular spaces of the superficial few layers of roots send signals that activate the host genome to initiate genetic cascades that lead to growth promotion and resistance to diseases caused by biotic and/or abiotic agents (Harman et al., 2004; John et al., 2010; Kleifeld and Chet, 1992; Quilliam and Jones, 2012; Vinale et al., 2014). Clearly, the complexity of the plant–*Trichoderma* interaction demonstrated by the most recently acquired scientific knowledge indicates that the *Trichoderma* genome provides the fungus with the ability to act not only as a soil colonizer or mycoparasitic agent but also to function in close association with the plant host to protect against pathogens or to withstand adverse environmental conditions such as drought. In fact, genome coevolution has been postulated for some plant and *Trichoderma* species, just as has been demonstrated in many plant-pathogen interactions (Harman et al., 2004; Lorito et al., 2010; Shoresh et al., 2010).

Effects of the application of *Trichoderma* species to increase seed germination and enhance growth of vegetable and horticultural crops and conifers began to appear in scientific articles in the 1980s and 1990s. Independent of their mycoparasitic and antibiotic activities, application of *Trichoderma* species resulted in increases in biomass, leaf area, photosynthetic ability, drought tolerance, and disease suppression (e.g., Chang et al., 1986; Hohmann et al., 2011; Inbar et al., 1994; Kleifeld and Chet, 1992; Mastouri et al., 2012; Regliński et al., 2012; see review in Bailey and Lumsden, 1998). The growth of these microbes in the rhizosphere is stimulated by sucrose that leaks from plant roots or possibly by as-yet-undiscovered signal molecules, and increased development of mycelium leads to a greater ability of the microbes to interact with plants. Similar to mycorrhizae, strains of *Trichoderma* species can influence plant growth by mobilizing minerals such as phosphate and iron, while other strains can produce the plant growth hormone indole-3-acetic acid (IAA), which leads to increased root production with an increased capacity for nutrient and water uptake (e.g., Altomare et al., 1999; Hoyos-Carvajal et al., 2009b; Saravanakumar et al., 2012; Vargas et al., 2009; Yedidia et al., 2001). The *Trichoderma* species metabolize sucrose in the roots, and this initiates a metabolic loop. Increased sucrose metabolism in the roots stimulates a resistance response in the leaves. When this resistance is triggered, the photosynthetic rate increases and more sucrose is translocated to the roots, resulting in increased root growth (Contreras-Cornejo et al., 2009; Vargas et al., 2009). The production of auxin by *Trichoderma* species further enhances root development and plant growth by modification of the plant's gene expression (Contreras-Cornejo et al., 2009; Druzhinina et al., 2011; Vargas et al., 2009). The application of *Trichoderma* species to seedlings subjected to drought and other stressors significantly improves their growth. In rice (Shukla et al., 2012) and cacao (Bae et al., 2009), respectively, *T. harzianum* and an endophytic strain of *T. hamatum* delayed the onset of drought symptoms by increasing root mass along with a net increase in photosynthesis and stomatal conductance. When cacao was inoculated with endophytic *Trichoderma* strains, there was repression of a member of the major intrinsic protein (MIP, microporins) superfamily of proteins that function as membrane channels and facilitate transport of water from cells (Bailey et al., 2006), and this delayed the effects of drought. Tomato seeds treated with *T. harzianum* T22 germinated and grew significantly better than the controls under abiotic stress conditions, including osmotic stress, heat and chilling, and high salinity, because of the ability of this strain to prevent the accumulation of reactive oxygen species in the plants, which is a normal response to abiotic stress (Mastouri et al., 2010, 2012). While for Mastouri et al. (2010, 2012), this growth stimulation in tomato occurred only in stressed seedlings, not in seedlings grown under optimal conditions where reactive oxygen species did not occur, stress was not obviously involved in other cases of growth stimulation (e.g., Chang et al., 1986; Inbar et al., 1994).

Harman et al. (2004) proposed that the main effect of *Trichoderma* species on host plants is the induction of localized or systemic resistance to attack by pathogens, and the path to induction of resistance was reviewed by Druzhinina et al. (2011), Hermosa et al. (2012), and Shoresh et al. (2010). The interaction between the microbe and the host plant in the development of resistance relies on, and responds to, the sequential production of metabolites by the host and the beneficial microbe and the activation of enzyme-mediated pathways, such as photosynthesis and respiration. Nonetheless, the several steps that lead to the induction of resistance by *Trichoderma* species can be exploited to achieve useful agricultural applications, whether by transformation of a host or other microbe using a gene that codes for a useful protein or by learning that some strains may induce resistance with less "cost" to the host than do others or may both induce resistance and promote growth. Genomic, transcriptomic, and metabolomic studies provide a window not only to the function of the beneficial microorganisms but also to stimulate the imagination of the researcher. Thus, for anybody considering utilizing a beneficial microorganism in an agricultural application, it is necessary to have a basic understanding of how these organisms affect responses in crop plants. Studies of structure and function of genomes of beneficial microorganisms, including *Trichoderma* species, can expand our understanding of basic sciences and host–microbial interactions. Certainly in this new era, the meaning of "biological control" must be expanded to include plant growth promotion and disease resistance along with classical antibiosis and mycoparasitism. Molecular genomic tools—knowing the expression patterns of genes and understanding the molecular mechanisms that underlie the production of enzymes or secondary metabolites—can lead to the selection of more effective strains and of strains that have novel abilities. The obvious result of the first -omics studies performed on *Trichoderma*–plant interactions is the astonishing level of metabolic changes that some fungal strains can cause in the plant upon contact, which involve hundreds of genes and metabolites—i.e., a plant fully colonized with effective *Trichoderma* strains may be substantially different from an uncolonized plant in terms of responses to the environment, expressed agronomic characters, yield, quality of produce, etc.

Research indicates that the *Trichoderma* species and the host plant interact, leading to the stimulation of the plant's immune system (see reviews in Bittel and Robatzek, 2007; Druzhinina et al., 2011; Harman et al., 2004; Shoresh et al., 2010; Van Wees, 2012; Van Wees et al., 2008). What is the nature of this interaction? In order to influence gene expression by the host, the *Trichoderma* species must enter its roots. To do this, small proteins known as hydrophobins produced on the surface of the *Trichoderma* hypha facilitate its attachment to host roots, while another small protein, the swollenin *TasSwo*, can recognize cellulose and modify the architecture of the plant root cells to enable entry of the *Trichoderma* species into the root tissue. Once inside the first few layers of host root cells, *Trichoderma* species can modify the plant's gene expression to activate its immune system to resist attack by pathogens. Plants have two basic immune systems. Nonhost disease resistance, or induced systemic resistance (ISR), renders plants resistant to most pathogens, while basal disease resistance, or systemic acquired resistance (SAR), reduces the severity of pathogenicity itself (Bittel and Robatzek, 2007; Druzhinina et al., 2011; Hermosa et al., 2012; Lorito et al., 2010; Pieterse et al., 2009; Shoresh et al., 2010). ISR results from many activities such as herbivory (Frost et al., 2008), the presence of proteins such as chitin that is associated with mycelium, or infection by beneficial microorganisms. The pathway of ISR is induced by the small-molecule hormones jasmonic acid and ethylene, which transmit a wound signal from roots to other parts of the plant (Frost et al., 2008; Pieterse et al., 2009). SAR is induced by pathogens and follows the salicylic acid pathway. In SAR, there is up-regulation of pathogenesis-related genes and the accumulation of pathogenesis-related (PR) proteins (e.g., chitinases and glucanases that degrade the walls of fungi and oomycetes) and phytoalexins. Stimulating ISR is not considered to have any "cost" to the plant in terms of energy expenditure. On the other hand, SAR does exact a "cost," for example, in productivity or growth. There is considerable cross-talk between the jasmonic and salicylic acid pathways (see reviews in Hermosa et al., 2012; Pieterse et al., 2009). ISR induced by *Trichoderma* species within plant roots is initiated by pattern recognition receptors in plants, localized in the plasma membrane of host cells. These receptors recognize potential pathogens by the presence of conserved molecules (microbe-associated molecular patterns [MAMPs], sometimes referred to as pathogen-associated molecular patterns [PAMPs]; see Bittel and Robatzek, 2007; Yun and Kwon, 2012). The MAMP molecules are essential for the life of the microbe, rather than for pathogenicity, and are not found in plants. Thus, the plant's sensors recognize them as being potentially dangerous. Several MAMPs, including proteins such as cellulases and chitinases as well as secondary metabolites such as endopolygalacturonase, peptaibiols, and 6-pentyl-α-pyrone, have been identified in *Trichoderma* species (summarized in Hermosa et al., 2012). Interestingly, the trichothecene toxin harzianum A, produced by *T. arundinaceum*, acted as a MAMP by activating both the salicylic and jasmonic pathways in tomato seedlings and primed tomato seedlings against *Botrytis cinerea* and *Rhizoctonia solani* (Malmierca et al., 2012). Activation of the receptor by its specific MAMP at the cell surface initiates an immune response in the plant (Yun and Kwon, 2012). The initial response by the host is commonly a hypersensitive reaction resulting from the release of reactive oxygen species, ethylene, callose, and antimicrobial compounds, which are produced in an effort to limit the perceived pathogen (Bittel and Robatzek, 2007; Shoresh et al., 2010). The MAMPs also modify the transcription of many host genes leading to the induction of SAR, often through the synthesis of salicylic acid (Shoresh et al., 2010) but also through the up-regulation of defense-related genes that are associated with biotic and abiotic stress and of RNA, DNA, and protein metabolism (Alfano et al., 2007; Bailey et al., 2006). Microbes that overcome the MAMP-triggered first immune response secrete effectors able to inhibit the defense reaction of the host, which in turn may develop the ability to specifically recognize such microbial compounds (which then became avirulence factors) and activate a strong effector-triggered immunity (ETI) often associated with the hypersensitive response (HR) or programmed cell death. The involvement of avirulence factors, possibly similar to pathogen avirulence proteins such as the small cysteine-rich Av4-like protein able to affect the activity of chitinases produced in defense by the host plant, has been suggested also in the ISR/SAR caused by *T. harzianum* and *T. atroviride* (Stergiopoulos and de Wit, 2009, review fungal effectors; see also Harman et al., 2004; Hermosa et al., 2012).

When necrotrophs such as *Trichoderma* species and other beneficial microorganisms invade the intercellular space of host plant roots, they initiate ISR, which "primes" or activates regulatory genes that are involved with defense or signaling pathways, so that the host plants can respond quickly to attack by a pathogen (Harman et al., 2004; Hermosa et al., 2012; Lorito et al., 2010; Shoresh et al., 2010; Van Wees et al., 2008) by invoking SAR and accumulating PR proteins designed to combat the pathogen. Basically, an effect of priming is to put plants into a state of low level stress, and plants in this condition can accelerate induced resistance in response to an attack by a pathogen (Frost et al., 2008). From the point of view of the host plant, inoculation with beneficial microorganisms is often an energy-efficient way to defend against pathogens. Not only do some microorganisms give a strong boost to plant growth, but also the induction of resistance via the priming effect is achieved at little "cost" to host fitness (Shoresh et al., 2010; van Hulten et al., 2006). Although, *Trichoderma* species are considered only to induce ISR rather than to activate the entire pathway to systemic resistance (e.g., Brotman et al., 2010; Hermosa et al., 2012), the exceptions (e.g., Segarra et al., 2007) suggest a less clear picture, with some strains inducing ISR, some inducing SAR, and still others inducing a combination of the two. This is not surprising given the ongoing communication among the signaling pathways regulated by the small-molecule hormones salicylic acid, jasmonic acid, and ethylene (Pieterse et al., 2009). Thus, the "cost" of using a given *Trichoderma* species to stimulate resistance must be considered.

Although little is known of the mechanisms, inoculation of plants using seeds treated with conidia of *Trichoderma* species can result in benefits far beyond the classical precepts of biological control, where a good microorganism kills a bad one. The biological effects once again include the result of intimate interaction at the genomic level among the host and the beneficial microorganisms. The effects also include direct effects of the *Trichoderma* species on pathogenic organisms and their environment, including parasitism and antibiosis through secretion of proteases, cellulases, and chitinases, all of which may be stimulated by molecules produced by the pathogen itself. The degree to which direct action on a parasite, induction of resistance, and growth promotion are observed is likely a function of the entire host–parasite–antagonist system. Parasitism might be the primary mechanism of biological control for some *Trichoderma* species. Genera such as *Hypomyces* and *Sphaerostilbella*, which include only mycoparasites, are the closest relatives to the genus *Trichoderma*, and indeed mycoparasitism and mycotrophy are considered the ancestral conditions for *Trichoderma* species (Druzhinina et al., 2011; Kubicek et al., 2011). Chaverri and Samuels (2013) demonstrated the existence of several clades, or phylogenetic groups, within the genus *Trichoderma*, the members of which are mycoparasites. For example, control of *Moniliophthora perniciosa*, the cause of witches'-broom of cacao, is most likely achieved by reduction of inoculum by *T. stromaticum*, which is a host-specific mycoparasite (see Fig. 3.2; Medeiros et al., 2010; Samuels et al., 2000), rather than through induction of resistance (De Souza et al., 2008). Some control of black pod disease of cacao in Africa might also result from direct parasitism by *T. asperellum* on *Phytophthora megakarya* (Tondje et al., 2007). Druzhinina et al. (2011) outlined a hunter–prey scenario in which the proteinases coming from the *Trichoderma* species stimulate the pathogen to release small molecules, possibly peptides, that are detected by sensors on the *Trichoderma* hypha and which then set in motion gene activity that results in the synthesis of cell-degrading enzymes, including cellulase, chitinase, and proteinases that initiate the attack on parasites (Seidl et al., 2009) and other pests such as nematodes (Sharon et al., 2001).

Induction of resistance and enhancement of plant growth are not mutually exclusive or even separate processes since immunity and plant development are interconnected in a network of hormone signaling pathways. However, growth-promoting benefits may have a wider application than the purely biocontrol benefits. For example, seed germination may be enhanced when they are inoculated with *Trichoderma* species, and the resulting seedlings develop more strongly. *Trichoderma* strains that grow among the first few layers of plant root cells initiate cascades of genetic activity in other plant tissues that enables crops such as rice and tomato to survive abiotic stress, including drought, salinity, and heavy metal and hydrocarbon contamination. The main effects of the *Trichoderma*–host partnership are probably increased photosynthesis and respiration, which results in increased root development, improved uptake of nutrients, and greater opportunity for beneficial microorganisms to associate with the plants, all ultimately leading to a healthier crop with increased production.

Specific *Trichoderma* strains and plant cultivar pairs can be matched on the basis of their molecular expressions. *Trichoderma* genes can be transformed into crop plants to confer desirable properties (Lorito, 1998), such as disease or drought resistance, or to promote growth through greater respiratory and/or photosynthetic capacity. Transgenic plants containing *Trichoderma* genes can confer resistance when transformed into plants, such as those coding for chitinase that make apple cultivars resistant to scab (e.g., Bolar et al., 2000), and conversely, the antifungal abilities of one strain of *T. harzianum* were enhanced by introducing a cellulose binding domain from an unrelated *Trichoderma* species (Limón et al., 2004). When considering genetic modification of antagonists or hosts, it is important to understand that modifying one process will no doubt affect other processes, and ultimately, the effect on the crop may not be worth the cost of the modification. This was illustrated by an *N*-acetyl-β-D-glucosaminidase-deficient mutant of *T. hamatum*. *N*-acetyl-β-D-glucosaminidase is a chitinolytic enzyme that releases nitrogen from nutrient-poor peat but also plays a key role in the biocontrol activities of *T. hamatum* against *Rhizoctonia solani* and *Sclerotinia sclerotiorum*. When this enzyme was disrupted, the strain lost its soil competitiveness and ability to protect against preemergence damping-off caused by *Sclerotinia sclerotiorum*, but on the other hand, its ability to promote plant growth was enhanced, possibly through the production of auxinlike compounds, such as was observed for *T. virens* growth promotion of *Arabidopsis* (Contreras-Cornejo et al., 2009; Ryder et al., 2012).

Applying *Trichoderma* Species in Agriculture

There are quite obvious reasons for the high interest shown in developing crop management programs that integrate beneficial microorganisms. The cost to the environment and the pocketbook can be much lower than the cost of applying synthetic pesticides. Biopesticides tend to be targeted on a pathogen or pathogens. Unlike many chemically based pesticides, they do not produce toxic secondary metabolites that adversely affect birds, insects, or mammals or linger in the crop or environment postharvest. They often are effective in small amounts and either disappear quickly or—a big advantage—may become established in the pathogen population, maintaining disease control over time. When used in combination with conventional pesticides such as copper, biopesticides can reduce the amount of conventional pesticides required to control a disease. Moreover, more recent research has shown a very clear advantage to applying selected *Trichoderma* strains to enhance seed germination and seedling development, especially under stress conditions that include drought, salinity, and soils contaminated by heavy metals or hydrocarbons. Each year, dozens of scientific articles describe beneficial effects of microorganisms such as *Trichoderma* species on crops, many of which originate in countries having low per capita incomes and economies based primarily on agriculture. As examples, recent publications summarize biocontrol programs in Cuba (Nalimova, 2007), the Philippines (Cumagun, 2012), Vietnam (Ha, 2010), and all of Africa (Babalola and Glick, 2012). In India and Vietnam, several *Trichoderma*-based products have been developed by universities and government institutions that may be commercially available (Table 2.1). In most cases, research and/or available products are based on native strains rather than on purchased cultures. For this reason, Okoth et al. (2013) emphasize the need to conserve the native soil microbiota, a key to crop health, and the pests, pathogens, and beneficial microorganisms associated with individual crops such as sweet potato and cassava.

The following sections review in detail some of the methodology that has been utilized in finding, evaluating, and applying *Trichoderma* species to control pathogens and to promote plant health. The examples are drawn mainly from the experimental literature and may not be applicable to large-scale crop production, but it is our hope that users of this book will be able to adapt experimental protocols to at least small-scale programs. While our primary desire would be to enable individuals or farmer cooperatives to develop their own *Trichoderma*-based biocontrol product, realistically, we are aiming at institutions, universities, governments, or nongovernmental organizations where there are facilities for sterile technique, storage of strains, their evaluation through in vitro and in vivo experiments, and investigating the mechanisms of observed pest control and/or plant growth promotion. The ideal situation would involve collaboration between such facilities and farmers, with the objective of educating farmers on biocontrol options and, where infrastructure exists, providing formulations using locally available agricultural by-products. At the same time, quality control and program improvement (screening for new, more effective strains; improving application techniques; resistant planting material; etc.) is left in the hands of the cooperating laboratory.

Integrating beneficial microbes into a crop management program is a multistep process requiring a coordinated approach. Fravel (2005) outlined the process of developing a biocontrol program, and the following steps for introducing a *Trichoderma* species into a crop management program are modified from Fravel. The steps, not necessarily in the order given below, are the following.

- Identify the problem: understand the crop and the target
- Find the right culture
- Assay for biological activity
- Identify the mode of action and ecology
- Determine methods of formulation and shelf-life studies
- Preserve cultures
- Register product

Identify the Problem: Understand the Crop and the Target

The first step in developing any crop management program is to understand the crop and diagnose its problems. Failure to understand this will frustrate any attempts to increase production through improved plant health and control of pathogens. Some evidence suggests that microbes such as *Trichoderma* species can be useful in promoting agricultural productivity from degraded lands by colonizing roots of crops such as cacao (Bae et al., 2011), rice (Rawat et al., 2012; Shukla et al., 2012), and maize (Shoresh and Harman, 2008) and by improving their root growth, shoot development, and plant dry weight. Traditionally, *Trichoderma* species have been used to control pathogens, and understanding the target pathogen and its niche is critical to developing any control program. What part of the plant is affected? When does infection occur? What environmental factors favor the host and the pathogen? Under what environmental conditions is the disease most and least severe? What environmental factors affect the antagonist? What are the sources of inoculum? Is the target pathogen soilborne or aerial? For example, a postharvest pathogen that appears in stored fruit might well originate from shelter belt trees surrounding the orchard, so control of that pathogen using a biocontrol agent might well be aimed at the pathogen in the shelter belt. The life cycle of the pathogen and its physiology, growth characteristics, cell wall structure, presence of resistant spore structures, and mode of infection all have important roles to play in the ability of any fungal pathogen to resist or succumb to actions of biocontrol *Trichoderma* species.

Find the Right Culture

Finding the "right" *Trichoderma* culture is vitally important to incorporating *Trichoderma* species into a crop management program. However, considering the great number of cultures cited in the extensive biocontrol literature and the very small number of species and cultures that have been formulated into commercially available products, the odds against finding a biologically useful, not to mention commercially viable, culture are great. Bear in mind, however, that most cultures used in a biological application were "local" in origin, isolated from a site that was most likely near home and discovered by some kind of winnowing process. An example of success is seen in recent work in Colombia (Santos et al., 2012; Smith et al., 2013), where 21 *Trichoderma* cultures were isolated from eight states in Colombia, and of the 21, two (*T. koningiopsis* Th003 and *T. asperellum* Th034) are now formulated for use in control of *Sclerotinia sclerotiorum* in tomato and potato. Similarly, local cultures of *T. asperellum* were developed to control black pod disease of cacao in Cameroon (Tondje et al, 2007).

There are several steps involved in selecting the right culture. These include not only isolating *Trichoderma* species from a substrate but also evaluating the potential of individual cultures for a specific biological activity in vivo and in vitro, improving on the abilities of individual cultures, and tracking the selected culture after it has been released into the environment.

Isolating *Trichoderma* species

Clearly, the first element is to find a useful culture. Because the evaluation process is neither easy nor inexpensive in terms of money and labor, it may be expedient to purchase cultures or products with proven capabilities, several of which are listed in Table 2.1, which is drawn in part from published lists (e.g., Cumagun, 2012; Fravel, 2005; Ha, 2010; Kaewchai et al., 2009; Punja and Utkhede, 2003; Woo et al., 2014). However, the cultures utilized in most published research reports are newly isolated, not purchased. In the following paragraphs, we describe how to obtain local *Trichoderma* cultures.

Trichoderma species are usually considered to be among the dominant components of the soil mycobiota

(Klein and Eveleigh, 1998). Indeed, despite the failure (Jumpponen and Jones, 2009) or near failure (Buée et al., 2009) of attempts of 454 pyrosequencing to detect *Trichoderma* species in some mycologically diverse forest soils, this would be a good first place to look for a *Trichoderma* species because reports of detecting them using different DNA primers (Friedl and Druzhinina, 2012) or direct isolation from all sorts of soils abound (a few important or recent examples taken from among many possible citations include Blaszczyk et al., 2011; Chaverri and Samuels, 2013; Domsch et al., 2007; Hoitink and Boehm, 1999; Hoyos-Carvajal et al., 2009a; Hujslová et al., 2010; Klein and Eveleigh, 1998; Kubicek et al., 2003; Kullnig et al., 2000; Lopes et al., 2012; Mulaw et al., 2010; Smith et al., 2013; Wuczkowski et al., 2003; Zachow et al., 2009). *Trichoderma* species can also sporulate on substrates in soils, including roots and diseased and/or healthy plant tissues (Fig. 2.1), and from fungal structures such as basidiocarps (mushrooms), sclerotia, and even mycelial mats (e.g., Chet, 1987; Tu, 1980). Typical greenish fungal growth on decaying plant tissue, wood, or target pathogen is a good indication of the presence of a *Trichoderma* species. If searching for an antagonist to a specific pathogen, burying tissue of the pathogen such as mycelial mats or sclerotia in soil can yield potential antagonists, including *Trichoderma* species. *Trichoderma* species are also common as endophytes living asymptomatically within plant roots (e.g., Ming et al., 2012; Quilliam and Jones, 2012; Ramão-Dumaresque et al., 2012) and sapwood and, less frequently, within leaves of tree species such as cacao, rubber, and *Cola* species (Chaverri and Samuels, 2013; Chaverri et al., 2011; Evans et al., 2003; Hanada et al., 2008, 2010). *Trichoderma* species are isolated with greater frequency and numbers of strains from soil or roots than from green plant tissues. Compared with the populations of soil bacteria (10^8–10^9 per gram of soil), fungal populations are relatively lower (10^4–10^5 per gram of soil). In soils from which *Trichoderma* species have been isolated, their populations range from 10^2 to 10^3 per gram of soil or per gram of root. Fields known to be relatively free of soilborne diseases may have suppressive soil, the disease suppression brought about by antibiosis or the effect of microbial antagonists on gene expression of the host plant (Alfano et al., 2007; Baker and Cook, 1982; Hoitink and Boehm, 1999; Hoitink et al., 2006; Howell, 1998). The apparent suppressiveness of soil to root-knot nematodes of banana in Guatemala actually resulted from the combined action of *Fusarium oxysporum* and *T. atroviride* endophytic within roots of the banana (Sikora et al., 2008).

Although *Trichoderma* species are often dominant members of the soil community, soil is not necessarily the location of the greatest *Trichoderma* species diversity. The *Trichoderma* species found in soils tend to be relatively few in number and belong to predictable communities of *Trichoderma* species that may be defined somewhat by latitude (Blaszczyk et al., 2011) and by their location in the soil and their associated species (Friedl and Druzhinina, 2012). On the other hand, many *Trichoderma*

Figure 2.1. **A,** The asexual and sexual states of *Trichoderma stromaticum* were originally found in Brazil on cacao tissues infected with *Moniliophthora perniciosa*, the cause of witches'-broom. **B,** Segments of cacao brooms infected with *Moniliophthora perniciosa* and artificially inoculated with *T. stromaticum. Trichoderma stromaticum* has only been found in association with cacao tissues infected with *Moniliophthora perniciosa* or *Moniliophthora roreri*.

species are rarely or never found in soils. The greatest diversity of *Trichoderma* species is actually found among sexually reproducing species that are not, or rarely, found in soil. *Trichoderma* stromata, formerly given names in the genus *Hypocrea*, are often found on wood and, less frequently, on other substrates such as sedges and on bracket fungi and other basidiomycetous fungi. Cultures can be obtained from stromata by germinating ascospores that form within. A large number of new species have been described for members of this group during the past decade (e.g., Chaverri and Samuels, 2003; Jaklitsch, 2009, 2011; Jaklitsch et al., 2006; Samuels et al., 2006a), most of which have not been evaluated for their potential in any agricultural application. Sapwood of woody plants is a good source of new species and of novel genotypes of described species (Chaverri and Samuels, 2013; Evans et al., 2003; Gazis and Chaverri, 2010; Hanada et al., 2010). Some of these species have been incorporated into experimental biocontrol programs (Krauss et al., 2010) or at least appear to have biocontrol and growth-promoting potential (Bailey et al., 2008; Hanada et al., 2010).

In classical biological control (Evans, 2008), antagonists that have coevolved with the host and its pathogens are sought in the center of origin of the crop. This is especially true of endophytic biocontrol strains (Evans et al., 2003; Mejía et al., 2008). In theory, the endophyte population of the crop is diminished when that crop is cultivated at a distance from its center of origin. A greater diversity of antagonists of any given coevolved pathogen will be found where the crop is most diverse, which is the center of its origin. There is a large and growing amount of literature describing the contribution of endophytic *Trichoderma* strains to the improvement of plant health and the induction of resistance to disease (e.g., Bae et al., 2011; Raghavendra and Newcombe, 2013). Of course, most crops are introduced from somewhere else and it is not possible to survey that distant area for biocontrol agents much less new, resistant, genetic material. Thus, most *Trichoderma* strains that are used in biological control are local in origin, isolated from farms having high disease pressure or by baiting soils with healthy host tissue or pathogen tissues. Another potential source of antagonistic microorganisms is in soils where crops exhibit heightened resistance to disease or pests.

Just because the diversity of *Trichoderma* species found in soil is lower than the usually wood-inhabiting sexual species, the potential for finding useful *Trichoderma* strains in soils has hardly been exhausted. New species are found in soil (e.g., Bissett et al., 2003; Hoyos-Carvajal et al., 2009a), and the physiological and biological variation among strains can be exploited. In the absence of sexual outcrossing and recombination, there is still considerable genetic variation in *Trichoderma* somatic hyphae resulting from mutations, transposons, plasmids, chromosomal polymorphisms, and the exchange of nuclei between strains by somatic hyphal fusion (see a review in Harman et al., 1998). The resulting heterokaryosis can lead to useful variation among strains of a species.

An important point to consider when isolating *Trichoderma* cultures is that it is easy to obtain a mixed culture. In nature, two or more *Trichoderma* species can be found in close proximity to each other, and in petri plate isolations, colonies may intermingle. Equally, it can never be assumed that a *Trichoderma* stroma and a closely juxtaposed *Hypocrea* stroma are parts of the same species. Most *Trichoderma* species cannot be distinguished on the basis of their macroscopic characters, and a mixed culture will not be obvious. Strain purity is essential both for identification and for developing a strain for an application. The only way to be absolutely certain of culture purity is to work with cultures that are derived from single conidia, ascospores, or hyphal tips. When isolating from soil or other "mass" substrates, a dilution series should be used to ensure that only single spores (conidia, chlamydospores) or hyphal tips are isolated. In these cases, it is also helpful to add to the medium a retardant such as chloramphenicol (0.5 g/L; Evans et al., 2003) that enables the recovery of slowly growing species on petri plates.

In order to isolate *Trichoderma* species from soil, serial dilutions of soil, root macerates, diseased or healthy plant tissues, fungal structures, etc. can be plated on several commercially available nonselective media such as potato dextrose agar (PDA) and cornmeal agar (CMA) or on selective *Trichoderma* medium E (TME) (see Table 4.3 for selective media; Davet, 1979; Elad and Chet, 1983; Papavizas and Lumsden, 1982). The objective of serial dilutions is to obtain the maximum number of discrete *Trichoderma* colonies on a 9-cm-diameter petri plate. In this case, we are not interested in quantifying the number of *Trichoderma* colony-forming units (CFU) in the sample. To achieve this objective, some trial and error may be required to find the optimum sample size. To start, suspend a sample weight of 1 g in 99 mL of sterile water; mix the sample in the water well in a closed container so as to prevent contamination. This will give a dilution of 1:100. Aseptically pipette 1 mL of this solution onto a selective or nonselective medium in a 9-cm-diameter petri dish. Using a sterile glass rod, spread the liquid over the surface of the agar. Using drier agar plates improves subsequent isolation of *Trichoderma* strains. Additional dilutions can be made by adding 1 mL of the 1:100 solution to 99 mL of sterile water in another container (10^4 dilution), adding 1 mL of the 10^4 dilution solution to another 99 mL of sterile water (10^6 dilution), and so on. Incubate the plates at room temperature for

3–7 days and observe for typical greenish colonies. Adjust a second round of dilutions according to the results of the first round by increasing or decreasing the initial sample or initial volume of sterile water.

Selective media have been developed for some species or species groups (see Table 4.3). Fujii et al. (2011) describe a cultivation method for the isolation of cellulose-producing microbes in which a distilled water suspension made from the substrate is placed on a solid mineral salts medium, lacking a carbon source, in petri plates and then overlain by sterile filter paper. The plates were incubated at 30°C and colonies appearing on the filter paper were isolated. In this way, they isolated several *Trichoderma* strains having high endocellulase and xylanase activity. Howell (1999) developed a medium for selecting *T. virens* and for distinguishing between "P" strains of the species that produce the antibiotic gliovirin and are effective against *Pythium* but not *Rhizoctonia* species and "Q" strains, that produce the antibiotic gliotoxin, and that are effective against *Rhizoctonia* but not *Pythium* species. Elad and Chet (1983) describe a selective medium for quantitative isolation of *Trichoderma* species from soil.

Isolating *Trichoderma* species from stems, leaves, and roots. When isolating endophytes from any tissue, rigorous surface sterilization is essential so as to ensure that the fungi appearing in the isolation medium truly emerged from within the host tissue and not from its surface. There are various published ways to achieve this, but the differences are in the type of plant and whether or not alcohol treatment is followed with hypochlorite. Huong et al. (2009) first freed rice roots of adhering soil by washing them in water; their surfaces were sterilized by immersing the roots in 1.5% NaHOCl for 3 min, followed by several rinses of sterile water to remove the bleach solution. After the final rinse, the sterility of the root surface was checked by imprinting the roots on 10% PDA plates emended with 150 parts per million (ppm) of streptomycin and penicillin to suppress bacterial growth. Root tips were removed with a sterile scalpel and discarded. Three randomly selected root sections were then cut into 15-mm-long sections and placed on 10% PDA plates. This was repeated three times for a total of nine samples per root system. Bailey et al. (2008) used 14% NaHOCl for 3 min to surface sterilize cacao roots. Carroll and Carroll (1978) briefly wet needles of Douglas-fir (*Pseudotsuga menziesii*) in 90% ethanol, followed by immersion for 10 min in 65% commercial Clorox (40% hypochlorite, bleach). Needles were cut into segments with a sterile scalpel and the segments were then placed on 2% malt extract agar. Ming et al. (2012) used a flame-sterilized scalpel to cut 15-mm-long segments of roots of the herb *Salvia miltiorrhiza*, which were then surface sterilized by sequential immersion in 75% ethanol for 30 sec, 1.3 M sodium hypochlorite (3–5% available chlorine) for 3 min, and 75% ethanol for 30 sec. These were then rinsed three times in sterile distilled water to remove excess sterilants and blotted on sterile filter paper. The root segments were cut into pieces approximately 2 mm thick and placed on PDA in petri plates amended with penicillin (100 mg/mL) to eliminate bacteria. The plates were incubated at 25°C and examined over 2–3 weeks for fungal growth. A protocol modified from Evans et al. (2003), who isolated *Trichoderma* species, *Clonostachys* species, and other endophytes from trunks (vascular cambium and phloem) of *Theobroma gileri*, is given here. An area of bark (approximately 6 × 8 cm) is removed from each tree bole (approximately 25–40 cm in diameter) at shoulder height (approximately 1.5 m), using a flamed knife. Immediately after removing the bark, a flamed No. 11 scalpel is used to remove 6 to 10 triangular slivers from the exposed wood, which are then placed individually onto 3-cm-diameter plastic petri plates containing a medium such as cornmeal dextrose agar (CMD), TME, or PDA or one-fifth strength PDA (20% PDA), but they are all supplemented with a 10-mg/L penicillin–streptomycin solution (for descriptions of media see Table 4.3). These are sealed immediately with Parafilm or electrical tape and stored in plastic food storage boxes. In order to isolate endophytes from leaves (Arnold et al., 2001; Chaverri and Gazis, 2010), squares measuring approximately 4 mm^2 are cut from various parts of apparently healthy leaves using a flamed No. 11 scalpel; these are surface sterilized in 2% sodium hypochlorite (chlorine bleach) for 3 min, and then in 70% ethanol for 2 min. The sterile leaf pieces are immersed in sterile water for a minimum of 1 min before being placed on TME or other media in 9-cm-diameter plastic petri plates; several pieces are placed onto a single petri plate. Regardless of whether isolations are made from wood or leaves, as the petri plates are incubated, they are observed frequently for fungal growth and newly emerging fungi are removed to fresh petri plates as they appear. Eventually, selected cultures are preserved for further study. *Trichoderma* species tend to emerge early, overgrowing and obscuring more slowly growing *Trichoderma* species or other fungi. Experience will determine the best age of a leaf for sampling. A smaller number of fast-growing endophytes dominate older leaves, while the youngest leaves have the smallest number of endophytes.

Strain improvement. Baker (1991) wrote of the need for diversity and for ways to increase diversity in biocontrol agents. With our new understanding of the genomic interactions between beneficial microorganisms and plants, it can still be seen that the desirability of obtaining new strains, or improving known strains, is no less important now than when finding biocontrol cultures was the main concern. The classical method of utilizing petri plate assays to evaluate antagonism of a *Trichoderma* species toward a pathogen is not very efficient. The artificiality of the technique can suggest a potential for a strain in vitro that can never be achieved in vivo; on the other hand, the process can miss strains that are good performers. Knowledge of the various "-omes"—the genome, the transcriptome, the metabolome—of *Trichoderma* species can provide clues as to whether and how a strain can have a beneficial impact, and several published correlative studies indicate the presence and regulation of important genes, proteins, and metabolites that relate to various stages of mycoparasitism, induction of resistance, or promotion of plant growth (see reviews in Druzhinina et al., 2012b; Hermosa et al., 2012; Lorito et al., 2010). Today, the complete genomes of only three *Trichoderma* species are known (*T. reesei*, *T. atroviride*, *T. virens*), but entries for many *Trichoderma* species, including nucleotide sequences, expressed sequence tag (EST) sequences, and genome survey sequences, are deposited in the Entrez Genome Project Database of the National Center for Biotechnology Information (NCBI, www.bioinfo.pte.hu/f2/GenPro/entrez.htm), and undoubtedly, the genomes of additional species will become available as the price for genome sequencing falls. With this will come an increased ability to examine the effects of individual genes and their products on biological activity. Genes that code for specific proteins that, in turn, lead to desirable activities can be cloned and subsequently inserted into strains so as to augment that activity by a strain (e.g., Limón et al., 2004; Reithner et al., 2011), or improve the performance of a crop plant (see review in Lorito et al., 1998).

One concern with utilizing transformed strains, even in a research phase of a project, that should be kept in mind is that the utilization of a genetically modified organism (GMO) at any stage in the research program can compromise any possibility of gaining organic certification. However, more classical approaches for improving strains are available, e.g., protoplast fusion and artificially induced mutation. *N*-methyl-*N'*-nitro-*N*-nitrosoguanidine-caused mutations rendered rhizosphere-incompetent strains rhizosphere competent, and this mutation was often accompanied by the development of resistance to benomyl (Baker, 1991). Papavizas and Lewis (1983) described the physiological and biocontrol characteristics of stable mutants of strains identified at *T. viride* that were resistant to methyl benzimidazole carbamate fungicides.

Protoplast fusion was described by Baker (1991) as a "reckless, random shot at genetic improvement;" but one of the most commonly cited *Trichoderma* strains, *T. harzianum* T22 (BioWorks; now *T. afroharzianum*; see Chaverri et al., 2015), is the result of the fusion of protoplasts of two native biocontrol strains identified as *T. harzianum* by a technique described in detail by Stasz et al. (1988). In order to generate protoplasts, treatment with specific lytic enzymes removes the cell walls, leaving the cytoplasmic membrane as the outer layer. When two or more protoplasts approach each other, their cytoplasmic membranes can fuse and whole contents, including organelles and genetic material, come together in a single cell. When protoplasts of organisms having desired and often complementary capabilities fuse, some of the resulting clones combine the characteristics of the parents. Methods for preparing *Trichoderma* protoplasts are described by Bae and Knudsen (2000), Besoain et al. (2007), Sivan et al. (1990, 1992), Stasz et al. (1988), and Thrane et al. (1995). Besoain et al. (2007) described the process by which they improved the performance of *T. harzianum* strains that had been isolated from tomato grown in suppressive soil and that already were shown to have good biocontrol properties by obtaining ultraviolet (UV) mutants and protoplast fusants of those cultures. Montealegre et al. (2010) found that treating plants with some of the mutants and one protoplast fusant generated by Besoain et al. (2007) improved the growth and yield of tomato over control and parent strains, but this improvement was dependent upon the tomato cultivar used.

One or more biological agents? Most applications using *Trichoderma* species are based on a single culture, but the formulation of some biocontrol products (e.g., the Binab line of products; Table 2.1) includes two or more species and/or strains, and in several applications, *Trichoderma* strains were combined with bacteria to control fungal pathogens. Krauss et al. (2001) argued that single-strain antagonists can be effective against single-strain pathogens under experimental conditions, but under field conditions, the host plant is more likely to be confronted by more than one strain of pathogen. Thus, results using a single strain of antagonist may be more inconsistent than a biocontrol treatment expressing greater genetic diversity with two or more strains of the antagonist. Krauss and Soberanis (2001) decreased the losses in cacao pods resulting from infection by *Moniliophthora perniciosa* and *Moniliophthora roreri* following treatments that mixed from two to six strains of the generalist mycoparasite *Clonostachys rosea* that had been isolated from the two cacao parasites. Similarly, mixtures of up to four strains of *Clonostachys rosea* improved efficiency in biological control of crown rot of banana caused by *Colletotrichum musae* (Krauss et al., 2001). In both cases, treatments that included the most strains were the most effective. In the banana study (Krauss et al., 2001), in addition to their effectiveness against the main pathogen, *Colletotrichum musae*, the mixtures had an effect on the secondary pathogens *Fusarium moniliforme* and *Lasiodiplodia theobromae*. A drawback to using multiple strains, or species, in a single treatment is the potential for antagonism among the strains, but the mixtures of *Clonostachys rosea* strains used by Krauss et al. (2001) in banana crown rot complemented each other in their respective mechanisms. In the cacao study, the absolute increase in healthy pods over untreated controls was not great (16.7%), but including this multistrain approach to biological control in an integrated crop management (ICM) program over a larger area would result in a much higher return for the farmer. Khan et al. (2011) successfully combined *T. harzianum* with the bacterium *Pseudomonas fluorescens* as a seed treatment to control fusarial wilt and root-knot nematode of pulses. The treatment reduced rhizosphere populations of both pathogens, an effect that persisted over 4 months. zum Felde et al. (2006) found that the numbers of nematodes in the roots of banana plants artificially inoculated with root-knot nematodes (*Radopholus similis*) were much lower in the plants that had been inoculated with endophytic strains of *Fusarium oxysporum* and *T. atroviride*, and the reduction was greater when two strains were simultaneously inoculated than when single strains or when two strains of each species were inoculated simultaneously. Mao et al. (1998) applied a combination of *T. virens* and the bacterium *Burkholderia cepacia* to pepper (*Capsicum* species) and tomato seeds, followed by drenches of the mixture, and found that in the greenhouse and in the field, plants in infested soil had greater fresh weight, and in the field, the disease severity index was much lower than in nontreated plots. Yobo et al. (2011) reported that combined inoculation with a strain of *T. atroviride* and a strain of *Bacillus subtilis* increased the biomass of cucumber and the effective control of *Rhizoctonia solani* damping-off in glasshouse trials, but while the combination enhanced growth, the increased growth did not translate to increased production. Control of damping-off was no better when the combination was used than when the *Trichoderma* species alone was applied to seeds. Paulitz et al. (1990) combined *Pythium nunn* and a soil-competent mutant of *T. harzianum*, T-95, to control Pythium damping-off of cucumber in two soils, a sandy Colorado soil and an Oregon soil that had been artificially infested with *Pythium ultimum*. The two antagonists did not interact negatively with each other, and combined, they suppressed damping-off in the Colorado soil but not in the Oregon soil. These examples demonstrate the utility of combining beneficial microbes to achieve disease control and/or plant growth enhancement.

Tracking cultures once they have been released into the environment. Once one or more cultures have been determined to be the "right" culture for a given application, it may be important to know whether that particular culture can survive in nature, invade host tissues, interact with other microbes, and so on. In short, how can you track the movement of one particular culture once it has been released into the environment? A tracking protocol must not only distinguish the applied microbe as a taxonomic entity (species, etc.), or even a genetically defined strain, but it must be able to discriminate a particular culture of a particular strain of a species from all other microbes in the site in which the microbe is applied. There is a subtle but fundamental difference between a culture—an entity in a tube or petri plate in a laboratory—that is developed for some biological activity, a strain of a species, and a species itself. For fungi, some of the genetic diversity that taxonomists accept within a species can be seen in any phylogenetic study: there is always genetic diversity. It is rather easy to identify *Trichoderma* species using a rapidly evolving genetic marker such as *tef*1, and even some of the finer genetic diversity of the species can be detected using this marker. A strain of a species, on the other hand, is a much finer subdivision than the lineages that are seen in phylogenetic analyses. A species typically occurs in nature as a complex of many genetically distinct strains, and when multiple cultures of a species from a given site are isolated, the possibility exists that each of those cultures represents a different strain (genetically distinct entity). Moreover, the possibility exists that the CFU (mycelium or other propagules) of those genetic entities (strains) that were isolated into pure culture are still present in the substrate. Several techniques, such as polymerase chain reaction (PCR), sequence-characterized amplified region (SCAR), microsatellites, single sequence repeats (SSRs), or other techniques, alone or in combination, have been described and are used to identify *Trichoderma* strains that are released into the nonsterile environment (e.g., Abbasi et al., 1999; Bahkali et al., 2012; Dodd et al., 2004; Feng et al., 2011; Hermosa et al., 2001; Miyazaki et al., 2009; Naef et al., 2005; Naeimi et al., 2011). While these techniques, especially microsatellites and SSRs, are highly specific and do a reasonably good job in picking out entities that are most likely genetically identical to the culture that was applied, they are unable to distinguish unequivocally the culture that was applied from what naturally occurs in the substrate. The following examples illustrate the difficulty in developing an unassailable tracking protocol.

Samuels et al. (2010) developed sequence polymorphism-derived (SPD) markers to track each of four strains of *T. asperellum* that were used in the control of black pod disease of cacao in Cameroon, only to find that the markers were widely spread in Cameroon and globally and, thus, of no use in tracking applied cultures. In another example, the strain of *T. atroviride* that Savazzini et al. (2008) used in microbial control applications in Italy was characterized by two mismatches in the sequence of its endochitinase (*ech*42) gene. The only other identical sequence they found in GenBank for this gene was in a Japanese strain, which suggests a reasonable degree of confidence in this particular marker. However, the culture used in the Italian study was isolated locally so that, even though the only identical sequence deposited in GenBank was for a Japanese strain, it could be argued that CFU carrying that mutation in *ech*42 are still present in the Italian study site, or that an Italian strain had been deposited in the Japanese culture collection. GenBank is not representative of nature; it is only representative of what individuals choose to deposit. In this case, Japan and Italy are indeed widely separated on the map, but to find this one mutation in such widely separated places says that the mutation may be actually quite widespread. It may or may not be common, but the utility of this mutation in tracking a particular culture is suspect. If the culture that was applied originated at some distance from the site to which it was applied, the chance of error in tracking the applied culture is lower than if the culture was isolated locally. But, however small, there is a risk.

The only way to be absolutely certain that the tagging protocol is specific to a particular culture (NOT strain) is to induce a distinctive change in that culture. Various methods have been described. For example, cultures can be cotransformed with the *Escherichia* β-glucuronidase (GUS) gene and the hygromycin B gene (Thrane et al., 1995) or they can be transformed by adding molecular markers for antibiotic resistance (Bowen et al., 1996; Lo et al., 1998) or for green fluorescent protein (Bae and Knudsen, 2000; Lu et al., 2004b) or by inducing mutants without transformation (e.g., Besoain et al., 2007; Tachapattaworakul et al., 2008). There is, however, a danger in losing the possibility for organic certification of a microbially based product if transformed cultures were utilized during the research phase of the project. Nonetheless, markers that do not depend on molecular transformations can be useful when used with the proviso that reasonable effort has been made to ascertain whether or not cultures having that marker are present in the area that is intended to receive the designated culture, as was done by Cordier et al. (2007).

Assay for Biological Activity

Finding the right culture obviously implies that the culture is capable of performing some kind of job, whether it be suppressing pathogens or enhancing drought tolerance or seed germination, and equally obviously implies that bioassays will be required to determine whether a given *Trichoderma* culture can perform as desired. In the following paragraphs, we describe some bioassay methods that have been used in culture selection.

Considerations in plant bioassays

Trichoderma species may be more effective against soilborne pathogens than against aerial pathogens. Although chlamydospore formulations are usually recommended for treating dry soil, potting media, and other substrates used for growing plants, conidium suspensions have been used for suppressing fast-growing pathogens such as *Pythium* and *Phytophthora* species (Papavizas et al., 1984; Widmer and Samuels, 2010). A bioassay incorporating either conidia or chlamydospores or both should be conducted to determine the effectiveness of the propagules for suppressing the target pathogen.

In in vivo plant bioassays, it is essential to standardize the effect of the pathogen on seedling emergence, plant stand, etc. before the potential of *Trichoderma* cultures can be evaluated further. Proper controls, with and without application of *Trichoderma* cultures and with and without pathogens, are certainly needed, but caution must be exercised in defining what constitutes an appropriate "control" in experiments that involve unknown biotic elements (Baker et al., 1984). Effectiveness parameters to be recorded include seedling emergence, plant stand, reduction in necrosis, and plant growth promotion including root growth and plant dry weight. Knowing baseline photosynthetic and respiration rates also provides information about growth promotion by beneficial microorganisms. However, clear differences between treatments cannot be observed if the disease incidence is low, and in screening trails, researchers often face the problem of low pathogen virulence. Therefore, standardization may also include additional evaluation of a pathogen's virulence and repeatedly observing differences, e.g., 50% reduction in seedling emergence, 70% seedling death, etc. Several rounds of bioassays are necessary to confirm the biocontrol ability of *Trichoderma* species. Once this baseline pathogenicity information is available, the possible antagonistic potential of *Trichoderma* cultures can be evaluated by incorporating their spore (conidia, chlamydospores) suspensions into the potting mixture or test medium. The minimum amount of *Trichoderma* conidia needed per gram of soil to demonstrate effectiveness of biocontrol activity should also be determined. The minimum amount of *Trichoderma* spores needed per seed per gram of soil to demonstrate a clear biological activity (e.g., 90% seedling emergence, 80% plant stand) should also be standardized. Adding more *Trichoderma* propagules may not always be beneficial. For example, when bioassays were being developed to evaluate *T. virens*, it was noticed that, over a certain concentration per gram of soil, the biocontrol agent was phytotoxic (Bailey and Lumsden, 1998; Howell, 2006). The biocontrol activity could be specific against a single pathogen or may be broad against several soilborne pathogens. While *T. virens* was shown to be more active against chitinaceous fungi such as *Rhizoctonia solani*, *T. asperellum* was more active against pathogens that contain cellulose in their cell walls, i.e., the oomycetes, including *Pythium* and *Phytophthora* species. Environmental factors can be critical to determining the success or failure of any crop management program and environmental variation can result in inconsistent results, especially of biological agents. In general, conditions should be recorded that affect performance one way or the other, such as soil moisture, temperature, humidity, pH, organic matter, and so on, including especially host cultivar, which can have a profound effect on disease control or growth promotion.

What are clues to predicting agricultural utility? One critical indicator of whether any culture will be useful in crop management, whether inhibiting development of pathogenicity, promoting growth, or both, is the ability of the selected culture to survive in the environment. Rhizosphere competence is one of the most important concepts in the application of any beneficial microbe to agriculture. Can the candidate outcompete other microbes in the soil in colonizing cellulose-rich substrates? Is the candidate culture rhizosphere competent? In *Trichoderma* species at least, rhizosphere competence is always associated with cellulase production (Baker, 1991). The earliest definition of rhizosphere competence, or competitive saprobic ability, was the ability to colonize cellulose-rich substrates. However, it is now known that beneficial microbes interact with host plant genomes and the interface occurs in the roots to promote growth and/or induce resistance to pathogens. Thus, another measure of rhizosphere competence is whether a beneficial microorganism can colonize the intercellular spaces of the surface layers of roots of the host (Ahmad and Baker, 1987; Harman et al., 2004). As the root elongates, the cells in the region of active growth, the mucigel region, are sloughed off and the remains of their cell walls form a gelatinous matrix that encloses this region. A beneficial microorganism such as a *Trichoderma* species would require cellulolytic enzymes to occupy this niche, and those that do can then be in a position to interact with the plant at a molecular level. Rhizosphere-incompetent strains of *Trichoderma* species have been mutated to rhizosphere competence and to benomyl resistance, using *N*-methyl-*N'*-nitro-*N*-nitrosoguanidine. This treatment most often resulted in rhizosphere colonization from seed treatments of up to 10^6 CFU per gram of soil 8 cm below the seeds, while the effect of the nonmutated parents was never found below 2 cm (Baker, 1991). Mutation to benomyl resistance was not always associated with rhizosphere competence, but as was stated above, rhizosphere competence is always associated with cellulase production. How can one assess the ability of a strain to metabolize cellulose? One way is to bury cellulosic materials, such as straw or cellophane disks, in field soil, inoculate the soil with a known quantity of CFU of the culture in question, and reisolate. Ahmad and Baker (1987) used benomyl tolerance as a marker and correlated colonization of polished straw and cellophane disks with cellulose production. The amount of cellulose produced was directly related to the strain's competitive saprobic ability and rhizosphere competence.

Assays for mycoparasitism

Despite the emerging literature that documents the abilities of *Trichoderma* cultures to influence the host at the genomic level and not simply through mycoparasitism, integrated pest management (IPM) using *Trichoderma* species to directly control pathogens remains popular. The classical method of determining whether a culture will kill a parasite, and thus be useful in biological control, has been to confront a pathogen and an antagonist in a petri plate. In this process, antagonists that grow over the pathogen, or that inhibit growth of the pathogen by a diffusing antibiotic, are selected for further study. However, while the petri plate, which was developed by Julius Richard Petri in the mid-nineteenth century, was a great boon to microbiology, it is a highly artificial system that can give misleading results to anybody wishing to discover new, biologically useful *Trichoderma* cultures. Nonetheless, we will describe a petri plate assay that does give some indication of mycoparasitic ability. We offer just this one caution: limiting one's thinking to mycoparasitism and petri plate assays is guaranteed not to reveal other possible uses for many cultures. So while petri plate assays or leaf disk or detached pod assays might give information about mycoparasitism and antibiosis, in planta assays for the activation of resistance, enhancement of seed germination, seedling growth, and ability to colonize host roots are strongly encouraged.

In vitro assays for mycoparasitism. Until facilities for examining genomes of potential strains become commonplace, the petri plate assays, despite their many drawbacks, will remain the first tool for assessing the potential for mycoparasitism of a large number of strains. One of the disadvantages of estimating mycoparasitic potential by confronting the antagonist and the target pathogen on agar in a petri plate is that the agar itself provides nutrient to the antagonist, thus possibly giving a misleading result with regard to the mycoparasitic capability of the antagonist. To avoid the possibility of the antagonist deriving its nutrients from the agar rather than from the parasite, the precolonized petri plate assay was developed (Bailey et al., 2008; Evans et al., 2003; Krauss et al., 1998; Widmer and Samuels, 2010). To assay mycoparasitic potential using this method, 9-cm-diameter petri plates containing a medium favorable to the target pathogen are inoculated at one edge with the pathogen. The plates are incubated under conditions that enable the pathogen to colonize the entire surface of the agar. At this time, a strip of inoculum, 0.5 × 2.5 cm, is excised from an actively growing colony of the antagonist and is placed at the edge of the plate opposite the inoculum of the target pathogen. Five replicate plates are prepared. The plates are incubated at 25°C in darkness or under black light (near UV [nUV]). After 5 weeks, 15 plugs, each measuring 3–5 mm in diameter, are removed from the colony in a line beginning with the inoculum strip of the antagonist and proceeding toward the inoculum of the parasite. These plugs are placed on 20% PDA in 9-cm-diameter petri plates and observed over 2 weeks at 25°C for the presence of the pathogen or the antagonist. The percentage of colonization of the disks by antagonist and pathogen is determined and comparisons are made by analysis of variance (ANOVA). Percentages are angular transformed prior to analysis to standardize the variance. The experiment is carried out for 5 weeks or until only *Trichoderma* species are recovered from the disks. A precolonized plate method was also used to assay the mycoparasitic ability of candidate fungi toward *Phytophthora megakarya* in Cameroon (Foley and Deacon, 1985; Krauss et al., 1998; Tondje et al., 2007). Sterile filter paper disks (5 mm in diameter; Whatman No. 1) were placed on top of the mycelium of *Trichoderma* isolates that had been grown on PDA for 5 days, and the fungal cultures were incubated for an additional 4 days. Filter paper disks overgrown with hyphae of individual *Trichoderma* isolates were placed on top of mycelium of 10-day-old cultures of *Phytophthora megakarya* on V8 juice agar. Growth of the *Trichoderma* culture on mycelium of the pathogen was observed daily following a 3-day incubation period at 25°C. The mycoparasitism was rated as follows: 3+, the entire culture of *Phytophthora megakarya* was colonized by the test isolate within 14 days; 2+, the entire culture of *Phytophthora megakarya* was colonized between 14 and 21 days; 1+, the entire culture of *Phytophthora megakarya* was colonized after 21 days; and –, when no growth on the *Phytophthora megakarya* mycelium was observed. Replicate plates of each isolate are required in each experiment, and the experiment should be repeated at least twice to be certain of the results obtained. Isolates that showed promise in the precolonized plate assay were screened for suppression of *Phytophthora megakarya* on detached cacao pods. The viability of the pathogen can be tested by plating agar pieces from precolonized plates on pathogen-specific medium or by reinfecting host plant tissue with the agar pieces. In the latter case, 1-cm-long strips of agar cut from plates containing the antagonist and *Phytophthora megakarya* were placed onto detached green cacao pods in closed containers lined with moist paper, and symptoms for black pod disease were rated. Mycoparasites from pods showing weak symptoms or no symptoms of black pod were then tested in the field (Tondje et al., 2007).

Naeimi et al. (2010) provide an example of the futility of using traditional petri dish assays for antagonism as an aid in selecting a useful strain. They wanted to know whether *Trichoderma* cultures isolated from rice fields would have any effect on growth rate and sclerotial production and germination of *Rhizoctonia solani*, the cause of rice sheath blight of rice. To test for the effect of a *Trichoderma* culture on the germination of sclerotia, sclerotia of uniform size were placed on 6-day-old cultures of the *Trichoderma* cultures in 9-cm-diameter petri plates. The cultures were incubated at 26°C in darkness for up to 30 days. Viability of the sclerotia was evaluated by placing them onto water agar plates and observing for hyphal development after 24 h; sclerotia that had not been exposed to the *Trichoderma* cultures were placed on water agar to serve as a control. To test for an effect on sclerotial production, they used a traditional petri plate confrontation assay and looked for the ability of a *Trichoderma* strain to overgrow *Rhizoctonia solani* and whether there was a diminution in the number of sclerotia produced in the petri plate. As a test of mycoparasitism, they spread molten PDA over the surface of glass microscope slides, leaving one end free to allow for manipulation, and then inoculated the slide with the *Trichoderma* cultures 1 cm from the *Rhizoctonia solani*. The slides were placed on 2% water agar in 9-cm-diameter petri plates and incubated at 28°C for 5 days. While observing with a compound microscope (unspecified magnification), they counted the amount of hyphal coiling and disintegration of cells of *Rhizoctonia solani* in three microscope fields. Finally, to determine whether volatile compounds produced by *Trichoderma* cultures affected the growth rate of *Rhizoctonia solani*, they individually inoculated PDA in the bottoms of petri plates with the *Trichoderma* cultures and *Rhizoctonia solani* and then positioned one plate over the other (they did not specify which fungus was on top) and incubated them at 26°C under a diurnal light regimen. They found that there was no correlation among the various tests; some cultures did indeed inhibit sclerotial production, some strains inhibited growth, some cultures were mycoparasitic, and so on, but cultures that were good at one activity were not good at another. None of the petri plate assays predicted success in limiting the development of rice sheath blight in in planta greenhouse trials in which conidia were sprayed directly onto rice plants. What was common to all seven of the cultures most effective in limiting the development of rice blast in a greenhouse trial was the ability to completely overgrow the *Rhizoctonia solani* in the petri plates, but about 55 of the 78 tested cultures were similarly capable of overgrowing the *Rhizoctonia solani*.

Krauss et al. (1998) used the preinoculated plate method to isolate mycoparasites from host tissue. In an effort to attract potential mycoparasites, they used the banana crown rot pathogens *Colletotrichum musae*, *Fusarium subglutinans*, *Fusarium moniliforme*, *Fusarium pallidoroseum*, *Botryodiplodia theobromae* (currently known as *Lasiodiplodia theobromae*), and *Nigrospora sphaerica* as bait. These were inoculated onto 9-cm-diameter petri plates as described above. Two or more strains of each pathogen were assayed. They then placed dry and green host material (i.e., flowers, leaves, banana peel) onto the cultures and observed for mycoparasitism and zones of clearing. There were differences among the bait pathogens in their ability to attract antagonists, with *Colletotrichum musae* attracting the most potential mycoparasites and the *Fusarium* species the fewest. Dry leaves provided more antagonists than did green plant material.

In planta assays for beneficial effect

In vitro assays can give an estimation of mycoparasitic potential but obviously the result is meaningless if the effect is not seen in host plants. Small-scale trials in the greenhouse under controlled conditions of the antagonist and the target pathogen are commonly a next step in evaluating the beneficial effects of the selected *Trichoderma* culture(s). Krauss and Soberanis (2001) studied the biological control of cacao diseases using mixtures of mycoparasites. For their preliminary assay using single antagonists, seeds of a susceptible cacao clone were planted in seedling bags under 50% shade and allowed to grow for 1 month. Ten replicate seedlings were simultaneously inoculated with conidia of the mycoparasites and spores of the pathogen, *Moniliophthora roreri*, a species that produces copious aerial spores. The hypocotyl was inoculated to runoff (2 mL per plant). Noninoculated plants and plants inoculated only with the pathogen served as negative and positive controls, respectively. Symptom development was observed over 3 months. Mbarga et al. (2012) sought to control Pythium root rot of cocoyam (*Xanthosoma sagittifolium*) using cultures of *T. asperellum* that had previously been shown to be effective against *Phytophthora megakarya*, the cause of black pod disease in cacao. They used the preinoculated petri plate assay described above to test for an in vitro effect against *Pythium myriotylum* and then used a solid-state fermentation to produce inoculum for an in vivo greenhouse assay for antagonism. For this, they grew cultures on PDA for 5 days to obtain conidia. Ten 5-mm-diameter agar disks of each culture were used to inoculate 100 g of sterile husked boiled rice in polyethylene fermentation bags (available from several sources). The inoculated bags were incubated in the laboratory for 10 days. When the substrate was completely colonized, conidia were extracted using a MycoHarvester (see "Selecting a fermentation (liquid, solid-state) method" below). After drying, 1 g of conidia of each *Trichoderma* culture was placed into 100 mL of sterile distilled water and the concentration was adjusted to 1×10^6 conidia per milliliter using a hemocytometer. For the pathogen, five petri plates containing PDA were inoculated with a *Pythium* strain known to be pathogenic to cocoyam. After 5 days, each culture was mixed in 200 mL of sterile distilled water and macerated with a blender. One milliliter of each suspension was placed into test tubes, mixed using a vortex, and diluted tenfold two times. One milliliter of both dilutions was plated onto petri plates containing 15 mL of PDA and incubated 24 h. The number of *Pythium* colonies appearing was used to calculate the concentration of CFU. The substrate for the assay was sterilized and fertilized local soil. Following sterilization, before planting, the soil was inoculated with the pathogen at a rate of 1 mL of the suspension described per gram of substrate. One kilogram of the inoculated substrate was placed into 15-cm-diameter plastic pots. The plants to be used were micropropagated and grown 6–8 weeks in a greenhouse. The plants were then soaked 24 h in conidial suspensions (10^6 conidia per milliliter) and replanted in the inoculated soil. Each treatment consisted of 10 pots, and the experiment was performed twice. Evaluations were made after 30 days.

When Naeimi et al. (2010) could not correlate the in vitro effects of a panel of *Trichoderma* cultures on mycoparasitism and sclerotial production in *Rhizoctonia solani*, which causes rice sheath blight (described above), they next utilized the following assay in a nonsterile greenhouse environment. This assay is typical of trials in rice using other pathogens. Twenty-day-old seedlings of a susceptible, high-yielding rice cultivar were planted in 20-cm-diameter plastic pots containing 4.5 kg of rice field soil. The plants were fertilized and watered regularly to keep them submerged and incubated at 26–34°C and 85–90% relative humidity. The plants were inoculated with *Thanatephorus cucumeris* by placing a 5-mm-diameter mycelial plug between the junction of the basal leaf sheath and the stem above the water line at the maximum tillering stage (five tillers per pot). Conidia were isolated from petri plates by washing the plates with sterile distilled water and scraping the surface with a sterile scalpel. The suspension was filtered through double-layer cheesecloth, and conidial concentrations were adjusted to 10^7 per milliliter. Tween 20 was added at 0.1% to facilitate adhesion of the *Trichoderma* conidia to the surface of the rice plants. Solutions of *Trichoderma* conidia (50 mL per replicate) were sprayed onto the rice plants using a hand sprayer twice, 24 h after inoculation with the pathogen and 7 days later. Untreated, inoculated plants served as a control. Watanabe et al. (2006) soaked 10 g of rice seeds that had been infected with *Gibberella fujikuroi*, the cause of bakanae disease, in 20 mL of conidial suspensions (10^6 conidia per milliliter) of selected *Trichoderma* strains at 15°C for 24 h. The seeds were decanted through gauze and incubated for 24 h at 32°C. Germinated seeds were sown in plastic pots and incubated for 3 days at 30°C and 100% relative humidity in darkness. The seedlings were then grown in a greenhouse for 18 days at 22°C and evaluated for development of bakanae disease. Three replicates were prepared for each treatment.

Chang et al. (1986) and Inbar et al. (1994) describe a typical method for evaluating plant growth in vegetable and floral crops in response to the application of *Trichoderma* cultures. Conidia of the *Trichoderma* candidate were obtained in two ways. In the first method, cultures were grown on yeast–malt–extract agar for 10 days at 25°C under illumination. Conidia were washed from the agar surface with sterile distilled water and the concentration of conidia was adjusted to 10^8 conidia per milliliter. The conidial suspension was sprayed onto unrooted chrysanthemum cuttings. Cuttings were also placed in a propagative medium containing *T. harzianum* growing in peat–bran (20%, v/v). After rooting (14 days), cuttings were transplanted to 9-cm-diameter pots (Chang et al., 1986). In the second method, cultures were grown in 500-mL Erlenmeyer flasks containing 50 mL of PDA for 10 days at 30°C under illumination. Conidia were harvested from the substrate, washed several times in sterile distilled water, and suspended in 0.001% Tween 20. This suspension was added to soil at a concentration of 5×10^6 conidia per gram of soil. The substrate for treatments consisted of a 1:1:1 mixture of peat, wheat bran, and water. This was autoclaved in polyethylene bags for 1 h on two successive days. The bran–peat mixture was incubated for 14 days at 28–30°C, air-dried, powdered by screening, and used immediately or stored in plastic containers for future use. The concentration of *Trichoderma* propagules in the mixture was between 10^7 and 10^9 CFU per gram. Inbar et al. (1994) combined this mixture with vegetable propagative material (peat and vermiculite, 1:1, v/v) at a concentration of 5% (v/v) and sowed cucumber and pepper seeds in seedling-type preformed trays. For floricultural experiments, Chang et al. (1986) propagated cuttings in a peat–perlite mix treated with rooting hormone and exposed to mist for 16 sec every 2 min during daylight hours until rooted. Container mix for growing plants was composed of soil–peat–perlite at 1:1:1. The propagative and container mixes were steamed at 83°C for at least 0.5 h.

To test the ability of *T. gamsii* to induce resistance to *Rhizoctonia solani*, Anees et al. (2010) added 15 mL of a suspension containing 4×10^6 conidia per milliliter of a culture of *T. gamsii* to 260 g of sand in each 8 × 8 × 7-cm square plastic pot (covered with 20 mL of calcined clay to hold the water). Twenty seeds of carrot (*Daucus carota*), a host susceptible to *Rhizoctonia solani*, were planted in each pot. Twelve days after sowing, 50 mL of natural or heat-sterilized soil was placed around the crown of each seedling. Two days later, four millet seeds infected by *Rhizoctonia solani* were placed in each pot, one at each corner. The millet seeds were infected by placing mycelial plugs of *Rhizoctonia solani* on seeds that had been sterilized at 105°C for 1 h on three successive days. To characterize direct interaction of *Trichoderma* isolates with *Rhizoctonia solani* in vivo, carrot seeds were sown as described above in sand without *Trichoderma* species. After 12 days, 15 mL of conidial suspensions were mixed with 50 mL of sterilized or natural soil and placed around the crown of the seedlings. Two days later, this soil was inoculated with millet seeds infected with *Rhizoctonia solani* as described above.

A simple way to inoculate plants such as maize, pepper, and tomato with potentially endophytic *Trichoderma* cultures is to insert wood toothpicks into stems (Sobowale et al., 2007), which is a common method of inoculating pathogens into host plants. The following method is adapted from that described by Drepper and Renfro (1990) and Pryor et al. (2000). Round or flat wooden toothpicks are washed and then sterilized in small glass jars. The toothpicks in the jars are moistened with a liquid medium such as potato dextrose broth or V8 juice and then inoculated with a suspension of conidia and incubated for a few days or until green conidia are visible on the toothpicks.

Inoculation with the *Trichoderma* conidia is performed by piercing the stem or roots to make a narrow channel into which the colonized toothpick is inserted with sterile forceps. They are left in the woody tissue to grow under optimal conditions. Evaluation of endophytic colonization can be done as was described above. Sobowale et al. (2011) could isolate the inoculated *Trichoderma* strain from points distant from the inoculation point up to 6 weeks after inoculation, although some strains moved further in the plant than others. An annual research report on the control of Armillaria root and crown rot of oak in California (Andris et al., 2002) describes inoculating trees with toothpicks infested by, respectively, two *Trichoderma* species. In this study, flat wood toothpicks were infested by incubating them in petri dishes with the *Trichoderma* cultures and then T-budding them into young trees at the mid-trunk level. The authors could reisolate the *Trichoderma* species from 100% of the trees 2 years after inoculation, although they did not describe how they knew that the reisolated *Trichoderma* culture was the same as that used in the inoculation. In this program, the researchers provided inoculated toothpicks to individual growers who could inoculate their own trees.

Establishing root colonization. Root colonization is essential for beneficial microorganisms to interact with the host genome, whether to influence growth and development or to induce resistance. The ultimate test of whether a strain is capable of forming an endophytic relationship with host roots is to reisolate the target strain from the roots of inoculated host seedlings. Bailey et al. (2008) described four methods for reestablishing an endophytic relationship in cacao seedlings by endophytic and nonendophytic cultures of several *Trichoderma* species. They examined the resistance of the newly established endophytic relationships to parasitism by the cacao pathogens *Moniliophthora roreri* and a *Colletotrichum* species. In all cases, pregerminated cacao seeds were planted in a soilless mix in Magenta boxes (Magenta LLC, Chicago, IL, U.S.A.) (Fig. 2.2), and the *Trichoderma* strains were inoculated into the mix as conidia before or after planting the seeds or as mycelial plugs after the seeds were planted. In a subsequent publication, Bae et al. (2011) described the ability of *Trichoderma* cultures that had been isolated as endophytes from trees to become established endophytically within pepper seedlings (*Piper anuum*) and to induce systemic resistance to *Phytophthora capsici* as follows. Two agar plugs (5 mm in diameter) of each *Trichoderma* culture taken from the leading edge of a 7-day-old sporulating culture grown on, for example, 20% PDA (4.8 g of potato dextrose broth powder per liter) were added to a sterilized soilless mix (PRO-MIX PGX; Premier Tech Horticulture Inc., Quakertown, PA, U.S.A.) in double Magenta boxes (77 × 77 × 194 mm). If a commercially made product is not available, soil can be prepared by mixing soil, perlite, and sand (1:1:1) in autoclave bags and autoclaving at 121°C for 20 min on two consecutive days; Magenta boxes are half filled with the soil mix and autoclaved a third time. The Magenta boxes contained 9 cm of sterile soilless mix and had four holes (5 mm in diameter) sealed with tape on the bottom. Sterile water (25 mL) was added to the soilless mix at the time of inoculation. The Magenta boxes were maintained in growth chambers for 14 days before being planted with pepper seeds. Seeds were surface sterilized in 30% sodium hypochlorite for 5 min, followed by five washes in sterile distilled water. Sterilized seeds were germinated on 1.5% water agar plates under fluorescent light for 4 days at 25°C. Three germinated seeds were planted in each Magenta box 3 cm deep into the sterile soilless mix with or

Figure 2.2. A Magenta box with a cacao seedling. The Magenta box is used in experiments to evaluate the ability of *Trichoderma* cultures to enter the roots of cacao seedlings and induce a biological effect. (Courtesy B. A. Bailey)

without *Trichoderma* cultures. Seedlings were grown in a 12-h-light/12-h-dark cycle at 25°C. After 35 days, the pepper seedlings (two per box) were dissected, and approximately 10-mm tissue sections of roots and stems were plated on cornmeal dextrose agar (CMA with 2% dextrose) and observed for presence of the *Trichoderma* species in the petri plates, determining success or failure to establish an endophytic relationship. A test for successful endophytic establishment by reisolating from host tissue was described above.

zum Felde et al. (2006) used a root dip method to establish root colonization by *Trichoderma* and *Fusarium* endophytes in banana. Infected plants were resistant to infestation by the nematode *Radopholus similis*. They sequentially dipped roots of banana plants that had been developed from tissue culture into 500-mL beakers containing 300 mL of a conidial suspension that included 1×10^6 CFU per milliliter of *T. atroviride* and/or nonpathogenic *Fusarium oxysporum* for 5 min. They considered the concentration to be of greater importance than the time, and additional time did not increase colonization. The conidial suspensions were prepared by growing the respective cultures on PDA to obtain good sporulation. Then 20 mL of sterile distilled water was pipetted onto the agar surface, conidia were freed using a metal bacterial cell spreader, and the liquid was decanted through sterile gauze into a sterile beaker, where conidial concentrations were adjusted using a hemocytometer. The plantlets were inoculated with nematodes at their base. After 2 months, the plants that had been inoculated with the endophytic fungi had healthier root systems and fewer nematodes than did the noninoculated plants, and those that were inoculated with both endophytes were healthier than those that had been inoculated with either one of the endophytes.

Identify the Mode of Action and Ecology

The mode of action, how and where the antagonist and target pathogen interact, is often viewed as a purely academic exercise, but actually the information obtained may be essential for the timing and placement of the product in the crop and, later, for product registration. Knowing how a culture acts to achieve an effect can certainly help in improving that culture and, thus, the result, or knowing how a culture acts can aid in finding more efficient strains. Several reviews of the mechanisms of action of beneficial microorganisms are cited above. Jeger et al. (2009) proposed a mathematical model to predict the likelihood of success of biological control of foliar pathogens such as *Botrytis cinerea* by *Trichoderma* species in relation to the mechanisms involved. Significant parameters include the colonization rate of diseased and healthy host tissue by the *Trichoderma* strain, how long the *Trichoderma* culture remains viable, and the host–pathogen infection rate. In general, four main mechanisms are cited as the mode of action: mycoparasitism, competition, antibiosis, and induced resistance. A single biocontrol agent can have more than one mode of action (Aneja et al., 2005; Bailey and Lumsden, 1998; Chet, 1987; Di Pietro et al., 1993; Harman et al., 2004; Jeger et al., 2009; Mejía et al., 2008). A specialized laboratory is required to conduct many mode of action studies, especially when studies involving isolation and sequencing DNA are involved, for example, real-time studies of gene activity during antagonist–pathogen–plant interactions. However, broad inferences can be made by observing the effect of applying a *Trichoderma* species. For example, increased germination, root growth and nodulation, and yield in soybean treated with a culture identified as *T. viride* was attributed to antagonism of the *Trichoderma* strain to the root parasites *Fusarium oxysporum* f. sp. *adzuki* and *Pythium arrhenomanes* (John et al., 2010). Antibiosis, including volatile and diffusible compounds, is an important mode of action of some *Trichoderma* species and can be observed by confronting pathogen and antagonist in petri plates (Lifschitz et al., 1986; Lumsden et al., 1992). However, one must always remember—as we have said elsewhere in this book—that a biocontrol effect observed in a petri plate is not necessarily indicative of the effect in nature. The lesson of the antibiotic effect of gliotoxin produced by *T. virens* in control of *Rhizoctonia solani* damping-off of cotton should not be forgotten: what was considered to be an antibiotic effect was exhibited by strains that were rendered incapable of producing the antibiotic gliotoxin. The biocontrol effect in these cases turned out not to have been antibiosis; biological control was actually achieved through induced resistance (Howell, 2003). Antibiotic effect has been attributed to nonanoic acid produced by a strain of *T. harzianum*, which inhibits the germination of two basidiomycete (*Moniliophthora* species) pathogens of cacao (Aneja et al., 2005), and the aromatic compound 6-pentyl-α-pyrone (6-PAP) produced by several *Trichoderma* species, which inhibits the germination of *Phytophthora* spores (Cutler et al., 1986). Mejía et al. (2008) found that two of the most common mechanisms in the antagonism of cacao endophytic fungi to major cacao pathogens were competition for substrate and antibiosis. The most common endophytes were the ones that were the best in competing for substrate and also the best at recolonizing cacao seedlings. The species that had antibiotic activity tended to be less common, slower growing, and less able to recolonize cacao seedlings.

Determine Methods of Formulation and Shelf-Life Studies

Selecting a fermentation (liquid, solid-state) method

While petri plates and agar media are sufficient to produce conidia for experiments, it is obvious that this technique could never produce the quantities of inoculum required for field application. In addition, unless the formulation containing *Trichoderma* propagules is intended to be used immediately ("just in time"), the formulation must provide for viability of the propagules, their shelf life, The longer the shelf life, the better. Fortunately, *Trichoderma* species can sporulate on a range of agricultural by-products, and this ability has been used successfully in developing growth substrates for the production of inoculum (Agosin and Aguillera, 1998; Miller and Churchill, 1986). By-products, such as bran, molasses, neem cake, soybean hull fiber, and cassava, have been used as substrates in solid-state fermentation. Extracts of the by-products in water have been used in liquid fermentation (Papavizas et al., 1984). Important factors to consider include local availability of the by-product, cost, type of biocontrol propagules (conidia, chlamydospores, mycelium) needed, target pathogen, and site of application (aerial plant parts or soil) (Lewis and Papavizas, 1983; Papavizas, 1985; Papavizas et al., 1984). Additional carriers (e.g., clay particles, sugars, talc) can be added to stabilize the formulations and improve the shelf life of propagules. When selecting a substrate for cultivating a biocontrol agent, it may be important to consider perception of the project by local inhabitants. For example, in a locality where rice is a dietary staple but is expensive, it might be well to consider the wisdom of letting rice become moldy for a scientific experiment if another, less expensive, commodity such as bran or cassava can be substituted.

In deciding which propagule to use (conidia or chlamydospores), consider that *Trichoderma* conidia germinate much faster than chlamydospores and are easier to harvest, formulate, and apply; however, unless refrigerated, their shelf life at room temperature is much shorter than that of chlamydospores. A cultivation substrate rich in carbon but low in nitrogen, such as rice grains, potato dextrose agar, or broth, will yield significantly more conidia per gram or per milliliter of substrate. However, low-carbon but nitrogen-rich substrates, such as molasses or soybean hull fiber, will yield conidia in lower numbers but with a higher number of desiccation-resistant, thick-walled chlamydospores, which can be used as inoculum. Simple hemocytometer counts can be used to visually determine concentrations of conidia and chlamydospores. Viability of the preparations can be tested by plating aliquots of serial dilutions on agar media such as PDA or CMA. A minimum concentration of 10^8 conidia and 10^6 chlamydospores per gram of solid substrate, or per milliliter of liquid substrate, is typically required for preparing inoculum intended for formulation as a biocontrol agent.

Either liquid or solid-state fermentation methods can be used to obtain the optimal concentration of a desired *Trichoderma* propagule (conidia, chlamydospore, mycelium). Liquid fermentation requires more specialized equipment (shakers, glassware, fermentation reactors, autoclaves) and space than does cheaper solid-state fermentation. For producing small (<10 kg) to large (>100 kg) quantities of *Trichoderma* spores, cheap locally available substrates, autoclavable plastic bags, simple pressure cookers, and shelf space in a clean incubation room are all that are needed.

A solid-state substrate fermentation with rice as a carbon source was used to produce *T. stromaticum* for field evaluations (Medeiros et al., 2010). Starter cultures with 250 mL of potato sugar broth (see Table 4.3) shaken at 200 rpm for 4 days at 25°C were used for inoculating (40 mL per bag) plastic bags containing 300 g of sterile rice grains. The rice grains were mixed with 210 mL of distilled water and autoclaved for 60 min at 121°C for two consecutive days before being inoculated with the starter culture. The inoculated bags were incubated for 48 h at 25°C and agitated daily to aerate and stimulate sporulation. After 2–3 days of incubation and when the fungal spores were visible on the rice grains, the rice was transferred from the bags to 30 × 50-cm sterile plastic trays (three bags of rice per tray) and left to sporulate further for an additional 3–4 days before being transferred to a clean room (if available, an air-conditioned room maintained at 16°C is preferable) for drying. The rice substrate was incubated for 7 or 8 days in the drying room until it reached a level of 12% moisture. The substrate was mixed regularly for the first 3 days to improve aeration. After the drying step, purity, spore counts, and percentage of spores able to germinate on water agar were checked. The rice substrate was stored in a refrigerator (7°C) until needed. The conidia were harvested by suspending and agitating the rice substrate in sterile water in a 5-L blender. The suspension was then sieved (#50 mesh) to separate the conidia and mycelium from the rice grains and used for field applications the same day. The concentration of conidia was determined using a hemocytometer. Germination rates greater than 90% were considered ideal for field application. The conidium preparations were transported to the field in polystyrene boxes and adjusted to a concentration of 10^7 conidia per milliliter before the suspension was sprayed onto the cacao trees.

An inexpensive liquid fermentation technology was developed by Papavizas et al. (1984) using 20-L autoclavable carboys to scale up liquid fermentation and obtain 7–10 L of *Trichoderma* conidium suspension. An air pump used for aerating home aquariums and a series of cotton filters and tubing supplied the required air to the carboy. An upright autoclave large enough to hold the carboy is essential for this technology.

A compact, benchtop device (MycoHarvester MH5; Fig. 2.3) has been successfully adopted in several laboratories to separate conidia of *T. stromaticum* from substrates such as rice grains (Bateman, 2004; www.mycoharvester.info; www.iparc.org.uk). The MycoHarvester has the advantage of eliminating particles that block sprayers, provides high-quality spore separation enabling development of physically stable formulations, draws dust into the machine that is potentially harmful to the operator, produces material quickly, and facilitates storage of conidia by concentrating them for subsequent drying. The MycoHarvester enables development of a formulation that has smaller particle size and is free of larger substrate debris, thus improving delivery of the conidia, especially in aerial applications. A larger version of the MycoHarvester (MH3) that enables two people to process 0.5 tonnes of grain substrate per day is also available (Fig. 2.3). Viability of the formulations stored at both room temperature (approximately 20–25°C) and at 4°C (refrigerator) can be tested by plating aliquots of serial dilutions on agar media such as PDA or CMA. A minimum of 6-month survival at room temperature with a minimum loss in viability is desirable.

Santos et al. (2012) developed two formulations for storing large volumes of conidia of *T. asperellum* and *T. koningiopsis* for up to 18 months. As above, conidia were developed on sterile cooked rice. To make dispersible granules, the sporulating substrate (20% of the final dry weight of the product) was mixed with industrial colloidal aluminum silicate (55%; Bell-Chem Internacional, S.A., Medillín, Colombia), starch to aid in dispersal (15%), and milk protein (10%; Colanta, Medillín, Colombia) to protect against desiccation. The mass was run through a sieve with 1-mm pores and dried in a stove at 25°C for 24 h. To make a dispersible powder, the sporulating rice mass was washed with a 1% solution of Tween 80 and the resulting solution was centrifuged at 3.075 × *g* for 30 min. The biomass was mixed with magnesium silicate USP pharmaceutical quality (Bell-Chem Internacional, S.A.), milk protein, and a red food dye to protect against UV radiation. The mixture was dried in a stove at 25°C for 24 h. The dry material was pulverized in a cutting mill. Before storing the material, the quality was tested by determining the percentage of germination in petri plates. Both the granules and the powder were vacuum sealed in aluminum foil packets, 1 g per packet, and stored at 8°C. The percentage of germination for the two storage methods remained fairly constant, at near 90% for 8 months, but at 9 months of germination, the percentage of the unformulated conidia dropped by more than 20%, whereas the germination rates of the granulated and powdered formulations dropped by less than 10%. After 1 year, germination percentages of the two formulations for the two species (*T. koningiopsis* and *T. asperellum*) remained high

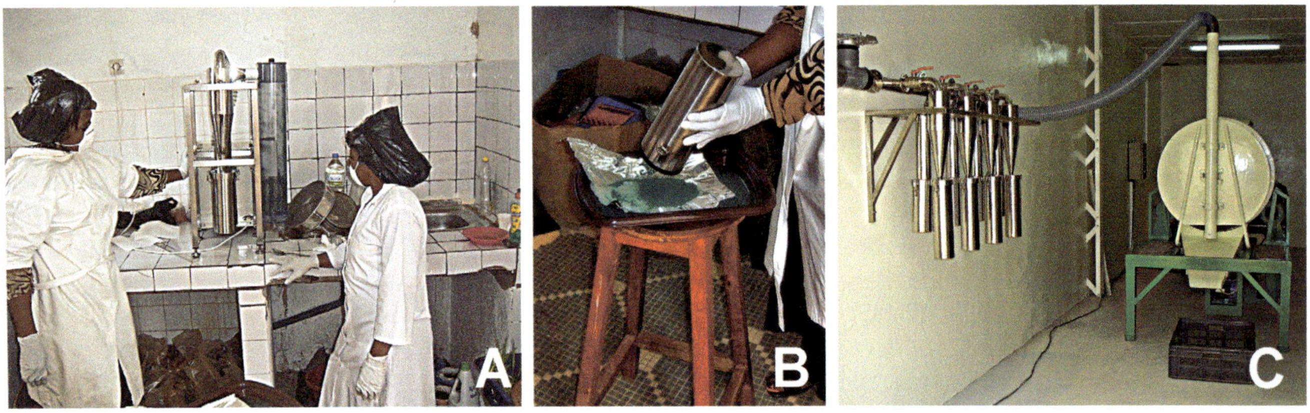

Figure 2.3. Quantities of pure *Trichoderma* conidia can be extracted from substrates such as rice or cassava through the use of a MycoHarvester. **A and B,** A benchtop model, MH5. Substrate is placed in the transparent tube, seen on the right in **A**, and through cyclonic action, conidia are drawn into the metal cylinder. **B,** Conidia are emptied from the metal cylinder. **C,** A MycoHarvester, MH3, which is designed for large-scale applications. (A and B, Courtesy D. Begoude; C, Courtesy R. Bateman)

for the powdered form (75% for *T. asperellum* and 83% for *T. koningiopsis*); the germination percentage of the granular form remained high for *T. koningiopsis* (83%) but was diminished to 30% for *T. asperellum*. Even after 18 months, the granular and powdered formulations supported 66 and 52%, respectively, for *T. koningiopsis* but only 10 and 33%, respectively, for *T. asperellum*.

Trichoderma conidia can be stored and applied to soil in pellet form. Following is the method for making alginate pellets, or prill, that was modified from Fravel et al. (1985) by Knudsen and Bin (1990). Slices of agar measuring 4 cm^2 were cut from a sporulating culture of the *Trichoderma* species. Each piece was placed in 500 mL of potato dextrose broth in a 1-L flask. Flasks were plugged with cotton and incubated at room temperature (approximately 22°C) on a rotary platform shaker at 120 rpm for 1 week. The mycelial mat was then strained, rinsed with sterile water, added to 100 mL of 1% aqueous sodium alginate solution, and stirred gently until evenly dispersed. The average wet weight of the mycelium was 37 g per flask. Ten subsamples were taken randomly from the flasks, oven-dried (100°C overnight), and weighed. Pellets were made by adding the mycelium–alginate mixture dropwise to 0.25 M aqueous $CaCl_2$, and in one treatment, 2 g of wheat bran was added to the mycelium–alginate mixture before pelletizing. Pellets forming in the $CaCl_2$ solution were removed within 10 min, air-dried on waxed paper, and stored in glass beakers at room temperature and ambient humidity. The mean oven-dried hyphal weight in each pellet was 1.8 mg, and the mean weight of the bran in the pellets was 2.5 mg per pellet. The *T. harzianum* remained viable after several months of storage. The bran used by Knudsen and Bin (1990) did not increase the radial extension of the *Trichoderma* colonies, but its addition did provide for an increase in hyphal biomass, which can also be considered to be proliferation. Fravel et al. (1985) used a clay bulking agent, Pyrax (pyrophyllite, hydrous aluminum silicate; Vanderbilt Minerals, LLC, Norwalk, CT, U.S.A.), in making the pellets. Ten grams of sodium alginate (available from various suppliers) and 100 g of Pyrax was added to 1 L of sterile distilled water and pulverized in a blender for 1 min. This material was inoculated with 20 mL of an aqueous suspension of the *Trichoderma* conidia. The amount of conidia in the suspension varied from 3.4×10^4 for *T. virens* to 6.8×10^6 for *T. asperelloides* (identified in the paper as *T. viride* T1-R9). The alginate–conidium–Pyrax suspension was stirred continuously as it was dripped through the tube and the pipettes into a solution of 25 M $CaCl_2$ or 0.1 M $C_{12}H_{22}CaO_{14}$ (calcium gluconate). The calcium displaces the sodium, resulting in the formation of a shell around the drop. Conidia of both *Trichoderma* species survived the pelleting process. However, while the rate of survival of the two species was greater when $CaCl_2$ was used as the gelling agent, the viability of both dropped much more rapidly with this reagent than with the calcium gluconate.

Preserve Cultures

Development of a program that aims to incorporate a beneficial microorganism into crop management is likely to be a protracted, multiyear process. Obviously, the culture or cultures that are eventually selected for development are going to have to be preserved in such a way that their biological properties are not altered by mutation and that the cultures are secured from contamination. It is important to not serial transfer cultures because of the strong possibility of contamination and/or mutation in the strains; rather, the working set should be large enough to provide inoculum for the duration of the research project. *Trichoderma* species vary in their ability to be preserved. The most certain method for long-term preservation is lyophilization, but this is expensive in terms of equipment and supplies and is labor intensive. Preservation on liquid nitrogen in 3-mL cryovials containing 10–20% (aq.) glycerin following pretreatment of a few hours at –80°C in an acetone bath is also a good method for preservation, but it requires a constant supply of liquid nitrogen and regular monitoring of the nitrogen tanks. Cryopreservation at –80°C in 3-mL vials containing 10–20% glycerin is a quick and effective method for storing most species but, again, requires an expensive low-temperature freezer and assumes a constant supply of electric power. Le et al. (2009) placed conidial suspensions onto Mast Cryobank tubes (Mast Group, Merseyside, U.K.) that contain specially coated beads, which are then stored at –80°C; a single bead is removed to initiate new colonies. Storage of *Trichoderma* conidia on silica gel has been successfully utilized for the preservation of bacteria and most spore-forming fungi but not for the water molds or downy mildews (Table 2.2). We have kept cultures on Difco CMA (Becton, Dickinson and Company, Franklin Lakes, NJ, U.S.A.) slants in 125-mm-long, glass tubes with screw caps at approximately 4°C in an incubator or refrigerator for several years without transferring them, and the strains remained viable. Cultures should not be kept on PDA because preservation on this rich medium can lead to mutation and loss of the culture. Cultures that sporulate heavily often can be revived from dry agar, but this is a last ditch effort. Cultures should be transferred before they dry!

Whatever method of storage is selected, it is a good idea to maintain two sets of important cultures: one set should not be touched and the second set is the "working set". It is also advisable to store the cultures using more than one method, because when working with newly isolated cultures, one cannot know at the outset what will be the best preservation method. For example, cultures of *T. stromaticum* do not survive well under liquid nitrogen or in cryovials at –80°C. From an academic and purely experimental point of view, it is very useful for cultures to be available to other researchers by depositing them in public culture collections, such as the American Type Culture Collection (ATCC) in the United States, the Centraalbureau voor Schimmelcultures (CBS) in the Netherlands, or the National Biological Resource Center (NBRC) in Japan. However, if there is a possibility that the strains will be part of a process or product that you will patent, you should discuss with technology-transfer specialists in your country the implications of making

Table 2.2. Silica Gel Technique for Preserving *Trichoderma* Cultures[a]

1. Grow the culture on an appropriate medium to obtain a healthy young population of conidia. Potato dextrose agar works well for most fungi.

2. Prepare silica gel tubes.
   A. In the fume hood, fill 8-mL glass vials (Wheaton #224744 with Teflon rubber-lined cap) about one-half full with desiccant activated silica gel (mesh size 6–12) and replace the cap. Make sure the rack is oven safe.
   B. Sterilize the capped tubes for 1 h (or overnight) at 180°C/350°F. The silica gel may become brown on top. These tubes may be made ahead of time and stored.

3. Prepare 10% instant milk solution.
   A. Add 10 g of instant milk to 100 mL of deionized water and autoclave for 20 min. The milk will turn a light brown after autoclaving and can be made ahead of time.
   B. Some people add 5% glycerol to the milk solution before autoclaving for storage of bacteria at very low temperatures.

4. Transfer of conidia.
   A. Use one sterile 1.5-mL Eppendorf tube for each strain. Transfer 1 mL of the 10% milk solution to each tube.
   B. Using a sterile toothpick, scrape the surface of the plate and transfer the conidia to the Eppendorf tube. Scrape the toothpick on the inside of the tube to remove all conidia.
   C. Vortex the tube to distribute the conidia.
   D. For each vial, add 1 mL of 10% milk to a 15-mL Falcon tube.
   E. Pipet the 1 mL from the Eppendorf tube into the larger tube and vortex or invert to mix well.

5. Transfer to vials.
   A. Place vials with silica gel in an ice bucket.
   B. Remove caps and add 1 mL of the mixed conidial suspension to each vial. Do not process more than five vials at once because the silica gel will clump as it sits.
   C. Working with one vial at a time, replace the cap and vortex the milk–silica gel mixture. Tip the vial upside down and tap the cap on the hood surface, if necessary, to loosen the silica gel granules.
   D. Vortex until the granules are mostly loose in the tube.
   E. Return the vials to the ice immediately. Repeat with the remaining vials.

6. Final steps.
   A. After 10–15 min, remove the vials from the ice and return to the rack.
   B. Allow the vials to sit at room temperature overnight.
   C. The next day, shake out a few granules from a few tubes onto the appropriate agar medium to check for viability. Immediately store tubes at –20°C or –70°C.
   D. The microbe should remain viable for several years if stored appropriately.

[a] Developed by G. E. Harman.

the strain or strains publically available, as this may hinder your ability to protect them with a patent. The international culture collections often have nonpublic options for preserving cultures.

Register Product

Eventually, small-scale projects can be taken up by commercial operations. Whether a product can be commercialized may depend upon the scope of the product, whether it is focused on a single disease or can be applied more widely, and whether intellectual property and national biodiversity laws limit the ability to distribute the product/organism across national lines. The researcher must become familiar with local regulations regarding the release of microbes, even if they are beneficial, and the registration of any product may be required.

Microbially based biological control of plant diseases differs fundamentally from chemical control, but pesticide regulations for microbial and chemical pesticides remain similar (Harman et al., 2004). At least in the United States, in general, any product that claims to be pesticidal must be registered, but the registration process is lengthy and often requires significant expense in assuring product safety through toxicological, environmental, mode of action, and often efficacy studies. For more details, refer to the Environmental Protection Agency (EPA) website (www.epa.gov/oppbppd1/biopesticides).

On the other hand, products that promote plant health and vigor but do not claim pesticidal action, such as compost or teas, rhizobia, and endomycorrhizal root endophytes, are largely unregulated. Harman et al. (2010) compared chemical and biologically based control of plant diseases and discussed various models: full registration and marketing of pesticidal claims; unregulated products promoting plant vigor, including biofertilizers; local production model; and state-supported production (Harman et al., 2010; Nalimova, 2007; Rossert, 1997). The state-supported model includes elements of the second and third models, but the materials are provided to farmers gratis or at low cost, such as in Cuba and Venezuela. Local production, or "cottage-scale" production, of biocontrol products is often found in developing countries. It offers the advantage of using local microorganisms, thus avoiding any intellectual property or introduced foreign agent issues, and at the same time is adapted to local conditions; locally produced substrates such as rice, cassava, or agricultural by-products are usually available at a low price and the scale of production does not need to be large, whether solid-state or liquid fermentation is used. Moreover, the product can be produced just in time, thus avoiding somewhat the problems of shelf life. Harman et al. (2010) describe several examples of "cottage-scale" biological control. A typical example is a project to control banana weevils in Ecuador using *Beauveria bassiana*. Researchers at a government institution (Instituto Nacional de Investigaciones Agropecuarias; INIAP) could demonstrate that their local culture could kill the weevils (mainly adults). Using the LUBILOSA project that employs *Metarhizium anisopliae* against locust in East Africa (Douthwaite et al., 2001) as a basis, they developed the skills and techniques needed to grow *Beauveria bassiana* in the laboratory and prepare inoculum. Finally, they participated in a small project, Biotechnological Tools for Farmers. Under this project, key young farmers and their wives and children were initially provided with minimum infrastructure (e.g., pressure cooker to sterilize substrate, rice, and inoculum or "seed") to scale up production of the fungus; later, only the "seed" was provided. INIAP monitored the project in the field and assured quality of the "seed" (Carmen Suarez-Capiello, *personal communication*; see also Norton et al., 2005).

Concluding Remarks

In this chapter, we addressed several key issues that researchers face when developing any fungus-based microbial product. We have seen that, as our understanding of the interrelationships among beneficial microbes, plants, and their pathogens has increased, the range of possibilities for exploiting beneficial microbes to improve crop management by more than biological control of pathogens has expanded. Successful crop management is a multidimensional process, and despite the bias in this book toward *Trichoderma* species, the need for crop rotation, fertilization, and sanitation can never be forgotten. In the absence of healthy and well-cultivated plants, no beneficial microorganism is going to deliver a profitable crop from plants that are stressed by environment, planted in exhausted or pesticide-saturated soils, choked by weeds, or beset with pathogens of all sorts. Microbially based disease control most likely will not, by itself, be enough to keep serious pathogens at bay. Thus, we emphasize the need to practice integrated crop management (as opposed to integrated pest management, which is rather unidimensional because it refers only to disease), in which microbial products, biofertilizers, ground covers, crop rotation, etc., may be incorporated along with some chemical agents to control disease. In order to develop a *Trichoderma*-based microbial agent, it is essential to foster close cooperation between research institutions and private or public entities and farmer groups. Quality control of the formulation is absolutely critical at every step. Poor quality control, loss of viability, contamination with competing microbes, poor storage conditions, improper application methods, lack of user education, etc., are reasons for the failure of microbial control products. The adoption of an IPM, or perhaps better phrased as ICM, strategy may be as much philosophical as it is physical. An important aspect of an ICM strategy is to obtain socioeconomic information about farmers and where they live, to review agronomic practices, and to identify all the reasons for low productivity. Through farmer schools and other means of information dissemination, IPM/ICM projects integrate farmer families and scientists. This has been shown to increase the personal investment of the whole family, and the community of families, in the crop as they become involved in the process of developing a plan for managing their crop, while at the same time, hopefully providing the farmer with the means for pest control at minimal cost. Under this model, whole families and communities become invested in the success of the project. Examples of developing participatory crop, or pest, management programs are given in the book *Globalizing Integrated Pest Management: A Participatory Research Process* (Norton et al., 2005). One of the main effects of an ICM program for the control of black weevils in plantain in Ecuador was to increase participation by women from zero to nearly 50%. Along with increasing participation by women, there were changes in the behavior of the men, including decreased drinking and smoking and a greater regard by men for the contributions made by their wives (Muniappan et al., 2009).

CHAPTER 3

Applications of *Trichoderma* Species in the Field and Greenhouse

So far, we have described some of the mechanics of introducing a *Trichoderma* strain into a crop management program. Once a desired effect is obtained from a selected culture, or cultures, at a small scale and under controlled conditions, the next step is to demonstrate whether the same effect can be obtained under natural conditions, during which more factors come into play than are seen in the laboratory, and on a larger scale. In short, can the strain that showed promise in the laboratory help to obtain a better harvest? In this chapter, we describe several scenarios in which *Trichoderma* cultures have been applied. Many of the examples include field application, but we have also included examples that, while still experimental in scope, might stimulate local researchers to apply the techniques to their own situations. *Trichoderma* cultures have been applied directly to seeds and soils, incorporated into potting and container mixes, added to hydroponic or other closed systems, and introduced into tissue-cultured plantlets. Following are examples of each of these applications.

A Few Comments on Testing

Strains should be tested under the conditions in which they will be used. If, for example, the *Trichoderma* strain is to be used to enhance seed germination, then the treated seeds should be planted following the normal procedure and into all the places where that crop is going to be planted. If it is intended for biological control, it should be tested in fields or sites that are infested with the pathogens. The testing should be undertaken over two or more seasons. Cooperation with farmers will provide "real world" experience while investing the farmers in the program. This was seen in field testing cultures of *T. asperellum* against *Phytophthora megakarya* in cacao in Cameroon, where Tondje et al. (2007) cooperated with farmers in two localities by establishing a randomized block design for the treatments in producing cacao plots that had not been sprayed.

Soil Application of *Trichoderma* Cultures

Most *Trichoderma* cultures are applied directly to the soil, and several commercial preparations are applied as soil drenches or mixed into the substrate/soil. For soil application to be successful, the *Trichoderma* hyphae must survive and proliferate in the soil or planting medium, either finding their prey or colonizing host roots. Often in soil applications, *Trichoderma* conidia are mixed into a nutrient base, which is usually an agricultural waste such as neem cake, mustard cake, groundnut cake, linseed cake, banana peel, wheat bran, pulse bran, rice bran, wheat straw compost, ground wheat straw, ground cotton straw, and, probably thc most common, wheat bran–peat mixtures (Bhagat and Pan, 2007; Lumsden et al., 1998; Mao et al., 1998; Sivan et al., 1984). Lewis and Papavizas (1984) found that by inoculating a sand–bran (1:1, w/w) mixture with a solution of conidia and then incorporating that mixture into soil after 1–3 days of incubation, the number of *Trichoderma* colony-forming units (CFU) in the soil achieved and maintained was higher over several months than when conidia in bran were added directly to the soil. Sivan et al. (1984) used a wheat–bran mixture

(1:1, v/v) adjusted to 40% moisture (w/w) that was sterilized in autoclavable polyethylene bags (50 × 50 cm) for 1 h at 121°C on three successive days. The substrate mixture was inoculated with a conidial suspension and incubated in an illuminated chamber for 14 days at 30°C. This was mixed with soil (5 g of soil per kilogram) or rooting mixture (20% preparation by volume) before sowing the seeds. When added directly to rooting flats in water or in a peat–bran mixture soil in greenhouse trials, *T. harzianum* T95 decreased the germination time for pepper (*Capsicum* species), accelerated the flowering of periwinkle (*Vinca minor*), increased the number of blooms on chrysanthemum, and generally increased weights and/or heights of various flowering crops and vegetables (Chang et al., 1986). However, the population density of *Trichoderma* CFU usually exceeded 10^5 CFU per gram of container medium or soil, making the liquid application economically unfeasible. The peat–bran–conidium mixture was formulated as follows: equal volumes of wheat bran, peat moss, and water were autoclaved 1 h on two successive days and then conidia (10^8 per milliliter) were added. The mixture was incubated for 14 days at 28–30°C, air-dried, and then powdered by grinding and screening. The mixture was kept in plastic bottles until needed. Lo et al. (1996) developed a granular application for *T. harzianum* by culturing a soil-competent strain of *T. harzianum* on a 1:1 (w/w) wheat bran–peat mixture that, after drying, contained 5×10^8 CFU per gram. This was effective in reducing the effects of various turfgrass diseases.

Some authors have added *Trichoderma* cultures to soil in sodium alginate pellets, also known as prill, which are biodegradable, uniform in size, convenient for storage, and compatible with agricultural machinery. The process of making alginate prill, or pellets, was described in Chapter 2. *Trichoderma virens* added as prill to a nonsterile soilless mix controlled damping-off, caused by *Pythium aphanidermatum* and *Rhizoctonia solani*, in cabbage, zinnia, and cotton. The antagonist that was used was selected from among 50 fungi as being able to control both pathogens (Lumsden and Locke, 1989). Once in the soil, the hyphae of *T. virens* radiate from the pellets, able to search for their prey or colonize root surfaces. Knudsen and Bin (1990) postulated that this ability of the *Trichoderma* hyphae to explore in the rhizosphere may be more important than their proliferation by spatially localized sporulation. Montealegre et al. (2005, 2010) added native *T. harzianum* and mutants of a native strain as sodium alginate pellets to planting holes of tomato and observed reduced mortality from *Rhizoctonia solani* with a consequent increase in plant health. This effect was amplified by summer solarization. Treatment with *T. harzianum* alone or with the bacterium *Paenibacillus lentimorbus* combined with solarization during the summer was proposed as an alternative to treatment with methyl bromide to control *Rhizoctonia solani*. Unfortunately, during the summer months, no biological treatment was as effective as methyl bromide in controlling this serious pathogen.

Suppressive Soils and Composts

Suppressive soils are those in which a pathogen is unable to become sufficiently established to cause more than low-level disease for several years. All natural soils suppress plant disease to some extent, in large part because of their diverse microbiota, which outpaces and outcompetes many plant parasites through various mechanisms. Intensive agriculture, which relies on monoculture, tillage, pesticides, and fungicides, tends to deplete populations of beneficial microorganisms living in the soil, and this can open the door to pathogens. Composting is the decomposition of organic material under controlled conditions, followed by sterilization and reintroduction of beneficial microorganisms. This is distinct from the uncontrolled decomposition of material rotting in open dumps, manure piles, or open fields (Hoitink, 1980). Some composts can be suppressive when mixed with organic or inorganic materials such as peat or perlite. Composts can be amended with beneficial microorganisms, such as *Trichoderma* species, to enhance their suppressiveness. Composts, usually mixed with peat, have become an important component in the container plant and nursery cropping systems. Suppressiveness in field soils can also be restored, but far less is known about this than about the use of composts in container cropping. Baker (1991), Craft and Nelson (1996), De Ceuster and Hoitink (1999), Harman and Björkman (1998), Stone et al. (2004), and Zinati (2005) reviewed the extensive literature about the effect of suppressive soils and composts on plant disease control in field soils. This literature should be studied and understood since the exploitation of suppressive soils and composts in crop management is a complex subject. In the following paragraphs, only a brief summary is provided as it relates to the application of *Trichoderma* species.

There are two types of disease suppression, and both have been discussed in detail by Stone et al. (2004). In general, suppression of the pathogen is held in check by the totality of the microbial inhabitants of the soil. In specific, suppression is a defined organism or set of organisms or conditions that reacts to specific activities or properties of a pathogen. General suppression applies to genera such as *Pythium* and *Phytophthora*, the species of which cannot sustain saprobic growth, are poor competitors, and do not produce long-lived resting structures such as sclerotia or chlamydospores. In contrast, specific suppression usually refers to pathogens such as *Rhizoctonia solani* and *Fusarium* species that form resistant resting structures and are capable of saprobic life apart from the host crop, thus requiring a more targeted and often complex effort for their control. The characteristics of general suppression of damping-off caused by *Pythium* species by the addition of organic matter in containers were enumerated by Stone et al. (2004) as follows.

- Many types and sources of organic amendments consistently generate suppression (e.g., hardwood and conifer bark, oat straw–chicken manure, vermicomposted cow manure).
- Suppression is generated immediately after high-rate organic amendment unless the organic substrate is raw, in which case suppression takes longer to develop.
- Suppression is of fairly short duration (typically weeks to 1 year).
- Suppression is positively related to microbial activity (specifically hydrolysis of fluorescein diacetate, a measure of activity of soil microbes; Boehm and Hoitink, 1992). As microbial activity declines, the *Pythium* population increases.

The characteristics of general suppression of damping-off caused by *Pythium* species and of diseases caused by *Phytophthora* species are essentially the same. The main difference between *Pythium* and *Phytophthora* species is that populations of the former do not decline, while hyphae and propagules of the latter are degraded by soil microbes, with the result that populations decline. The nursery and container plant industry takes advantage of general suppression to control damping-off and wilts caused by *Pythium* and *Phytophthora* species through the use of planting mixes that incorporate composted organic materials. The composts reestablish the diversity of beneficial microorganisms. Craft and Nelson (1996) discussed the microbial properties of composts that suppress *Pythium*-induced damping-off and root rot of creeping bentgrass. The use of composts in greenhouse and container cropping systems can eliminate or at least reduce the need for soil sterilization, including sterilization by methyl bromide, and for fungicides to control common soilborne pathogens (De Ceuster and Hoitink, 1999).

Hardwood composts are generally considered to be more effective in disease suppression than are composts made from conifers. The process of making hardwood compost was reviewed by Harman and Björkman (1998), Hoitink (1980), and Hoitink et al. (1991). The formation of compost under controlled conditions consists of three phases. (1) In the first few days, easily degraded materials decompose. (2) Next is a thermophilic phase that can last as long as 3 months, during which temperatures reach 40–80°C and cellulose is degraded and most organisms, including plant pathogens and beneficial organisms, are killed. This thermophilic phase proceeds more quickly when the cellulose content of the wood is higher, and hardwoods have a higher cellulose content than conifers. (3) Finally, there is a stabilization period, during which temperatures decline and the compost is recolonized by mesophilic organisms. During this third stage, a free film of water must be present on the organic material in the compost (≥40%, w/w) to allow bacteria to recolonize (Hoitink et al., 1991), some of which are beneficial to plants either through their control of pathogens, their influence on plant health, or both. During the final phase, recolonization by beneficial microorganisms is random, but composts produced in the open near a forest are colonized by more suppressive fungi, such as *Trichoderma* species, than are those produced at a distance from a forest or in enclosed chambers (Hoitink et al., 1991; Kuter et al., 1983). *Trichoderma hamatum* and *T. harzianum* are the predominant fungal parasites isolated from lignocellulosic wastes (Zinati, 2005). Although hardwood is usually more efficient in disease suppression than is conifer wood, mixes containing ≥20% composted pine bark support suppression of Pythium damping-off (Hoitink et al., 1991).

The use of composts in container systems is very well documented and is an economically successful strategy of crop management. Composts and fortified composts are particularly suited for use in greenhouse applications, where conditions are more easily defined and controlled than in the field. Rather early in the development of container media, it was realized that peat mixes were not very effective in suppressing root diseases caused by *Pythium* and *Phytophthora* species and were not at all effective in controlling foliar diseases. But composted bark of various tree species was a good substitute for peat in container mixes, and "aged" or composted bark media were able to naturally suppress Pythium and Phytophthora root rots, even though fungicides were still required for foliar pathogens (Hoitink et al., 2006). However, the inoculation of container media with specific biocontrol agents, such as *T. hamatum* T382, can induce resistance to root and foliar diseases over the natural suppression inherent to the medium (Hoitink et al., 2006). In nursery and container cropping, hardwood bark, the most commonly composted material, is mixed in a ratio of approximately one part "light" sphagnum peat to four parts composted bark

(Hoitink, 1980). "Light" peat is taken from the top of the peat bog. It is less composted than peat taken from deeper in the bog and is suppressive to damping-off caused by *Pythium* species, whereas the more heavily composted peat—taken from deeper in the peat bog—is conducive to disease (Boehm and Hoitink, 1992; Stone et al., 2004). To control Fusarium wilt (*Fusarium oxysporum* f. sp. *melonis*) of muskmelon (*Cucumis melo*) in the greenhouse, Martínez-Medina et al. (2009) prepared an inoculum of *T. harzianum* (strain CECT 20714) by adding 10 mL of a suspension containing 10^8 conidia per milliliter to a sterilized mixture containing 20 g of oat, 50 mL of bentonite (a type of clay), 100 mL of vermiculite, and 60 mL of water. This was incubated at 28°C for 8 days to reach a population of 10^8 conidia per gram. The inoculated mixture was then mixed with commercial peat, reaching a population density of 10^6 conidia per gram of peat. Melon seeds were planted into this mixture in plastic cells. Compared with adding conidia to water, conidial numbers in the vermiculite–bentonite mixture increased and remained viable for far longer. Moreover, Fusarium wilt was suppressed and plant fresh and dry weights increased.

There has been far less work done on the use of amendments with organic matter to suppress disease in field soil. However, it might be possible to restore the natural balance of organisms in the soil by applying organic residues (crop residues, cover crops, organic wastes). An extensive amount of literature discussing this was reviewed by Stone et al. (2004). One large study compared treatments in potato fields in Maine that were managed organically with those that were managed by conventional methods (Bernard et al., 2012). The treatments included the addition of compost, rotation with rapeseed (*Brassica napa*), and the addition of biocontrol agents, including *Bacillus subtilis*, *T. virens* (SoilGard; Certis, Columbia, MD, U.S.A.), and a hypovirulent strain of *Rhizoctonia solani*. While the study did not evaluate the effects of the treatments on crop production, some treatments increased the soil microbiota, thereby enhancing the potential for disease suppression. Individual treatments had different effects. For example, rotation with rapeseed increased the fatty acid methyl ester (FAME) profiles indicative of gram-negative bacteria, and the addition of compost increased the FAME profiles indicative of gram-positive bacteria and fungi. The combined treatments were additive. Of interest is that the amendment with *T. virens* was directly responsible for increased population numbers of fungi in general and *T. virens* in particular. The increases in microbial populations were more marked in the conventionally managed site than they were at the organic site, presumably because of a lesser initial microbial component in the conventionally managed site. This emphasizes the importance of the history of a site when planning a crop management program. We have no examples of successfully augmenting field soils with microorganisms to achieve disease control. However, Leandro et al. (2007) amended field soil with organic compost that had been fortified with *T. hamatum* T382, but they did not attribute any biological effect on the strawberries planted into the soil to the T382 fortification, and fewer than 10% of the sampled roots were infected with the strain. Nonetheless, the fact that the population of that particular strain was maintained at 10^3 CFU per gram of soil indicates that augmenting field soil with organic compost inoculated with a *Trichoderma* strain, and maintaining that strain in the rhizosphere of a host, is a distinct possibility. Lewis and Papavizas (1984) found that adding young mycelium of either of two *Trichoderma* strains (identified by the authors as *T. viride* T-1-R4, but which is probably actually *T. asperelloides*, and *T. harzianum* WT-6-24) in a sterile sand–bran–water (1:2:1, w/w/v) mixture was associated with proliferation of the microbes in the soil, as measured by an increase in the numbers of CFU of the strains in soil over 3 weeks. The numbers were sustained for up to 36 weeks. However, the addition of conidia into the mixture did not support an increase in *Trichoderma* populations in the soil. Age of the inoculated mycelial inoculum was critical. CFU increased over incubations of 1–3 weeks, but although composts incubated for nearly 6 weeks had abundant conidia, soil CFU decreased after its application. Lewis and Papavizas concluded that young hyphae were able to proliferate in the soil while conidia failed to germinate.

Paulitz and Bélanger (2001) reviewed several biocontrol agents, including commercially available *Trichoderma* species and strains that have been used in greenhouse systems. Planting media inoculated with one strain, *T. hamatum* T382, can induce systemic resistance to stramenopile and ascomycete pathogens in a wide range of crops and reduce damage from foliar (*Phytophthora*) and dieback (*Botryosphaeria*) pathogens (Hoitink et al., 2006). There is a note of caution in the fortification of composts by single microbes: when Khan et al. (2004) added *T. hamatum* T382 to a heated composted cow manure mix, the heating caused a loss of suppression of Phytophthora damping-off, and the postheating addition of the *Trichoderma* culture resulted in the partial restoration of suppression but not to the same high level observed for naturally colonized compost. This means that suppression is a complicated matter involving the interaction of the entire microbial community of the rhizosphere and the suppression caused by the community cannot be attributed to, or be duplicated by, a single microbe. Zinati (2005) listed seven methods, alone or in combination, to intensify the activity of antagonists already in a soil: (1) crop rotation; (2) adding amendments to stimulate antagonists; (3) altering soil pH to favor the antagonist, inhibit the pathogen, or both; (4) employing tillage methods that modify soil structure and aeration; (5) selection of planting date to reduce disease incidence; (6) applying organic amendments in a way that reduces the level of available nitrogen at the infection site; and (7) managing irrigation to maintain the soil water potential favorable to antagonists at the infection sites (moist for bacteria, drier for actinomycetes) and still ensure that water is available for the plant.

While control of *Pythium* and *Phytophthora* pathogens can be achieved by the general suppressiveness of most composts, control of *Rhizoctonia solani* and *Fusarium* species requires a more specific approach. Aggressive strains of *Rhizoctonia solani* are notoriously difficult to control for several reasons (reviewed in Papavizas, 1970; summarized in Stone et al., 2004). These include a wide host range, large sclerotia that are insensitive to fungistasis and resistant to decomposition, the rapid colonization of fresh organic matter, extensive mycelial growth, mycelium that is relatively resistant to degradation, hyperparasitic potential, and the capacity to escape soil competition under humid conditions by growing on surface organic matter or by aerial plant-to-plant spread. Stone et al. (2004) summarized the extensive literature that should be considered for the management of crops for diseases caused by *Rhizoctonia solani*. While some soil amendments suppress this pathogen, most are conducive, that is, they favor development of the disease. Because of the tremendous variability of *Rhizoctonia solani*, Stone et al. (2004) did not recommend attempts to control it using organic amendment of the soil under field conditions. Although some suppression of *Rhizoctonia solani* may be achieved through organic amendments in container-grown systems, it is not at a commercially viable level (Stone et al., 2004). Chet and Baker (1980) induced suppressiveness in soil heavily infested by *Rhizoctonia solani* by adding an aqueous suspension containing 10^6 conidia per gram of a *T. harzianum* strain and replanting with radish, alfalfa, and sugar beet at weekly intervals, but suppression of the disease depended on the acidification of the soil. Kuter et al. (1983) associated suppression of *Rhizoctonia solani* by media in containers amended with hardwood bark that had been composted in the field, in windrows, and in closed bins with high numbers of three *Trichoderma* species, *T. harzianum*, *T. hamatum*, and *T. virens*, which were the taxa most commonly isolated from the mix. However, they could not attribute the suppression to a single species. The incidence of wilt disease in muskmelon (*Cucumis melo*) in the greenhouse caused by *Fusarium oxysporum* f. sp. *melonis* was reduced when it was grown in composted vineyard prunings that had been fortified with a *Trichoderma* strain (Bernal-Vicente et al., 2012). To make the compost, Bernal-Vicente et al. (2012) formed piles (1 m^3) of vineyard wastes in the open air; the moisture content of the piles was maintained at 40–50% by watering, and the composting process lasted 120 days. The piles were inoculated with suspensions containing 10^5 CFU per gram of a *T. harzianum* strain that had been isolated from diseased muskmelon at the beginning of the composting process and after 60 days, at the beginning of the maturation process. In greenhouse trials, they planted muskmelon seedlings into the mature compost resulting from the two different treatments, and as a third treatment, they inoculated untreated compost with the *T. harzianum* culture at the time of planting muskmelon seedlings. They found that the incidence of *Fusarium* species and severity of disease were reduced in all treatments, but the greatest effect was seen when the *T. harzianum* culture was inoculated as the compost entered its maturation stage. Interestingly, when the *T. harzianum* culture was inoculated at the time of planting, the beneficial effect was lower than in the other two treatments. The authors concluded that the number of CFU of the *T. harzianum* strain was higher in this treatment, which could have resulted in increased competition for nutrients between the melon and the *T. harzianum* strain. Thus, the number of CFU was important.

Container media amended with composts made from municipal wastes is conducive to *Rhizoctonia solani* because the composting process needed to kill fecal pathogens also kills all the beneficial microorganisms. Those composts may need to be allowed to stabilize for several months before they can be colonized by beneficial fungi and bacteria (Hoitink et al., 1991). A commercial product, Metrocompost, made from vegetable and animal wastes, sewage sludge and yard wastes by a proprietary process (Castelldefels, Barcelona, Spain), contains naturally occurring strains of *T. asperellum* that are suppressive to Fusarium wilt. Among the microorganisms directly isolated from it (as opposed to taxa revealed by Illumina sequencing, which tends to show a greater diversity) were five strains of *T. asperellum* (Cotxarrera et al., 2002). The sterilized compost was not suppressive, but suppressiveness was renewed by the addition of one of the *Trichoderma* strains (CECT 20417). Thus, this strain of *T. asperellum* was viewed as the principle agent of suppression of Fusarium wilt of tomato in this particular compost.

The mechanisms of suppression in composts include mycoparasitism, competition for nutrients and niches, antibiosis, and systemic acquired resistance (SAR) (Hoitink et al., 1997a,b, 2006; Yogev et al., 2010). Composts are a source of organic matter that, when added to a soil, stimulates the growth of microbes and the competition between beneficial and pathogenic microorganisms. This can reduce the level of nutrients and number of infection sites available to pathogens, which often are less competitive (Bernard et al., 2012). Interestingly, a compost made of tomato residues and the coarse fractions of cow manure and then fortified with *T. asperellum* T 203 not only suppressed Fusarium wilt in melon (*Cucumis melo*) but also suppressed Botrytis foliar blight in melon and cucumber plants (Yogev et al., 2010). All the evidence pointed to induced systemic resistance by this well-known strain of *T. asperellum* (e.g., Shoresh et al., 2005) as the cause. Similarly, fortification of a composted cow manure mix with a strain of *T. hamatum* (T382) that is known to enhance the suppression of soilborne pathogens in container mixes reduced the severity of Phytophthora leaf blight in cucumber (*Cucumis sativus*) through the induction of systemic resistance (Khan et al., 2004).

Perhaps not surprisingly, composts fortified with *Trichoderma* species can also positively influence plant health. In one greenhouse study (MacKenzie et al., 1995), the addition of a strain of *T. harzianum* (T-12) to a peat–perlite (1:1, v/v) soilless mix increased root and shoot fresh weight in chrysanthemum cultivars that were considered easy or difficult to root during 21 days of rooting. When cuttings were then placed into soilless mix that had been treated with the *T. harzianum* strain, increases in height, shoot dry weight, and root fresh and dry weights of the hard-to-root cultivar were significantly increased over the control. In this study, the *Trichoderma* inoculum was prepared by growing on a peat–bran mixture for 2 weeks, grinding to a powder, and thoroughly mixing with the peat–perlite medium. The concentration of *Trichoderma* conidia ranged from 10^5 to 10^6 conidia per gram of heat-dried medium, and conidia were applied at a rate of 5 or 25 g per kilogram of medium. Most of the cultivars responded equally to the two applications, but the hard-to-root cultivar responded to the lower rate. In another example, an increase in growth and yield of Chinese radish (*Raphanus sativus* var. *longipinnatus*) was obtained in pot experiments when the planting media, containing either sand and organic fertilizer (10:1, v/v) (Phuwiwat and Soytong, 1999) or sand, coconut coir, and organic compost (1:1:1, v/v/v) (Phuwiwat et al., 2001), were amended with aqueous solutions (20 mL per pot) of conidia (53×10^8 and 53×10^9 conidia per milliliter, respectively) of a strain of *T. harzianum*. The strain, *T. harzianum* T-22, when added at a rate of 4 g per liter to a mixture of peat (60%) and compost (40%) that was heavily infested with *Rhizoctonia solani*, suppressed damping-off and increased the dry weight of bean in the greenhouse. However, other commercially available *Trichoderma* products that were based on *T. asperellum* and *T. gamsii* had no effect on damping-off (Puglese et al., 2011). This example is a demonstration that biological activity is strongly a function of the strain rather than of the species.

The Use of *Trichoderma* Species in Soilless and Hydroponic Systems

In hydroponic and soilless systems, plants are either cultivated in water, without soil, and minerals and other nutrients are supplied in aqueous solutions or cultivated in inert material, such as perlite, sand, rockwool, sawdust, coir, or peat, and the nutrients are supplied in solution (reviewed by Douglas, 1985). The systems can be open ("run to waste"), wherein the added water–nutrient solution is not recycled, or closed and the water–nutrient solution is recirculated. Soilless mixes are commonly used for the production of flower, foliage, bedding, and vegetable crops. Since the conditions can be exactly controlled and can be tailored to a specific crop, this is popular in greenhouse cropping for a wide variety of plants (Paulitz, 1997). The closed system is ideal in situations in which the availability of water is a limiting factor, and hydroponic production is particularly well suited to tropical or desert regions where temperatures are too high for field cropping (Jensen, 2002). However, soilless substrates are essentially sterile and thus lack the complement of beneficial microorganisms, including *Trichoderma* species, and bacteria that can influence plant growth and suppress pathogens. A risk of using hydroponic systems is that nutrients are supplied in water, and of course, pathogens such as *Pythium* species thrive in aquatic systems. Pathogens can be introduced with planting material, tools, and the recycling of water and nutrient solutions, and in the absence of any antagonism from the planting substrate, they do very well (Gravel et al., 2006). Damping-off and root and crown rots are serious drawbacks in greenhouse cropping. Since few chemical pesticides are approved for greenhouse use, methyl bromide has been used to sterilize soil and mixes, but because of its effect on the environment, the industrialized countries phased out its use in 2005. Thus, the possibility of replacing methyl bromide in crop management by manipulating the beneficial microorganisms present in the nutrient solutions is attractive (van Os et al., 2004). Paulitz (1997) reviewed the biological control of root pathogens in soilless and hydroponic systems but *Trichoderma* species were not mentioned, although rhizobacteria such as *Pseudomonas* and *Bacillus* species are able to control species of *Fusarium*, *Pythium*, and other genera. The ability of any *Trichoderma* species to have an effect in a soilless system no doubt depends upon the strain and the crop. van Os et al. (2004) did not observe any effect against *Pythium aphanidermatum* or *Phytophthora cryptogea* when they added *T. harzianum* T50 (Vitalin; Vitalin Pflanzengesundheit GmbH, Ober-Ramstadt, Germany) to cucumber and tomato, but there was a remarkable reduction in disease incidence when this strain was used with gerbera. *Trichoderma* species can be established in seedlings through seed treatment before they are planted into a hydroponic system, and in a few publications, conidia of *Trichoderma* species have been added to the soilless planting substrate, demonstrating a potential for the use of *Trichoderma* species in hydroponic systems. Suppression of tomato root rot caused by *Pythium ultimum* was combined with increased root development and yield when 200 mL of an aqueous solution containing 10^6 conidia of a culture of *T. atroviride* was applied to three 10-cm-tall plants that had been transplanted to rockwool slabs in the greenhouse or to two 10-cm-tall plants that had been transplanted to an organic medium consisting of peat, pine sawdust, and compost (60:30:10) in 9-L plastic containers. One week later, the plants were inoculated with a solution containing 10^6 spores of *Pythium ultimum*. The plants were grown under greenhouse conditions and drip irrigated with nitrogen, phosphorus, and potassium (N-P-K, 6:11:31) over a period of 11 weeks (Gravel et al., 2006). Gravel et al. applied cultures of bacteria and fungi, including *T. atroviride*, which had been isolated from organic composts and had earlier been shown to be antagonistic to *Pythium ultimum* in in vitro assays. Several of the bacteria and fungi, including *T. atroviride*, suppressed the *Pythium ultimum* and promoted growth to varying degrees. Ozbay et al. (2004) evaluated commercially available strains of *T. harzianum* (T-22, T-95) to control crown and root rot of tomato caused by *Fusarium oxysporum* f. sp. *radicis-lycopersici*. The antagonists were applied at two inoculation densities (10^6 and 10^7 conidia per milliliter) by soaking rockwool or coir blocks (7.5 × 18 × 91, encased in opaque plastic film) in the conidial solution either at the time of sowing or at the time of transplanting when seedlings were 5 weeks old. Three seedlings were planted in each block. *Fusarium oxysporum* f. sp. *radicis-lycopersici* was added as a conidial suspension (10^6 conidia per milliliter) 2 weeks after transplanting. Treatment with both antagonists resulted in reduced disease incidence (73–79% over control) and increased yield (25–37% over control). There was no difference in yield between coir and rockwool, leading the authors to suggest coir as a replacement for the commonly used rockwool. Although the authors attributed the increased yield to disease suppression, T-22 is known to positively influence yields as it interacts with the plant, suggesting the possibility that more than disease suppression was responsible for the increased yield. van Os et al. (2004) explored the optimization of microbes in nutrient solutions as a means to replace methyl bromide. When they added conidia of *T. harzianum* T50 (Vitalin) to gerbera grown in a closed hydroponic system, where the water and nutrients are recycled, conidial numbers increased and the incidence of disease caused by *Pythium aphanidermatum* and *Phytophthora cryptogea* was reduced by 90%. Although this is a preliminary result, the dramatic reduction in disease caused by these waterborne pathogens was an optimistic sign that, at least in some hydroponic systems, the introduction of an adapted *Trichoderma* strain can provide an alternative to methyl bromide fumigation of soil in the greenhouse.

Seed Treatments

Treating seeds with *Trichoderma* species before planting, either through seed dressing or seed coating, can have three effects, all of which may occur simultaneously. These effects are protection, rhizosphere sanitation, and stimulation of germination. There are several reports of the beneficial effects of *Trichoderma*-treated seeds, and seeds pretreated with *Trichoderma* species are available for some crops. Some of those reports are summarized below, along with the protocols followed for coating the seeds. Seeds coated with *Trichoderma* species can be protected against pathogens, giving the plant an opportunity to develop before it can be infected by a soilborne pathogen. The *Trichoderma* species that coat the seeds can also provide a longer term beneficial "cleaning" of the rhizosphere by increasing in abundance in the soil because they are able to grow from the surface of the seeds. More recently, certain *Trichoderma* strains have been shown to increase the percentage of germinating seeds and the performance of the seedlings. A critical first element is to use strains known to be rhizosphere competent; they must be able to colonize the seed surface and survive in the rhizosphere. A second consideration is the size of the seeds. In order for *Trichoderma* species to colonize the roots and initiate the *Trichoderma*–plant interaction at the molecular level, the *Trichoderma* species need to have an initial nutrient base, and those nutrients are leached from the seeds. Crops that have small seeds, such as onion and carrot, may not be able to provide the needed nutrients. This may have contributed to the equivocal results reported by Bennett et al. (2009) for seed treatments following an extensive series of studies that compared the seeds of onion and carrot that had been primed with *T. harzianum* T-22, *Clonostachys rosea*, and *Pseudomonas chlororaphis* with and without pesticides in the greenhouse and with different soil types. The fact is that commercial-scale success in the growth promotion of several crops following seed treatment with *Trichoderma* species, alone or combined with rhizobia, has been achieved once the "correct" cultures have been obtained (i.e., rhizosphere competent and able to interact with the genome of the host plant). In the case of seed treatment, as with all applications of beneficial microorganisms, finding the "right" strain, or combination of strains, is a key to success.

Applying *Trichoderma* Conidia to Seeds

The simplest method for applying *Trichoderma* conidia to seeds is to immerse the seeds in dry masses of conidia or in liquid suspensions of conidia. In either case, the seeds and the masses of conidia can be mixed in a pot and, in the case of liquid conidial suspensions, spread on a plastic sheet to dry. To coat seeds, a binder or sticker is often used; this process may require special equipment. In seed pelleting, inert materials are added to the seeds to change their shape and increase their bulk. This is the most expensive method of seed treatment because of the requirement for equipment and additional pelletizing materials, but use of this technique results in increased seed bulk, which can facilitate the accurate placement of small seeds such as onion and carrot. The Plant Protection Information Network (PPIN) gives recommendations for the rates of application of *Trichoderma* conidia to "rabi" crops (crops that are planted in the winter and harvested in the spring) (http://ppqs.gov.in/PDF/Seed Treatment Rabi123.htm). The seeds are mixed with conidia in pots or polythene bags. The list includes sugarcane (root rot, wilt; 4–6 g of conidia per kilogram of seeds), rice (5–10 g of conidia per kilogram of seeds), chilies (anthracnose, damping-off; 4 g of conidia per kilogram of seeds), pigeon pea (wilt, blight, root rot; 4 g of conidia per kilogram of seeds), tomato (soilborne pathogens, early blight, damping-off; 2 g of conidia per kilogram of seeds + captan 75% water soluble [w.s.] at 1.52 g of active ingredient per liter), coriander (wilt; 4 g of conidia per kilogram of seeds), leguminous vegetables (soilborne pathogens; 2 g of conidia per 100 g of seeds), sunflower (seed rot; 6 g of conidia per kilogram of seeds), crucifers (damping-off; 2 g of conidia per kilogram of seeds), and gram (chickpea) (wilt, damping-off; 1% wettable powder [w.p.] at 9 g of conidia per kilogram of seeds). Sivan et al. (1984) coated seeds with *T. harzianum* to protect against damping-off induced by *Pythium aphanidermatum*, *Rhizoctonia solani*, and *Sclerotinia sclerotiorum* in pea, cucumber, tomato, *Capsicum* species, and *Gypsophila* species. Conidia were prepared in 200 mL of potato dextrose broth in 500-mL Erlenmeyer flasks. The conidia were separated by adding 100 mL of 0.001% Tween 20 solution in sterile tap water and 10 3-mm-diameter glass beads to each Erlenmeyer flask and then shaking the flasks on a rotary shaker at 400 rpm for 10 min. Conidial suspensions were adjusted to 5×10^9 conidia per milliliter and supplemented with 0.2 g of the polysaccharide Pelgel (Nitragin Sales Corp., Milwaukee, WI, U.S.A.). Three milliliters of this suspension were used to coat 150 g of pea, 50 g of cucumber, or 10 g of tomato seeds and immediately dried by warm ventilation.

To coat seeds of pea (*Pisum sativum*), snap beans (*Phaseolus vulgaris*), and cucumber (*Cucumis sativus*), Hadar et al. (1984) prepared a conidial suspension containing 10^8 conidia per milliliter by scraping conidia from 9- to 14-day-old potato dextrose agar (PDA) petri dish cultures. Aqueous conidial suspensions contained various "stickers" to keep the conidia attached to the seeds (2% Methocel A4C Premium, Dow Chemical Company, Midland, MI, U.S.A.; 20% w/v polyvinyl alcohol; or 2% Polytran N, also known as Actigum CS6, Cargill, Inc., Minneapolis, MN, U.S.A.). Aliquots (0.4 mL) of conidial suspensions were added to 100 bean or pea seeds, while 0.3 mL was applied to 100 cucumber seeds. These mixtures were shaken in Erlenmeyer flasks to distribute the conidia over the surface of the seeds. After shaking the coated seeds in water and plating them on PDA, 10^4–10^5 *Trichoderma* propagules were recovered. These seeds were planted in the field in soil infested with *Pythium* species. Lifshitz et al. (1986) treated pea seeds with an aqueous suspension that contained 10% of the sticker Pelgel and 3×10^8 conidia of proven biocontrol strains of two *Trichoderma* species. Aliquots (0.75 mL) of the conidial suspension were added to 50 g of pea seeds three times 24 h before sowing in raw soil in plastic pots in the greenhouse. The incidence of preemergence damping-off of pea caused by *Pythium* species was reduced by the application of *Trichoderma* conidia to the pea seeds. Nelson et al. (1988) found that adding organic acids, such as carboxylic acid, at a rate of 1 mg/mL to an aqueous conidial suspension (approximately 10^8 conidia per milliliter) and 2% (w/v) of the sticker Methocel A4C Premium increased the biocontrol activity of a strain of *T. harzianum* but not *T. koningii*.

Most of the reports of seeds treated with *Trichoderma* cultures refer to laboratory experiments and not field trials. In the limited field trials performed by Hadar et al. (1984) with pea, snap beans, and cucumber, seeds treated with *T. koningii* T-8 performed better than the controls but not as well as seeds treated with captan. Dubey and Singh (2013) adopted an integrated approach to control diseases of mung bean (*Vigna radiata*), including wet root rot (*Rhizoctonia solani*), Cercospora leaf spot (*Cercospora canescens* and *Pseudocercospora cruenta*), and yellow mosaic (transmitted by the whitefly *Bemisia tabaci*), and combined coated seeds and foliar sprays of *T. virens* (as the product Pusa 5 SD [Indian Council of Agricultural Research, India], with *T. virens* at 10^8 CFU per gram) in combination with the neonicotinoid insecticide thiamethoxam and/or the fungicide carbendazim. Field trials were carried out over 2 years. Compared with the untreated controls, combined treatments of foliar sprays and seed dressing with the insecticide, fungicide, and *T. virens* resulted in higher germination, better developed seedlings, and a lower incidence of wet root rot, Cercospora leaf spot, and yellow mosaic. The seed treatment consisted of thoroughly mixing a known quantity of seeds contained in polythene bags, seed dressing drums, or pots with *Trichoderma* conidia at a rate of 10^8 conidia per gram of seeds. Hohmann et al. (2011) found that coating seeds of *Pinus radiata* with *T. hamatum* LU592 (4×10^5 conidia per seed, 10^7 conidia per gram of seeds, applied to the seed surface using 1% methyl cellulose) improved seedling growth and plant health in nursery situations. These authors also applied conidia directly to the drill hole at planting in one series of experiments, but there was no difference in biological effect between adding conidia to the soil at planting and coating the seeds. Thus, rhizosphere competence was a critical factor in the biological effect of LU592 in pine seedlings. This rhizosphere-competent strain of *T. hamatum* colonized the rhizosphere of the pine. In contrast, the application of a culture of *T. atroviride* that was a poor root colonizer did not provide the growth-promoting effects. Both methods of inoculation with LU592, spray and seed coat, were suitable for container-grown *Pinus radiata* seedlings. Mastouri et al. (2010) reported that *T. harzianum* T-22 could improve the performance of tomato seeds under biotic (*Pythium* species) and abiotic (poor quality seeds, salinity, heat, osmotic) stress conditions, but the effect was not seen in the absence of stress. They suggested that the *Trichoderma* species induced physiological responses in the plant that protected them against oxidative damage. To coat tomato seeds, conidia of *T. harzianum* T-22 were applied with cellulose power and then encapsulated with tapioca dextran (Crystal-Tex 627; Ingredion Incorporated, Westchester, IL, U.S.A.), which provides wettability to the conidium surface, and then they were suspended in ultra-pure water (Mastouri et al., 2010, 2012). Seeds were treated with the conidial suspension at a rate of 2 μL per gram to deposit 2×10^7 CFU per gram of seeds. Treated seeds were used for planting without air-drying.

In field trials in soil naturally infected with *Rhizoctonia solani*, cotton seeds were treated with *T. harzianum* and *T. hamatum* in a Pelgel adhesive solution (10%) containing 5×10^8 conidia per milliliter. In greenhouse trials, cotton plants issuing from the seeds treated with either *Trichoderma* species were resistant to *Rhizoctonia solani*, but *T. hamatum* was more efficient at low temperatures (approximately 20°C), whereas *T. harzianum* was more efficient at higher temperatures (approximately 27°C). In field trials, both species significantly reduced damping-off from *Rhizoctonia solani* and increased plant stand. The incidence of damping-off was greater when seeds were planted to a depth of 3 cm than when planted at 1 cm. The differences in species response to temperature and the differences in protection in relation to planting depth highlight the importance of considering conditions when applying a biocontrol agent (Elad et al., 1982).

Potato is not grown from seeds, per se, but rather from seed pieces, that is, pieces of tubers that have buds. Tubers are collected during the previous season and stored whole. The potato seeds are cut sometime before planting and stored until planting. Seeds infected with late blight (*Phytophthora infestans*), dry rot (*Fusarium sambucinum*), and stem canker and black scurf (*Rhizoctonia solani*) can be transferred from the seeds to sprouts, stems, and foliage. This infection can result in the loss of production or even epidemics. Treatment with a fungicide can increase the emergence of the sprouts and reduce the incidence of disease caused by these pathogens. For the control of late blight and dry rot, seeds treated with *T. harzianum* T-22 (as T22 Planter Box, BioWorks, Inc., Victor, NY, U.S.A.) gave results equivalent to the control given by a fungicide (fludioxonil 5 g per kilogram + mancozeb 96 g per kilogram) in decreasing seed decay and disease incidence when the seeds were stored under optimal conditions (Wharton and Kirk, 2014; Wharton et al., 2012). To treat the seeds, tubers that had been stored at 3°C since the preceding year were gradually brought up to 12°C over 12 days. The tubers were then cut in half lengthwise, ensuring that each half had viable sprouts, and then they were either not inoculated, inoculated with either a virulent strain of *Phytophthora infestans* or of *Fusarium sambucinum*, or inoculated with the pathogen and one of three treatments: a commercial seed protectant (Maxim MZ; Syngenta Crop Protection, Greensboro, NC, U.S.A.), *Bacillus subtilis* (Serenade Max; Bayer CropScience Canada, Calgary, Manitoba, Canada), or *T. harzianum* T-22 at a rate of 125 g per 100 kg of potato seeds. An uninoculated control was sprayed with water. The cut and treated tubers were stored according to guidelines provided in a Michigan State University extension bulletin (Wharton and Kirk, 2007). They were placed in well-ventilated boxes in controlled environment chambers and held at 8°C for 2 days. Over the next 10 days, the temperature was increased every 2 days by 2°C intervals up to 18°C in a 14-h/10-h photoperiod and 95% relative humidity with forced air ventilation at 5,950 L per minute and then held at 18°C until planting. A key element in the success of the *Trichoderma* in preventing late blight and dry rot is storing the potatoes under proper conditions. When tubers were stored under less than optimal conditions, *T. harzianum* T-22 did not afford any protection. However, under less than optimal conditions, performance of even the fungicide was diminished. When proper storage conditions are observed, control of late blight and dry rot seed treatment with *T. harzianum* T-22 can be as successful as applications with fungicide; moreover, in recent years, resistance in *Fusarium sambucinum* to fludioxonil + mancozeb has developed, making the use of biological control even more attractive. Like *Phytophthora infestans* and *Fusarium solani*, *Rhizoctonia solani* can initiate a complex of diseases (stem canker, black scurf) from infected potato seeds. Briefly dipping tubers in a 1% solution of Trianum-G (*T. harzianum* T-22; Koppert B.V., Berkel en Rodenrijs, The Netherlands) containing 1.5×10^8 conidia per gram of preparation did not control *Rhizoctonia solani* (Wilson et al., 2008). However, when seeds were treated with an aqueous solution of flutolanil (673 mg of active ingredient per liter) and then planted in rows to which the *T. harzianum* T-22 was added (at a rate of 50 g per meter), damage to the sprouts and the severity of stem canker was reduced over 30 days. When *T. harzianum* T-22 was applied to furrows, alone or in combination with flutolanil-treated seeds, the yield of marketable-sized tubers increased and the incidence of black scurf on progeny tubers decreased, a result not seen when the fungicide was used alone. The explanation for the longer term positive effect of the *T. harzianum* T-22 was explained by the possibility of its having colonized the developing potato plants, thereby inducing the cascade of protective events described above. However, dipping the potato seeds in a solution of *T. harzianum* T-22 was, by itself, ineffective in the control of *Rhizoctonia solani* (Wilson et al., 2008), whereas treating the seeds with a powdered preparation of *T. harzianum* T-22 (but in a different product) (Wharton and Kirk, 2014; Wharton et al., 2012) did control *Fusarium solani* and *Phytophthora infestans*.

Topical Applications

Most often, *Trichoderma* species are applied at the root level in soil, in soilless mixes and composts, or hydroponically. Reports of topical applications to control stem and leaf pathogens are less common in the literature, but many or most of the available *Trichoderma*-based products can be formulated in water and applied as sprays to aboveground plant parts or to soil. Usually, conidial concentrations of 10^7–10^8 are used. Difficulties encountered when using cultures of *Trichoderma* species as a spray application include the survivability of the propagules in the harsh environment of leaves or fruit, the ability of the propagules to resist being washed from the host surface, and the likelihood of the biocide coming into contact with a fungicide and being rendered ineffective if it is used as part of an integrated crop management program. Mbarga et al. (2014) developed an oil-based formulation that was composed of locally available 74% soybean oil, 15% Tensiofix NTM (an emulsifying/dispersing agent; S.A. Ajinomoto OmniChem N.V., Louvain-la-Neuve, Belgium), 5% Tensiofix 869 (a bentonite clay structural agent), 4% glucose, and 2.7×10^7 conidia of *T. asperellum* per milliliter that considerably extended the life of conidia in solution while providing protection of cacao pods against highly virulent *Phytophthora megakarya*. This formulation, when applied to cacao under experimental field conditions, improved persistence and time of protection over a formulation based on water and over the fungicide Kocide 2000 (DuPont Crop Protection, Wilmington, DE, U.S.A.). As always, control is highly dependent upon the culture or cultures of biocontrol agents used, the virulence of the pathogen, and the susceptibility of the host cultivar.

Formulations of *Trichoderma* species can be delivered quite effectively to target pathogens on plant surfaces or in the soil using sprayers, but in developing a spray program, many factors must be considered. Those factors include precisely defining the target (i.e., the site at which the efficiency of the biocontrol agent can be optimized); the ability of the spray nozzle to deliver the product to the site; the most efficient size of the spray droplets; and whether to apply in water alone or with adjuvants such as oil, a sticker, or a wetting agent. The International Pesticide Application Research Consortium (IPARC) has developed a website (www.dropdata.org/index.htm) to promote the selection and safe application of pesticides, including microbial agents. Essentially, microbial agents are applied as particles that must be delivered to the correct site in appropriate numbers. For this, the concentration in tank mixtures and drops, the numbers of colony-forming units (CFU) of the microbial agent in the tank and in the drop, and what happens to them once they are applied must be considered. Although a farmer may have little latitude to change parameters for spraying using existing equipment, in general, for aerial applications (on leaves, flowers, and vegetable and fruit surfaces), formulations having smaller droplet volumes that contain a minimum number of CFU are desirable because this ensures uniform spread of the inoculum, with each droplet containing at least one CFU (Bateman, 2004, 2014; Bateman et al., 2005). Bateman et al. (2005) described the process of spray application of *T. asperellum* to control cacao black pod (*Phytophthora megakarya*) in Cameroon. The formulation of *T. asperellum* that was developed by Mbarga et al. (2014), a dispersion of conidia in oil as opposed to the more usual water-in-oil type, or reverse emulsion, left conidia uniformly suspended and thus was better adapted to the side level backpack sprayers commonly used by small-scale farmers than to mist sprayers or foggers (Fig. 3.1).

The mechanism underlying the application of *Trichoderma* species directly to pathogens on plant surfaces is usually thought to be mycoparasitism, and intuitively one would think that direct application of a mycoparasite to the plant surfaces should control the disease. Of course, this is the principle behind aerial spray applications. However, inhibition of these aerial pathogens can sometimes be achieved by applications to the soil where interaction between *Trichoderma* species and host roots can lead to the induction of resistance to foliar pathogens in the absence of evidence for the colonization of aboveground plant parts (e.g., Alfano et al., 2007; Elad, 2000b; Hoitink et al., 2006; Horst et al., 2005). It is likely that both mycoparasitism and the induction of resistance play significant roles but to differing degrees, depending on the crop and pathogen at hand, and when developing a program, both mechanisms must be considered. Lee et al. (2006) used a strain of *T. harzianum* (YC459) as a foliar spray to control Botrytis blight on cucumber but achieved greater suppression of the pathogen when the *Trichoderma* conidia were applied as a soil drench in combination with a foliar spray. In a plastic tunnel experiment, the authors sprayed an aqueous solution containing a surfactant and 250 g of a wettable powder that contained 10^8 conidia per gram, and the protective effect on cucumber leaves lasted 1 week after the last application. Freeman et al. (2004) significantly reduced Botrytis gray mold and Colletotrichum anthracnose in strawberry with applications of aqueous solutions of *Trichoderma* strains containing 10^5–10^6 conidia per milliliter, alone or in combination, to runoff at 7- and 10-day intervals at a rate of about 10^7 conidia per plant. Lo et al. (1997) controlled turf diseases by applying *T. harzianum* directly to the soil in a granular form and then applying the same strain in a spray formulation to the cut leaves.

Figure 3.1. Application of *Trichoderma asperellum* to cacao pods using a side level backpack sprayer. (Courtesy D. Begoude)

The production of desiccation-tolerant conidia of *T. harzianum* in small-scale liquid cultures has been obtained through the use of osmoticants such as polyethylene glycol (PEG), 2% sucrose or glycerol (Jin and Custis, 2011; Jin et al., 1991, 1996). A formulation with a conidium concentration of approximately 10^8 per gram not only improves the ability of selected *Trichoderma* species to suppress a given pathogen but also reduces the amount needed to conduct bioassays and improves its eventual commercial application. Krauss et al. (2010) utilized a sticker (Break-Thru [Evonik Industries, Hopewell, VA, U.S.A.] or Emulgator NP-7 [Bayer SA, CropScience, São Paulo, SP, Brazil]) in the application of the systemic fungicide flutolanil and *T. ovalisporum* to control frosty pod rot of cacao and black pod of cacao (*Phytophthora palmivora*). Increases in the yields of cacao over 3 years were obtained when the application of *T. ovalisporum* or flutolanil, which is aimed at basidiomycetes such as *Moniliophthora* species, was switched to a broad-spectrum contact copper hydroxide fungicide in water (Kocide). The success of the endophytic *T. ovalisporum* in the control of frosty pod rot (*Moniliophthora roreri*) was attributed to the fact that the antagonist and this basidiomycete cacao pathogen are thought to have coevolved along with cacao in the upper Amazon region of South America (Fig. 3.2). Compared with the systemic flutolanil applied in a sticker, the behavior of *T. ovalisporum* was more erratic, being effective in the control of frosty pod rot in only two treatments, but in those two cases, *T. ovalisporum* outperformed the fungicide in controlling black pod caused by *Phytophthora palmivora* in absolute and relative yields of cacao pods most likely because the main efficacy of flutolanil is against basidiomycetes (Krauss et al., 2010). In the study cited above, Lo et al. (1997) found that adding the detergent Triton X-100 (Sigma-Aldrich Co., St. Louis, MO, U.S.A.) at 0.1% to the spray formulation of *T. harzianum* gave greater control of turfgrass diseases than when no detergent was used or when additives such as Pelgel or Tween 20 were added. The sprayed conidia persisted on the cut grass leaves and provided control for more than 1 month. Aqueous sprays containing single or combined biocontrol strains (conidial concentrations of $2.5–8 \times 10^7$) were tested for the control of anthracnose of strawberry (*Colletotrichum acutatum*) and strawberry gray mold (*Botrytis cinerea*). All combinations controlled anthracnose, the higher concentrations being more effective. Rather few single or combined biocontrol agents had any effect on gray mold, and in these cases, a lower concentration and timing of the applications were critical (Freeman et al., 2004). The lower concentration of conidia was also more effective in the control of tomato gray mold but not in the control of cucumber white mold (*Sclerotinia sclerotiorum*) and gray mold (Elad, 2000a).

Medeiros et al. (2010) describe a successful integrated management program for cacao that greatly reduced pod losses resulting from infection by *Moniliophthora perniciosa*, cause of the destructive witches'-broom disease of cacao (Fig. 3.2). This program combined important elements of farm sanitation and timed, alternating spray applications of the microbial control agent *T. stromaticum* and copper hydroxide fungicide. In this case, *T. stromaticum* is a specific mycoparasite of species in the mushroom genus *Moniliophthora* and it controls witches'-broom disease by killing the small mushrooms that are the fruit bodies of *Moniliophthora perniciosa*, thus reducing inoculum.

Figure 3.2. Diseases of cacao caused by *Moniliophthora* species. **A,** Frosty pod rot, caused by *Moniliophthora roreri*. **B–D,** Witches'-broom disease caused by *Moniliophthora perniciosa*. **B,** An infected and distorted pod (cherimoya) with *T. stromaticum* sporulating on it. **C,** The fruit body of the witches'-broom pathogen, *Moniliophthora perniciosa*, arising from a stem. **D,** A witches'-broom resulting from infection by *Moniliophthora perniciosa*. (Courtesy H. C. Evans)

Using *Trichoderma* Species to Enhance Plants Developed by Tissue Culture

Nematodes and soilborne pathogens are major constraints to the production of banana in countries where cooking bananas are a staple crop and dessert bananas are a source of income for many people. Because many pests are associated with planting material, it is essential that the planting material is clean. Tissue culture is used as a way to produce banana clones that are free of nematodes. The use of clean planting material of course has the benefit of giving the plantlets a head start before they are affected by one or more of these ubiquitous pests. However, there is a downside to this in that the plantlets are not only free of pests but they lack the source of nutrients that is provided by the corm, and they are also free of beneficial microorganisms that, we now know, enhance growth and root development and provide protection against pests. In short, these essentially sterile little plants are at a serious disadvantage in the world outside of a tissue culture plate. However, in Central America and East Africa, endophytic fungi isolated from banana tissue have been reintroduced into tissue-cultured plantlets in an effort to induce resistance to pathogens. Dubois et al. (2006) described the process of developing endophyte-enhanced, tissue-cultured banana in Kenya and Uganda through a public–private partnership. In Kenya, Waweru et al. (2013) described lower population densities of the root lesion nematode *Pratylenchus goodeyi* and reduced root necrosis in pot-grown tissue-culture–derived banana plantlets that they infected with endophytic, nonpathogenic *Fusarium oxysporum*, thus demonstrating the potential for using this technology. In order to establish beneficial microorganisms in the developing banana plants, endophytic fungi, including *Trichoderma* and *Fusarium* species, vesicular-arbuscular mycorrhiza (VAM) fungi, and bacteria isolated from the roots of banana have been introduced to the plantlets at the nursery stage. Sikora et al. (2008) reviewed the use of endophytic fungi in developing resistance to nematodes in banana. Kavoo-Mwangi et al. (2013) applied several biocontrol products, including mycorrhizae (*Glomus* species), bacteria (*Bacillus* species, *Paenibacillus* species), and *T. harzianum*, to banana plantlets that had been produced by tissue culture. When applied to the plants in natural planting soils, there was improved survival in the nursery and in the field but results depended greatly on soil type. Chaves et al. (2009) reported that combinations of endophytic *T. atroviride*, nonpathogenic *Fusarium oxysporum*, and *Bacillus* and *Pseudomonas* species inhibited root penetration by the nematode *Radopholus similis*. Inoculation with the biocontrol agents was achieved when the roots of the plants that had been developed in tissue culture were dipped into suspensions containing the biocontrol agents. Although combinations did not inhibit penetration of *Radopholus similis* any better than singly inoculated bacteria, the potential exists for an effect against other pathogens by the combined inoculation. zum Felde et al. (2006) reported a significant reduction in the number of root-knot nematodes in the field after banana plantlets developed by tissue culture were exposed to solutions of propagules of *T. atroviride* and nonpathogenic *Fusarium oxysporum*, both isolated from banana soil suppressive to plant-parasitic nematodes. Although the single inoculation with *Fusarium oxysporum* was more effective than the single inoculation with *T. atroviride*, combination inoculations that included two strains of *T. atroviride*, or two of *Fusarium oxysporum*, or all four were better than any of the single inoculations. The protective effect appears to have been a direct result of the interaction between the microbes and the plants through the induction of resistance (Sikora et al., 2008).

CHAPTER 4

Systematics of the Genus *Trichoderma*

Trichoderma is a genus of the ascomycete order Hypocreales, family Hypocreaceae (Ascomycota, Pezizomycotina, Sordariomycetes, Hypocreales, Hypocreaceae). Perithecia, where known, are formed within stromata. Sexual stromata (the teleomorph) are brightly colored, usually in a shade of yellow or brown, less commonly green, black, or red. Stromata of most species are discrete and 1 mm or less in diameter, others are erect and clavate, sometimes several centimeters tall, while others are broadly effused over the substrate and up to several centimeters in extent. Excellent illustrations of stromata of *Trichoderma* species can be seen in several publications, including Chaverri and Samuels (2003), Jaklitsch (2009, 2011), Jaklitsch and Voglmayr (2015), Jaklitsch et al. (2006), Overton et al. (2006a,b), and Samuels et al. (2006a). The centrum development is of the Nectria type (Canham, 1969; Hanlin, 1965; Luttrell, 1951), in which apical paraphyses initiated by a meristem located in the top of the developing centrum grow downward and dissolve at ascospore maturity, leaving a centrum with no obvious sterile filaments among the asci. Eight bicellular ascospores form in each ascus but each of these disarticulates into two unequal part ascospores early in development. Usually, the upper part of the ascospore is conical to subglobose and the basal part is longer and wedge shaped, although in some species, all but parts of the top and bottom ascospores in the ascus are globose. In some species, the eight ascospores remain unicellular and these species have been classified in the genus *Sarawakus*, which is now considered to be *Trichoderma* (Jaklitsch et al., 2014). Ascospores are colorless or green (classified in the past in the genera *Creopus* and *Chromocrea*) and usually roughened. Teleomorphs are commonly found on very rotten or firmer wood, bark, and other fungi; rarely, they completely deform the fruiting surface of bird's nest fungi or agarics. Some species that have clavate stromata (formerly placed in the genera *Podostroma* and then *Hypocrea*) arise from soil or wood or capsules of tree fruits in the soil or, in one case, the basidioma of a stink horn fungus. *Trichoderma* teleomorphs have been classified in the genera *Hypocrea*, *Aphysiostroma*, *Creopus*, *Chromocrea*, *Podostroma*, and *Sarawakus*. These genera are now considered to be synonyms of *Trichoderma* (Jaklitsch and Voglmayr, 2013, 2015; Rossman et al., 2013). Approximately 400 species have been described in *Hypocrea* and other teleomorph-typified genera, and approximately 300 have been described in the genus *Trichoderma*. Currently, approximately 250 species of *Trichoderma* are considered to be "in use", that is, they are represented by diagnostic DNA sequences either taken from living cultures or extracted directly from sexual stromata. While many *Trichoderma* species are linked to a named or unnamed teleomorph (e.g., *T. viride* and *H. rufa*, *T. polysporum* and *H. pachybasioides*, *T. floccosum* and a *Hypocrea* species), the complete life cycles of only a few of the *Hypocrea* species described since early in the eighteenth century are known. Until 2013, the International Code of Botanical Nomenclature (ICBN) allowed for the separate naming of anamorphs (asexual stages, e.g., *Trichoderma* species) and teleomorphs, and the name of the whole species (anamorph + teleomorph), the holomorph, was the *Hypocrea* name. However, beginning in 2013, the name International Code of Botanical Nomenclature was changed to International Code of Nomenclature for algae, fungi, and plants (Melbourne Code) (ICN; www.iapt-taxon.org/nomen/main.php), and this new code ended the dual nomenclature for pleomorphic fungi. The correct name for the holomorph (teleomorph + all anamorphs) is now determined, in most cases, by which of the two or more names for a genus is older. In the case of *Trichoderma*, which was described in 1791, and *Hypocrea*, which was described later in 1825 (all other generic synonyms are younger), *Trichoderma* takes priority over *Hypocrea* (Rossman et al., 2013). Therefore, in principle, all names that were proposed for the genus *Hypocrea* should now be transferred to the genus *Trichoderma*, but in practice, each species must be evaluated. There are many anomalous cases that must be resolved. Fortunately, this nomenclatural revolution will have little impact on the names of the *Trichoderma* species that are applied in ecology, agriculture, and biotechnology. In a few cases, the principle of priority would dictate that the older name in the genus *Hypocrea* replace a well-known name in the genus *Trichoderma*. Perhaps the most significant of these cases involves *T. reesei* (1977) and its older teleomorph, *H. jecorina* (1873). But in this case, the name *T. reesei* has been proposed for conservation over the infrequently used name *H. jecorina*. Conservation has been proposed for other such cases (Samuels, 2014).

The new nomenclatural rules will pose conceptual difficulties for some users, at least in the short term. The biggest impact of the new rules will be in the taxonomic/monographic studies that emphasize collections of the teleomorph and where the teleomorph name has traditionally been used. In several cases, the *Trichoderma* species seems to form only a teleomorph or, despite its phylogenetic position, does not form an anamorph that is recognizable as a species of *Trichoderma* on the basis of its morphology. For example, no obvious anamorph is found in *H. peltata* (Samuels and Ismaiel, 2011), which is now *T. peltatum* (Jaklitsch and Voglmayr, 2015); *H. spinulosa* (Chaverri and Samuels, 2003), which is now *T. spinulosum* (Jaklitsch and Voglmayr, 2015); or *Aphysiostroma stercorarium* (Barrasa et al., 1985), which is now *T. stercorarium* (Jaklitsch and Voglmayr, 2015). Anamorphs of *H. avellanea* (Carey and Rogerson, 1976), which is now *T. avellaneum* (Jaklitsch and Voglmayr, 2015), and species in the Hypocreanum clade (Bissett, 1991a; Jaklitsch, 2011; Overton et al., 2006a,b) are *Acremonium*- or *Verticillium*-like and have hyaline conidia and thus are not recognizable as *Trichoderma* species.

A tremendous amount of literature about the systematics and taxonomy of the genus *Trichoderma* is summarized by Domsch et al. (2007), Rifai (1969), Samuels (1996, 2006), and Seifert et al. (2011). The systematic arrangement of *Trichoderma* species has gone through three phases. From the first recognition of the genus by Persoon in 1794 (Persoon, 1794) until 1969, the genus was viewed as an intractable mess and few species were described or recognized, and in some cases, species that we now can easily separate by their appearance were lumped together (Bisby, 1939). The second phase of *Trichoderma* systematics is characterized by critical morphological observation. When Joan Dingley monographed the genus *Hypocrea* in New Zealand, she described and illustrated what appear to be different *Trichoderma* anamorphs but identified them all as *T. viride* (Dingley, 1957). Yoshimichi Doi described life cycles of *Hypocrea* species for Japan (Doi, 1969, 1972), Latin America (Doi, 1975, 1976), and elsewhere. He elegantly illustrated but did not name the anamorphs. John Webster, alone or with Mein Rifai, fully described some *Trichoderma* species and their teleomorphs for Britain (Rifai and Webster, 1966a,b; Webster, 1964; Webster and Rifai, 1968). Rifai's monograph of the genus *Trichoderma*, where he recognized nine aggregate species, or species complexes, was the standard taxonomic reference for *Trichoderma* for nearly 15 years (Rifai, 1969). John Bissett's highly perceptive and now classical taxonomic revisions of the genus *Trichoderma* (Bissett, 1984, 1991a,b) trebled the number of species in the genus to nearly 35; these were divided among five morphologically based sections. Bissett brought *Trichoderma* systematics to the door of molecular phylogenetics, which opened in the mid-1990s. Druzhinina et al. (2006) reviewed the first molecular taxonomic studies of the genus *Trichoderma*. The first taxonomic revision of part of the genus *Trichoderma* to be based on DNA sequences and phylogenetic inference focused on *Trichoderma* section *Longibrachiatum* (Kuhls et al., 1997; Samuels et al., 1998). That first study was based on a single gene, the internal transcribed spacer region of rDNA (ITS 1+2), a gene that is now recognized to be too highly conserved to distinguish closely related species in this genus. One major lesson from the adoption of molecular phylogenetics was confirmation of a reality that was voiced by Rifai in 1969—morphologically distinctive species, such as the aggregate species that he recognized, are often complexes that include two or more morphologically cryptic phylogenetic species. Closely related, or even not so closely related, species can be morphologically indistinguishable. Since that time, fungal systematics has adopted the use of multiple genes as the preferred basis for recognizing fungal species. This approach is known as Genealogical Concordance Phylogenetic Species Recognition (GCPSR; Taylor et al., 2000), and species correspond to clades. According to Dettman et al. (2003), a clade is recognized as an independent evolutionary lineage if it satisfies either of two criteria. (1) Under the genealogical concordance criterion, a clade must be present in the majority of the single-locus trees, regardless of the level of support. (2) Under the genealogical nonconcordance criterion, a lineage is considered to be independent if it is well supported in at least one single-locus tree and is not contradicted in any other single-locus tree as judged by both maximum parsimony bootstrap values and Bayesian posterior probability (Huelsenbeck and Ronquist, 2001). The morphologically defined sections of the genus *Trichoderma* have now largely been abandoned in favor of clades because of the high degree of morphological convergence manifested among phylogenetically distant species or because of a more precise defining of the meaning of morphology in light of phylogenetic analysis.

There is no generally agreed upon grouping of *Trichoderma* species. There is little phylogenetic, morphological, or biological reason for taxonomic subdivision of the genus into sections, which was done by Bissett in his work. As can be seen from Figure 4.1, too many species do not cluster with other species to impose a formal sectional taxonomy on the genus. Moreover, new species of *Trichoderma* are being described at a fast rate (e.g., Zhu and Zhuang, 2015), and as new species are added to the analyses, new relationships are revealed. The system that we have adopted here (Fig. 4.1; Table 4.1) is modified somewhat from the phylogenetic, clade-based arrangement used by Jaklitsch (2011) in his monograph of European *Hypocrea* species.

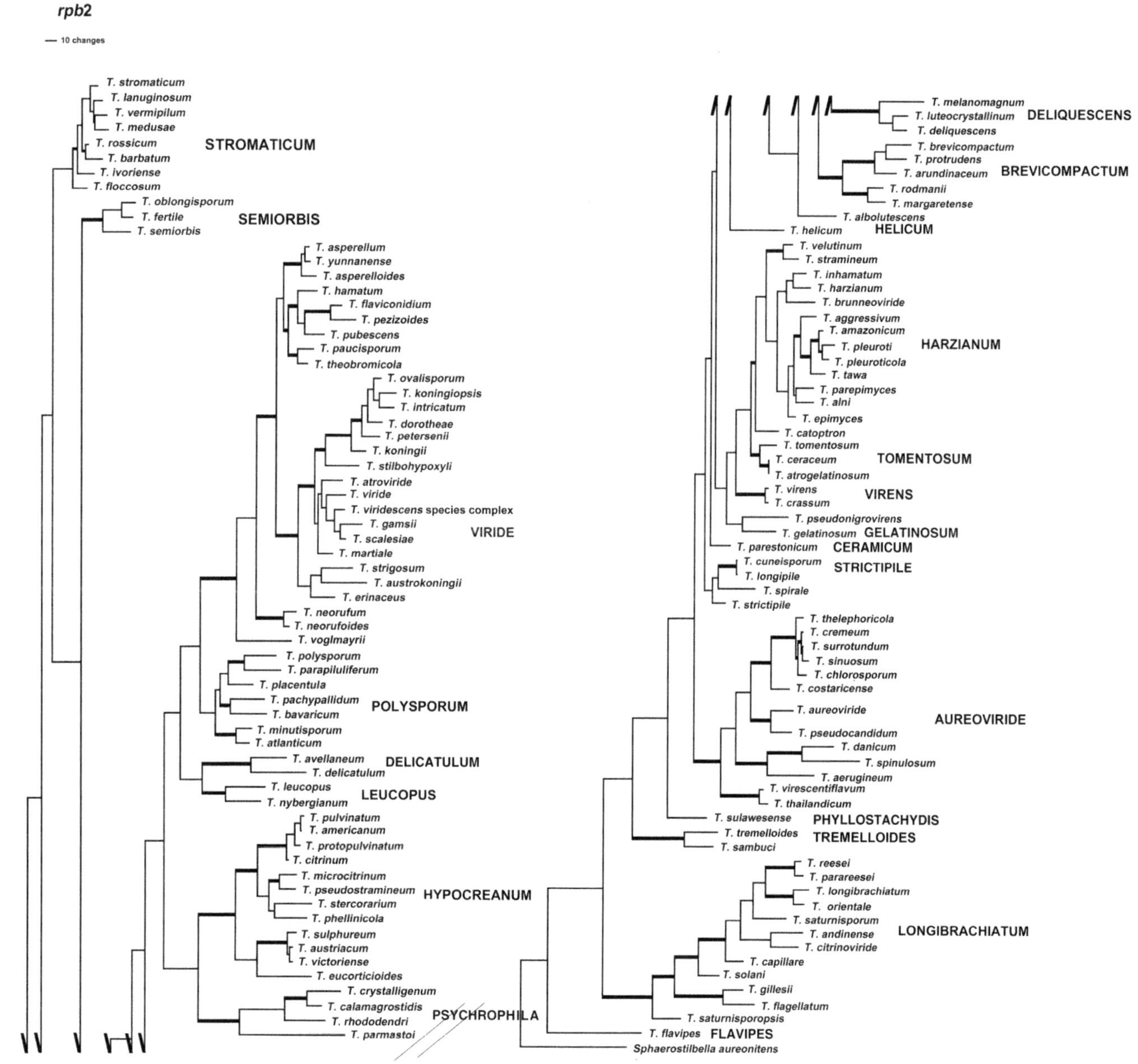

Figure 4.1. A phylogram based on *rpb2* showing the relationships of *Trichoderma* species. The heavy lines indicate bootstrap support of 70% or greater.

Table 4.1. Grouping of *Trichoderma* Species[a]

Clade	Species
AUREOVIRIDE CLADE	
	T. aerugineum Jaklitsch**[b]
	*[c] *T. aureoviride* Rifai**
	T. chlorosporum P. Chaverri & Samuels**
	T. costaricense (P. Chaverri & Samuels) P. Chaverri, Jaklitsch & Voglmayr
	T. cremeum P. Chaverri & Samuels**
	T. danicum (Jaklitsch) Jaklitsch & Voglmayr
	T. pseudocandidum P. Chaverri, Samuels & Minnis**
	T. sinuosum P. Chaverri & Samuels**
	T. spinulosum (Fuckel) Jaklitsch & Voglmayr**
	T. surrotundum P. Chaverri & Samuels**
	T. thailandicum P. Chaverri & Samuels**
	T. thelephoricola P. Chaverri & Samuels**
	T. virescentiflavum (Speg.) Jaklitsch & Voglmayr
BREVICOMPACTUM CLADE	
	T. arundinaceum Zafari, Gräfenhan & Samuels
	T. auranteffusum Jaklitsch**
	* *T. brevicompactum* G. F. Kraus, C. P. Kubicek & W. Gams
	T. margaretense Jaklitsch**
	T. protrudens Samuels & P. Chaverri
	T. rodmanii (Samuels & P. Chaverri) Jaklitsch & Voglmayr**
	T. turrialbense Samuels, Degenkolb, K. F. Nielsen & Gräfenhan
CERAMICUM CLADE	
	T. ceramicum P. Chaverri & Samuels**
	T. estonicum P. Chaverri & Samuels**
	T. parestonicum Jaklitsch**
DELICATULUM CLADE	
	T. avellaneum (Rogerson & S. T. Carey) Jaklitsch & Voglmayr**
	T. delicatulum Jaklitsch**
DELIQUESCENS CLADE	
	* *T. deliquescens* (Sopp) Jaklitsch**
	T. luteocrystallinum Jaklitsch**
	T. melanomagnum P. Chaverri & Samuels**
FLAVIPES CLADE	
	T. alcalifuscescens (Overton) Jaklitsch & Voglmayr**
	T. flavipes (Peck) Seifert, Jaklitsch & Voglmayr**
GELATINOSUM CLADE	
	T. chromospermum P. Chaverri & Samuels**
	T. gelatinosum P. Chaverri & Samuels**
HARZIANUM CLADE	
	T. afarasinum P. Chaverri & F. B. Rocha
	T. afroharzianum P. Chaverri, F. B. Rocha & Druzhin.
	* *T. aggressivum* Samuels & W. Gams
	T. alni Jaklitsch**
	* *T. amazonicum* P. Chaverri & Gazis
	T. atrobrunneum F. B. Rocha, P. Chaverri & Jaklitsch**
	T. brunneoviride Jaklitsch**
	T. camerunense P. Chaverri & Samuels
	T. cinnamomeum P. Chaverri & Samuels**
	T. compactum Z. F. Yu & K. Q. Zhang
	T. dacrymycellum Jaklitsch**
	T. endophyticum F. B. Rocha, Samuels & P. Chaverri
	T. epimyces Jaklitsch**
	T. guizhouense Q. R. Li, McKenzie & Yong Wang bis
	* *T. harzianum* Rifai
	T. inhamatum Veerkamp & W. Gams
	T. lentiforme (Rehm) P. Chaverri, Samuels & F. B. Rocha**
	T. lixii (Pat.) P. Chaverri**
	T. neotropicale P. Chaverri & F. B. Rocha
	T. parepimyces Jaklitsch**
	* *T. pleuroti* S. H. Yu & M. S. Park
	* *T. pleuroticola* S. H. Yu & M. S. Park
	T. pseudogelatinosum (M. Komatsu & Yoshim. Doi) C. S. Kim
	T. pyramidale Jaklitsch & P. Chaverri
	T. rifaii F. B. Rocha, P. Chaverri & Samuels
	T. simmonsii P. Chaverri, F. B. Rocha, Samuels & Jaklitsch**
	T. stramineum P. Chaverri & Samuels**
	T. tawa P. Chaverri & Samuels**
HELICUM CLADE	
	T. helicum Bissett, C. P. Kubicek & Szakács
	T. silvae-virgineae Jaklitsch**

[a] Details of publication of the species can be found in MycoBank (www.mycobank.org).
[b] Species known to have a teleomorph are indicated by **.
[c] Species preceded with an * are included within this book.

(continued on next page)

Table 4.1. Grouping of *Trichoderma* Species *(continued from previous page)*

HYPOCREANUM CLADE	**PHYLLOSTACHYDIS CLADE**
T. americanum (Canham) Jaklitsch & Voglmayr**	*T. phyllostachydis* P. Chaverri & Samuels**
T. austriacum Jaklitsch**	*T. sulawesense* (Yoshim. Doi) Jaklitsch & Voglmayr**
T. citrinum (Pers.:Fr.) Jaklitsch, W. Gams & Voglmayr**	**POLYSPORUM CLADE**
T. decipiens (Jaklitsch, K. Põldmaa & Samuels) Jaklitsch & Voglmayr**	*T. alutaceum* Jaklitsch**
T. eucorticioides (Overton) Jaklitsch & Voglmayr**	*T. atlanticum* Jaklitsch**
T. microcitrinum (Yoshim. Doi) Jaklitsch & Voglmayr**	*T. bavaricum* Jaklitsch**
T. phellinicola Jaklitsch**	*T. europaeum* Jaklitsch & Voglmayr
T. protopulvinatum (Yoshim. Doi) Jaklitsch & Voglmayr**	*T. luteffusum* Jaklitsch**
T. pseudostramineum (Yoshim. Doi) C. S. Kim**	* *T. minutisporum* Bissett**
T. pulvinatum (Fuckel) Jaklitsch & Voglmayr**	*T. pachypallidum* Jaklitsch**
T. stercorarium (Barrasa, A. T. Martínez & G. Moreno) Jaklitsch & Voglmayr**	*T. parapiluliferum* (B. S. Lu, Druzhin. & Samuels) Jaklitsch & Voglmayr**
T. sulphureum (Schwein.) Jaklitsch & Voglmayr**	*T. piluliferum* J. Webster & Rifai**
T. victoriense (Overton) Jaklitsch & Voglmayr**	*T. placentula* Jaklitsch**
LEUCOPUS CLADE	* *T. polysporum* (Link) Rifai**
T. leucopus Jaklitsch**	*T. stellatum* (B. S. Lu, Druzhin. & Samuels) Jaklitsch & Voglmayr
T. nybergianum (T. Ulvinen & H. L. Chamb.) Jaklitsch & Voglmayr**	**PSYCHROPHILA CLADE**
T. seppoi Jaklitsch**	*T. calamagrostidis* Jaklitsch**
LONGIBRACHIATUM CLADE	*T. crystalligenum* Jaklitsch**
T. aethiopicum Mulaw, C. P. Kubicek & Samuels	*T. megalocitrinum* (Yoshim. Doi) Jaklitsch & Voglmayr**
T. andinense (Samuels & Petrini) Samuels, Jaklitsch & Voglmayr**	*T. parmastoi* (Overton) Jaklitsch & Voglmayr**
T. bissettii Sandoval-Denis & Guarro	*T. psychrophilum* Jaklitsch**
* *T. capillare* Samuels & C. P. Kubicek	*T. rhododendri* (Jaklitsch & Voglmayr) Jaklitsch & Voglmayr**
* *T. citrinoviride* Bissett**	**SEMIORBIS CLADE**
* *T. effusum* Bissett, C. P. Kubicek & Szakács	*T. fertile* Bissett
* *T. flagellatum* Mulaw, C. P. Kubicek & Samuels	*T. fomiticola* Jaklitsch**
* *T. ghanense* Yoshim. Doi, Y. Abe & Sugiy.	*T. mienum* C. S. Kim, Nakagiri & N. Maek.
T. gillesii Samuels**	*T. moravicum* Jaklitsch**
T. gracile Samuels & Szakács	*T. oblongisporum* Bissett
T. konilangbra Samuels, Petrini & C. P. Kubicek	*T. semiorbis* (Berk.) Jaklitsch & Voglmayr**
* *T. longibrachiatum* Rifai	**STRICTIPILE CLADE**
T. novae-zelandiae (Samuels & Petrini) Jaklitsch & Voglmayr**	*T. cuneisporum* P. Chaverri & Samuels**
* *T. orientale* (Samuels & Petrini) Jaklitsch & Samuels**	* *T. longipile* Bissett**
T. parareesei Jaklitsch, Druzhin. & Atan.	* *T. spirale* Bissett
T. parceramosum Bissett	*T. strictipile* Bissett**
T. pinnatum Samuels	**STROMATICUM CLADE**
* *T. pseudokoningii* Rifai	*T. barbatum* Samuels
* *T. reesei* E. G. Simmons**	*T. caesareum* Samuels**
* *T. saturnisporopsis* Samuels & Jaklitsch	*T. floccosum* Samuels**
* *T. saturnisporum* Hammill	*T. ivoriense* Samuels
T. solani Samuels, V. Doyle & V. S. Lopez	*T. lanuginosum* Samuels**

(continued on next page)

Table 4.1. Grouping of *Trichoderma* Species *(continued from previous page)*

STROMATICUM CLADE *(continued)*

T. medusae Samuels**

* *T. rossicum* Bissett, C. P. Kubicek & Szakács

* *T. stromaticum* Samuels & Pardo-Schulth.**

T. vermipilum Samuels

TOMENTOSUM CLADE

T. atrogelatinosum (Dingley) Jaklitsch & Voglmayr**

T. ceraceum P. Chaverri & Samuels**

T. cerinum Bissett, C. P. Kubicek & Szakács

* *T. tomentosum* Bissett

TREMELLOIDES CLADE

T. sambuci (Jaklitsch & Voglmayr) Jaklitsch & Voglmayr**

T. tremelloides Jaklitsch**

VIRENS CLADE

* *T. crassum* Bissett

* *T. virens* (J. H. Mill., Giddens & A. A. Foster) Arx**

VIRIDE CLADE

T. aeroaquaticum K. Yamag., Tsurumi, Chuaseehar. & Nakagiri

T. appalachiense Samuels & Jaklitsch**

T. asperelloides Samuels**

* *T. asperellum* Samuels, Lieckf. & Nirenberg

* *T. atroviride* P. Karst.**

T. austrokoningii Samuels & Druzhin.**

T. caerulescens (Jaklitsch & Voglmayr) Jaklitsch & Voglmayr

T. composticola Samuels & Jaklitsch

T. dingleyae Samuels & Dodd**

T. dorotheae Samuels & Dodd**

* *T. erinaceus* Bissett, C. P. Kubicek & Szakács

* *T. evansii* Samuels

T. flaviconidium (P. Chaverri, Druzhin. & Samuels) Jaklitsch & Voglmayr**

* *T. gamsii* Samuels & Druzhin.

* *T. hamatum* (Bonord.) Bainier**

T. hispanicum (Jaklitsch & Voglmayr) Jaklitsch & Voglmayr**

T. intricatum Samuels & Dodd**

T. junci Jaklitsch**

* *T. koningii* Oudem.**

* *T. koningiopsis* Samuels, C. Suárez & H. C. Evans**

* *T. lieckfeldtiae* Samuels

T. martiale Samuels

T. neokoningii Samuels & Soberanis

T. neorufoides Jaklitsch**

T. neorufum (Samuels, Dodd & Lieckf.) Jaklitsch & Voglmayr**

T. neosinense Samuels & Jaklitsch**

T. ochroleucum (Berk. & Ravenel) Jaklitsch & Voglmayr**

T. olivascens Jaklitsch, Samuels & Voglmayr**

* *T. ovalisporum* Samuels & Schroers

T. paratroviride Jaklitsch & Voglmayr

T. paraviridescens Jaklitsch, Samuels & Voglmayr**

* *T. paucisporum* Samuels, C. Suárez & K. Solis

T. petersenii Samuels, Dodd & Schroers**

T. pezizoides (Berk. & Broome) Samuels, Jaklitsch & Voglmayr**

* *T. pubescens* Bissett

T. rogersonii Samuels**

T. samuelsii Jaklitsch & Voglmayr**

T. scalesiae Samuels & H. C. Evans

* *T. stilbohypoxyli* Samuels & Schroers**

T. strigosellum C. A. López-Quint., W. Gams, Boekhout & Druzhin.

* *T. strigosum* Bissett

T. subeffusum Jaklitsch**

T. taiwanense Samuels & M. L. Wu**

T. theobromicola Samuels & H. C. Evans

T. trixiae Samuels & Jaklitsch**

T. valdunense Jaklitsch**

T. vinosum Samuels**

T. viridarium Jaklitsch, Samuels & Voglmayr**

* *T. viride* Pers.**

* *T. viridescens* (A. S. Horne & H. S. Will.) Jaklitsch & Samuels**

T. viridialbum Jaklitsch, Samuels & Voglmayr**

T. virilente Jaklitsch & Voglmayr**

T. voglmayrii Jaklitsch**

T. yunnanense Z. F. Yu & K. Q. Zhang

LONE LINEAGES

T. albolutescens Jaklitsch**

T. catoptron P. Chaverri & Samuels**

T. patella (Cooke & Peck) Jaklitsch & Voglmayr**

T. peltatum (Berk.) Samuels, Jaklitsch & Voglmayr**

T. pseudonigrovirens P. Chaverri, Samuels & Minnis**

T. subalpinum Jaklitsch**

T. taxi Chu L. Zhang, F. C. Lin & C. P. Kubicek

* *T. velutinum* Bissett, C. P. Kubicek & Szakács

The species that do not have any obvious close relatives in the present analysis are referred to here as Lone Lineages. However, as additional species are included in phylogenetic analysis, the interrelationships among species may change, new clades will be revealed, and the Lone Lineages will find their closest relatives. The tree presented in Figure 4.1 is based on sequences of *rpb*2 because this gene is able to give some support for phylogeny of the entire genus while being sufficiently conserved to distinguish closely related species. The *tef*1 gene is probably the best gene for the diagnosis of *Trichoderma* species, but as always, an analysis that includes multiple characters, whether it is classical and morphological or it is molecular phylogenetic, is always more robust than an analysis that includes a single gene or character. A more detailed analysis that includes multiple genes may show more, or fewer, groups than we recognize here. We are confident that the major *Trichoderma* groups are outlined here. We include most of the species that we know today in this analysis, but the most complete visual arrangement of species of *Trichoderma* as of 2015 was published by Jaklitsch and Voglmayr (2015), based on partial sequences of the *rpb*2 gene. In these analyses, *Trichoderma* species are distributed among well-supported clades (dark lines in Fig. 4.1). The Polysporum clade (comprising most of the Pachybasium Core Group of Jaklitsch [2011]) is an exception because it is only weakly supported in the *rpb*2 phylogram presented here. However, in Jaklitsch's combined *tef*1/*rpb*2 phylogram (Jaklitsch, 2011), this group received reasonable support, which is further bolstered by the morphological homogeneity presented by the members of this clade.

Trichoderma vs. *Gliocladium*

Trichoderma and *Gliocladium* are distinct genera in the Hypocreales. The well-known and commonly encountered soil fungi *T. virens*, *T. deliquescens*, and *Clonostachys rosea* were originally described in the genus *Gliocladium*, and there is a large amount of literature for each of these species under their *Gliocladium* names. *Gliocladium virens* and *Gliocladium deliquescens* are now known to be *Trichoderma* species (Jaklitsch, 2011; Rehner and Samuels, 1994, 1995). *Gliocladium roseum* is now *Clonostachys rosea* (Rehner and Samuels, 1994; Schroers and Samuels, 1998), in the family Bionectriaceae (Hypocreales); its teleomorph is described as *Bionectria ochroleuca*. True *Gliocladium* (type: *Gliocladium penicillioides*) is a distinctive genus of fungi; its species grow on bracket fungi (species of *Trametes* and *Stereum*) and their teleomorphs are described as species of the genus *Protocrea* (Jaklitsch et al., 2008).

Are There Biocontrol Species of *Trichoderma*?

One often reads about biocontrol species as though some species are more proficient than others at controlling pathogens. Are there really biocontrol species? In this case, "biocontrol" most likely refers to a mycoparasitic or otherwise antagonistic ability. However, to really answer this question, one must first answer another question. What does the term "biocontrol" mean? We have seen that the term "biocontrol" can refer to direct parasitism of a pathogen; production of a wide diversity of chemicals that variously affect pathogens; stimulation of the host immune response to resist a pathogen; or enhancement of plant health and vigor to enable the host plant to resist a pathogen, compete for nutrients or space, and so on. Can we predict a biological ability based on phylogenetic relationships?

The first problem is one of taxonomic precision. To be blunt, many, or at least most, of the cultures that are reported to have some desirable biological property have been misidentified (Table 1.1). The species most often cited in the nontaxonomic literature are *T. harzianum* and *T. viride*, but these two species are actually rather uncommon in the environment and have a Northern Hemisphere distribution (Chaverri et al., 2015; Jaklitsch et al., 2006; Lieckfeldt et al., 1999). It does not appear that either, in the restricted sense in which we see each of them today, is utilized in agricultural applications. Nonetheless, from a phylogenetic perspective, most nontaxonomic reports of species of *Trichoderma* refer to species belonging to only a few main groups (Fig. 4.1; Table 4.1): the Viride clade (e.g., *T. asperellum*, *T. atroviride*, *T. hamatum*, *T. ovalisporum*) and the distantly related Harzianum (*T. afroharzianum*) and Virens clades. Is this because there is a phylogenetic basis for a wide range of metabolic activities or is it that, once again, these species happen to be the ones most commonly isolated in surveys?

Trichoderma is indeed a hyperdiverse genus. Chaverri and Samuels (2013) combined diversification and multilocus phylogenetic analyses of 143 *Trichoderma* species in a study of the evolution of nutritional modes within the genus. Within the genus *Trichoderma*, switches between nutritional modes (e.g., mycoparasites to plant saprotrophs) and habitat preferences (e.g., fungi, dead plant material, living plant material as endophytes, soil) have been numerous and appear to be related to the evolution of new taxa. Because recently evolved species tend to be ecologically similar, biological ability may be inferred by reference to the closest relatives. For example, exclusively endophytic clades might be more efficient in inducing resistance than nonendophytic taxa (Bailey et al., 2008; Harman et al., 2004), and species belonging to the several major mycoparasitic clades in the genus may have a greater potential for biological control that is mediated by the direct action on the pathogen. In support of this, Kubicek et al. (2011) found genes associated with mycoparasitism (e.g., genes from the glycosyl hydrolase family that hydrolyze the wall-building materials chitin/chitosan and β-1,3-glucan, and polyketide synthases and nonribosomal synthetases that are involved in the formation of secondary metabolites) to be well developed in genomes of "biocontrol" species *T. atroviride* and *T. virens* but less well developed in the genomes of the supposed saprotrophic species *T. reesei*, and in non-hypocrealean fungi.

Of the approximately 500 total described *Trichoderma* species (including those species described as *Hypocrea*, which as mentioned above is now considered to be a synonym of *Trichoderma*), very few have been grown in pure culture and the names of fewer still enter the literature. Jaklitsch (2011) correctly pointed out that most *Trichoderma* species are actually not known apart from their *Hypocrea* stromata and are not encountered in soil surveys or isolated directly from substrates. The biological potential of these species has never been considered. Most of the cultures reported in the scientific literature were isolated originally from soil, and the number of *Trichoderma* species isolated from soil is small compared with the total (Blaszczyk et al., 2011; Friedl and Druzhinina, 2012). In addition, *Trichoderma* species have only recently been discovered to be a dominant component of the endophytic mycota of some trees (Evans et al., 2003), and cultures isolated as endophytes are now showing promise in biocontrol applications (e.g., Bae et al., 2009; Krauss et al., 2010). The scheme provided by Chaverri and Samuels (2013) provides at least a hypothesis for the exploration for novel agricultural applications for strains of *Trichoderma* species.

A most important point to understand is that a species is the sum of its strains, and each strain is genetically unique. The point is that each hyphal tip or conidium potentially is a representative of a genetically distinct population, and in an isolation plate, it cannot be assumed that two or more colonies of any given species are going to behave in the same way. We cite as an example a study in which multiple cultures of *T. gamsii* cultures were isolated from fields where sugar beet was affected by *Rhizoctonia solani*, but in patches, not over the entire field. The authors found that the cultures—strains—most effective in controlling *Rhizoctonia solani* were isolated from areas of the field having the highest disease incidence. They concluded that the antagonistic activity was a property of a population, which we interpret here to be a strain, not of the species (Anees et al., 2010).

To return to the question as to whether there are biocontrol species in the genus *Trichoderma*, the answer is probably no. However, the use of genomic tools to discover relevant genes, combined with multilocus phylogenetic studies to provide taxonomic precision and reveal interrelationships, can help to direct the search for biologically useful strains. Species in some groups of species may have greater potential than species in other groups, and some groups may find applications not found in other groups. Moreover, the many species awaiting evaluation may well provide cultures for agricultural applications.

Identification of *Trichoderma* Species

Correct identification is critical to the effective communication of results. Unfortunately *Trichoderma* species have classically been notoriously difficult to identify. Or, putting it another way, it seems to be quite easy to **misidentify** *Trichoderma* cultures. The literature on the genus *Trichoderma*, especially prior to the year 2000, when molecular diagnostic tools were developed, is replete with reports based on misidentified strains (Kullnig et al., 2001; Table 1.1). An important point to be aware of is that identification of *Trichoderma* species based on morphology alone is not nearly as precise as identification based on molecular markers. Molecular phylogenetic studies undertaken since the late 1990s have left us with a taxonomy that overlaps morphological boundaries. Every effort should be made to obtain a correct identification as part of any publication of research results or application. For the identification of strains, it may be necessary to collaborate with an expert taxonomist. The easiest way to provide cultures for identification is to touch 10-mm^2 squares of sterile filter or other paper to a heavily sporulating culture, one or a few for each culture, and wrap those squares in small glassine bags, or paper packets made from clean, smooth paper. Several such packets can be mailed in a first class envelop. However, before sending cultures, permits to send cultures out of the country and the permission of the receiving person must be obtained if necessary. If a DNA lab is available, a polymerase chain reaction (PCR) product may be sent to a laboratory for sequencing.

Identification of *Trichoderma* Species Using DNA Sequences

The most effective way to identify *Trichoderma* species is through DNA sequences, and with few exceptions, all described *Trichoderma* species are represented by partial sequences of two or more gene regions in GenBank (National Center of Biotechnology Information (NCBI); www.ncbi.nlm.nih.gov) and the International Subcommission on *Trichoderma* and *Hypocrea* Taxonomy site (www.ISTH.info). From the point of view of identifying species using DNA databases, with no more than a small handful of exceptions, all of the species described as *Trichoderma* are characterized by cultures, modern descriptions, and perhaps most importantly, DNA sequences. Today, approximately 250 *Trichoderma* species have been characterized and are represented by cultures in international culture collections and in GenBank (Table 4.1).

Several genes are utilized for the identification of *Trichoderma* species, most notably the ITS region of a nuclear rDNA gene cluster (ITS 1+2 and the 5.8S rDNA gene), the third and fourth introns of the translation-elongation factor gene (*tef*1), and the section encoding conserved amino acid motifs numbers 5 and 7 of RNA polymerase II (*rpb*2). In addition, other genes are used in phylogenetic reconstruction; these include the actin (*act*) gene section that covers partial sequences of exons 5 and 6 and the complete sequence of intron 5 based on the complete sequence of the actin gene of *T. reesei* (GenBank accession no. X75421), a fragment of the calmodulin gene (*cal*1), an 0.8-kilobase (kb) fragment of the GH18 chitinase 85 gene (*chi*18-5, formerly called *ech*42), and part of the larger subunit of ATP citrate lyase (*acl*1).

An approximation of species identity can be obtained using ITS sequences. This cluster has been recommended as a "bar code" for *Trichoderma* species (Druzhinina et al., 2006; www.isth.info), but as more species are recognized, this gene cluster is proving to be too highly conserved to separate species that are closely related. Moreover, although not common, nonorthologous copies of the ITS cluster are sometimes found and these give misleading results (Hoyos-Carvajal et al., 2009a). The most useful gene for *Trichoderma* species identification is *tef*1, and sequences of this gene have been deposited in GenBank for all species.

Primers for the most commonly used gene regions are given in Table 4.2.

Table 4.2. Primers Commonly Used in Phylogenetic Studies of *Trichoderma* and Identification of Its Species

Primer	Sequence
Internal transcribed spacer (ITS) region rDNA[a]	
ITS 5	5′-GGA AGT AAA AGT CGT AAC AAG G-3′
ITS 4	5′-TCC TCC GCT TAT TGA TAT GC-3′
Actin (*act*)[b]	
TRI-ACT1	5′-TGG CAC CAC ACC TTC TAC AAT GA-3′
TRI-ACT2	5′-TCT CCT TCT GCA TAC GGT CGG A-3′
ACT-500 F	5′-ATT CCG TGC TCC TGA G-3′
ACT-511 R	5′-CTC AGG AGC ACG GAA T-3′
Calmodulin (*cal*)[b]	
CL1	5′-GAR TWC AAG GAG GCC TTC TC-3′
CAL737RM	5′-CAT YTT TCK KGC CAT CAT GG-3′
ATP citrate lyase (*acl*1)[c]	
acl1-230up	5′-AGC CCG ATC AGC TCA TCA AG-3′
acl1-1220low	5′-CCT GGC AGC AAG ATC VAG GAA GT-3′
Endochitinase (*chi*18-5)[d]	
chi18-5-1a	5′-GCT (CT)TC CAT CGG TGG CTG GAC-3′
chi18-5-2a	5′-GGA GTT GGG GTA GCT CAG C-3′
RNA polymerase 2 (*rpb*2)[e]	
fRPB2-5f	5′-GA(T/C)GA(T/C)(A/C)G(A/T)GATCA(T/C) TT(T/C)GG-3′
fRPB2-7cr	5′-CCCAT(A/G)GCTTG(T/C)TT(A/G) CCCAT-3′
Translation-elongation factor 1-alpha (*tef*1)	
EF1-728[f]	5′-CATCGAGAAGTTCGAGAAGG-3′
TEF1R[g]	5′-GCCATCCTTGGGAGATACCAGC-3′

[a] White et al., 1990.
[b] Samuels et al., [2011 online] 2012.
[c] Gräfenhan et al., 2011; Jaklitsch et al., 2013.
[d] Kullnig-Gradinger et al., 2002; Lieckfeldt et al., 2000.
[e] Liu et al., 1999.
[f] Forward primer. Carbone and Kohn, 1999.
[g] Reverse primer. Samuels et al., 2002.

Identification of *Trichoderma* Species Using Chemical Methods

Biolog FF MicroPlates

Biolog FF MicroPlates (Biolog, Hayward, CA, U.S.A.) is a commercially available system for the identification of bacteria and some molds. In this system, species identification is based upon carbon assimilation and mitochondrial activity. In two *Trichoderma* publications that incorporated these metabolic data (Hoyos-Carvajal et al., 2009a; Kubicek et al., 2003), groupings of species was evident, but there was also significant overlapping among species and also some species that were grouped but are phylogenetically unrelated. While the taxonomic power of the Biolog system is limited, the profiles of substrate utilization provide ecological insights for some species.

MALDI-TOF mass spectrometry

Matrix-assisted laser desorption/ionization time-of-flight intact cell mass spectrometry (MALDI-TOF ICMS) has had some success in the identification of *Trichoderma* species (De Respinis et al., 2010; Neuhof et al., 2007; Santos et al., 2010). De Respinis et al. (2010) observed spectral mass peaks in two ranges, *m/z* 2,000–5,000 and *m/z* 6,000–8,000, and considered these spectra to be characteristic of *Trichoderma* species. Those in the *m/z* 6,000–8,000 range most likely were hydrophobins, as was previously observed in a smaller study by Neuhof et al. (2007), who did not observe the peaks in the lower range in their study. The resolving power for species is roughly equivalent to that of ITS sequences (De Respinis et al., 2010), but at that level of precision, the MALDI-TOF MS technique provided results more quickly than did sequencing. Another significant result of the technique is that each of the studied strains possesses a unique spectral mass for hydrophobins and other peptides (De Respinis et al., 2010) and, thus, MALDI-TOF MS may provide a way to track individual strains in biocontrol or other applications where it is important to know the fate of a strain.

Identification of *Trichoderma* Species Using Cultures and Morphology

In the following discussion of characters, illustrations of the characters can be seen in the figures of the mentioned species that appear in this book.

Media and incubation

Some media reported in the morphological observations described in this book are taken largely from colonies grown on Spezieller Nährstoffarmer agar (SNA) and potato dextrose agar (PDA); occasional references to characters observed on cornmeal dextrose agar (CMD) or cornmeal agar (CMA) are reported. These media are described in Table 4.3. Micromorphology is taken predominantly from SNA and less frequently from CMD or CMA. Inoculum is taken from cultures actively growing on SNA or CMD. Petri plates are sealed with Parafilm and incubated in a 25°C incubator under a regimen of 12-h cool white fluorescent light/12-h darkness. Colony morphology, pigmentation, and odor are taken from Difco PDA (Becton, Dickinson and Company, Franklin Lakes, NJ, U.S.A.) incubated as above. According to Jaklitsch (2011), there are considerable differences among brands of PDA. In the present work, we have used PDA from Difco; 20 mL of freshly made medium is poured into 9-cm-diameter petri plates.

Most *Trichoderma* species develop their conidia within 1 week at 25°C on media such as PDA, CMD, malt and malt extract agar, and SNA; the conidia develop more slowly in species that form their conidia in discrete pustules, such as *T. polysporum* and *T. hamatum*. An unusual characteristic of *Trichoderma* species is that, although their conidiophores develop quickly, they just as quickly become overmature and phialides are no longer visible. In this condition, conidiophores are useless for the observation of morphology. The best time to observe conidial development is when there is a mixture of conidia that are still uncolored with green conidia in pustules. On a rich medium, such as PDA, conidia tend to form in a dense lawn where individual pustules are not apparent. In the present work, our observations are primarily made from cultures grown on SNA, less frequently on CMD. In *T. spirale*, which typically produces both pustules and a diffused *Verticillium*-like synanamorph, the ability to produce pustules may be lost and only diffused conidiophores appear in cultures.

Macroscopic characters: Characters observed in the petri plate directly or with use of a stereomicroscope

Conidial color. Conidia of most *Trichoderma* species are green at maturity; conidia of a small number of species are white. The shade of green varies from yellow-green through gray-green to very dark green. In this work, we have not attempted to specify a shade of green. In the present treatment, only one species, *T. polysporum*, has white conidia. Additional species having white conidia include *T. parapiluliferum*, *T. piluliferum*, *T. placentula*, and *T. stellatum* (Jaklitsch, 2011; Lu et al., 2004a). Conidia of most members of the Hypocreanum clade, which are not morphologically recognizable as *Trichoderma* species, are held in colorless drops of liquid (Bissett, 1991b; Overton, 2006a,b). Conidia of *T. stromaticum* (Samuels et al., [2011 online] 2012) and some members of the Viride clade (e.g., *T. flaviconidium*, *T. koningiopsis*, *T. paucisporum*, *T. viridescens*) (Druzhinina et al., 2004; Jaklitsch, 2011; Jaklitsch et al., 2006; Samuels et al., 2006b) may be yellow as they develop when growing on PDA, often linked to a diffusing yellow pigment and/or a coconut odor.

Conidial matrix. In most species, conidia appear to be dry or, as in *T. brevicompactum* and its relatives, moist and somewhat slimy, but in some species, such as *T. deliquescens*, *T. gelatinosum*, and *T. virens*, conidia are conspicuously held in drops of clear green liquid. Many

Table 4.3. Media Used in Isolating and Cultivating *Trichoderma* Species

Name	Description/Preparation
Media for general purposes, including morphological characterization	
Spezieller Nährstoffarmer agar (SNA) without filter paper (modified from Nirenberg, 1976)	For use in characterizing morphology and growth rates. The medium was originally described for characterizing *Fusarium* species and included placing a piece (approximately 1 cm^2) of sterile filter paper on the surface of the solidified medium in petri plates. The filter paper is omitted here. **Preparation:** KH_2PO_4 1.0 g KNO_3 1.0 g $MgSO_4 \cdot 7H_2O$ 0.5 g KCl 0.5 g Glucose 0.2 g Sucrose 0.2 g Agar 20 g Distilled water 1 L Sterilize 15 min.
Potato dextrose agar (PDA) (various manufacturers)	According to Jaklitsch (2009), different brands of commercially available PDA give different results. This medium is not acceptable for storing *Trichoderma* cultures. **Preparation:** Made following manufacturer's instructions.
Cornmeal agar (CMA) (various manufacturers)	This medium is useful for storing cultures in screw-capped tubes. **Preparation:** Made following manufacturer's instructions.
Cornmeal dextrose agar (CMD)	This medium is useful for characterizing morphology and for documenting cultural characteristics. **Preparation:** Prepare CMA following manufacturer's instructions. Add 20 g of dextrose per liter. Sterilize 20 min.
Media for producing inoculum	
Potato dextrose agar (PDA) (various manufacturers)	Cultures cultivated on PDA usually produce abundant conidia, and this medium is often cited in the production of inoculum.
Potato sugar broth (Medeiros et al., 2010)	This is useful for obtaining conidia to inoculate solid substrate. **Preparation:** Boil 1 kg of peeled potatoes in 2 L of distilled water. Strain through cheesecloth. Add 1 L of presterilized (autoclaved twice at 121°C for 30 min to reduce contamination) fresh sugar cane juice. Add sterile distilled water to bring the final volume to 3 L.
Yeast–malt–extract agar for producing inoculum (Inbar et al., 1994)	**Preparation:** Yeast extract 2 g Malt extract 20 g Agar 20 g Distilled water 1 L

(*continued on next page*)

Table 4.3. Media Used in Isolating and Cultivating *Trichoderma* Species (continued from previous page)

Name	Description/Preparation
Selective media	Media used for the selective isolation of *Trichoderma* cultures.
Modified PDA (PDAm) (Gil et al., 2009)	The authors compared several selective media used for the quantitative isolation of *Trichoderma* species from soil (including TME and TSM described below) and found this medium to be superior. **Preparation:** Start with PDA medium. Add the following per liter of cooled medium. Rose bengal 0.02 g Chloramphenicol 0.3 g Streptomycin sulfate 0.1 g
***Trichoderma* medium E (TME)** (Papavizas and Lumsden, 1982)	This medium was developed for the isolation of *Trichoderma* species from soils infested with Mucorales (e.g., *Rhizopus* spp., *Mucor* spp.). The medium with sodium propionate but without alkylaryl polyether alcohol was effective, but the medium containing both was most effective. Benomyl was added (10 µg/mL) for the recovery of benomyl-tolerant *Trichoderma* strains. **Preparation:** I. Basal medium V8 juice 200 mL Glucose 1 g Agar 15 g Distilled water 800 mL Autoclave. II. Antimicrobials added to 1 L of cooled basal medium Neomycin sulfate (1%) 0.1 g Bacitracin (1%) 0.1 g Penicillin G (1%) 0.1 g Chloroneb 0.1 g Chlortetracycline hydrochloride 0.025 g Nystatin (0.25%) 0.020 g Sodium propionate 0.5 g Alkylaryl polyether alcohol 2.0 mL
Mineral salts medium for isolation of cellulose-producing microbes (Fujii et al., 2011)	**Preparation:** $(NH_4)2SO_4$ 5.0 g KH_2PO_4 1.0 g $MgSO_4·7H_2O$ 500 mg NaCl 100 mg $CaCl_2·2H_2O$ 100 mg H_3BO_3 0.5 mg $MnSO_4·5H_2O$ 0.4 mg $ZnSO_4·7H_2O$ 0.4 mg $FeCl_3$ 0.2 mg $Na_2MoO_4·2H_2O$ 0.2 mg KI 0.1 mg $CuSO_4·5H_2O$ 0.04 mg pH 5.3 Inositol 2.0 mg Calcium–pantothenate 0.4 mg Niacin 0.4 mg Thiamine–HCl 0.4 mg Pyridoxine–HCl 0.4 mg Riboflavin 0.2 mg Para-aminobenzoic acid 0.2 mg Folate 2 mg Biotin 2 mg Bacto agar 20 g Distilled water 1 L

(continued on next page)

Table 4.3. Media Used in Isolating and Cultivating *Trichoderma* Species *(continued from previous page)*

Name	Description/Preparation
Selective media *(continued)*	
***Trichoderma* selective media (TSM and TSMC)** (Elad and Chet, 1983)	TSM was developed as a medium for the selective isolation of *Trichoderma* species, but it was discovered that *Fusarium* species were recovered as well. The medium TSM with the addition of captan (50%) at a rate of 2 µm/mL (TSMC) was effective in isolating *Trichoderma* species in soils with a high concentration of *Fusarium* species. It should be noted that fungicides containing fenaminosulf may be moderately toxic (PAN Pesticide Database; www.pesticideinfo.org) and unavailable. Versions of this often-cited medium that omit fenaminosulf are given below (Askew and Laing, 1993; Williams et al., 2003). **Preparation:** Basal medium (TSM) $MgSO_4 \cdot 7H_2O$ 0.2 g K_2HPO_4 0.9 g KCl 0.15 g NH_4NO_3 1.0 g Glucose 3 g Chloramphenicol 0.25 g (Chloromycetin; Sigma-Aldrich Co., St. Louis, MO, U.S.A.) Fenaminosulf 0.3 g (sodium 4-dimethylamino benzene sulfonate; 60% wettable powder [w.p.]) Quintozene 0.2 g (pentachloronitrobenzene [PCNB]; Terraclor 75% w.p.; Amvac Chemical Corporation, Los Angeles, CA, U.S.A.) Rose bengal 0.15 g Agar 20 g Water 1 L TSMC: After autoclaving the basal medium, add 0.2 g of captan.
Selective medium for quantifying *T. harzianum* and *T. aggressivum* in mushroom compost (Williams et al., 2003)	This medium is essentially the same as TSM (Elad and Chet, 1983), but fenaminosulf is replaced with propamocarb, and captan is omitted. **Preparation:** I. Basal medium $MgSO_4 \cdot 7H_2O$ 0.2 g K_2HPO_4 0.9 g NH_4NO_3 1.0 g KCl 0.15 g Rose bengal 0.15 g Glucose 3 g Agar 20 g Distilled water 950 mL Autoclave at 121°C for 15 min. II. Antimicrobial and fungicidal ingredients Chloramphenicol 0.25 g Streptomycin 9.0 mL of a stock solution (1%, w/v) per liter Quintozene 0.2 g Propamocarb 1.2 mL (772 g of active ingredient per liter) Sterile distilled water 40 mL Mixture II is added to cooled mixture I.

(continued on next page)

Table 4.3. Media Used in Isolating and Cultivating *Trichoderma* Species *(continued from previous page)*

Name	Description/Preparation
Selective media *(continued)*	
Selective medium for quantifying *Trichoderma* species in oomycete-infested soil (Askew and Laing, 1993)	This medium is identical to TSM (Elad and Chet 1983), except that fenaminosulf is replaced by either metalaxyl or propamocarb. Both biocides were ineffective in suppressing *Pythium ultimum* in soil, but propamocarb was better. **Preparation:** I. Basal medium $MgSO_4 \cdot 7H_2O$ 0.2 g K_2HPO_4 0.9 g NH_4NO_3 1.0 g KCl 0.15 g Rose bengal 0.15 g Glucose 3 g Agar 20 g Distilled water 950 mL Autoclave at 121°C for 15 min. II. Biocidal ingredients Chloramphenicol (crystallized) 0.25 g Quintozene (PCNB, w.p., 750 g of active ingredient per kilogram) 0.2 g Captan (w.p., 500 g of active ingredient per kilogram) 0.2 g ADD EITHER/OR: Propamocarb hydrochloride (722 g of active ingredient per liter) 1.2 mL –OR– Metalaxyl (50 g of active ingredient per liter) 1.6 g Sterile distilled water 50 mL Mixture II is added to cooled mixture I.
***T. virens* selective medium** (Howell, 1999)	Both of these media contain antibiotics that suppress the bacterial and mycological biotas of field soil. "Q" strains of *T. virens* that produce gliotoxin develop on medium #1; these strains are able to clear *Bacillus subtilis* inoculated onto their colonies. "P" strains of *T. virens* that produce gliovirin develop on medium #2 and are unable to clear *Bacillus subtilis.* **Preparation:** Basal medium: PDA Medium #1: PDA with the following Rifampicin 50 µg/mL Gliotoxin 50 µg/mL Chlorothalonil 0.75 µg/mL Medium #2: PDA with the following Rifampicin 50 µg/mL Chlorothalonil 1.25 µg/mL Thiabendazole 0.6 µg/mL The antibiotics and fungicides are in acetone and are added to the medium at 60°C following autoclaving.

species produce conidia in more-or-less well-developed pustules as well as more effused, morphologically distinct conidiophores. Conidia produced from the effused conidiophores are usually held in drops of clear green liquid. In this work, these effused conidiophores are referred to as synanamorphs (e.g., *T. brevicompactum*, Fig. 6.10J).

Colony radius. The temperature optimum for most *Trichoderma* species is 25–30°C, and most species do not grow at 35°C or higher. Growth rate is determined from colonies grown on 9-cm-diameter plastic petri plates containing 20 mL of freshly made PDA or SNA. In the present work, we report diagnostic growth rates at 25 and 35°C after 72 h of incubation in darkness. Because *Trichoderma* species cover the surface of a petri plate rather quickly, we report here the radius of colonies while they are still young. The radius, reported in millimeters, is the greatest length of the colony measured from the outside of the inoculum plug to the most distant point of growth. For the inoculum, 3- to 5-mm-diameter disks are cut using a cork borer from the edge of an actively growing, 3- to 4-day-old colony on CMD, CMA, or SNA before conidiation. One disk is placed mycelium-side down (if possible) 10–15 mm from the edge of the petri plate. The unsealed petri plates are incubated in darkness with the colony surface facing up. In order to determine a growth rate curve for a species, we include all the strains of the species that we have in our collection and grow them in darkness at 15, 20, 25, 30, and 35°C for 96 h and measure the radius at intervals of 24 h. This is repeated three times and the measurements are averaged. Of course, there is variation among strains of a species at any temperature, but the standard error among the strains of any species is typically less than 10 mm. Chaverri and Samuels (2003) have an example of how growth rates are determined.

Colony characters. The physical appearance of the colony is a minor character of a species. Colony characters and patterns of conidiation are determined from colonies prepared as described above but incubated 7–10 days in an incubator at 25°C under cool white fluorescent light with a regimen of 12-h on/12-h off. Digital images of colonies are taken at the end of the incubation period and the images kept for future reference. Characters include distribution of conidia in colonies on PDA and SNA, presence of pigmentation in colony reverse (usually yellow when present), color of conidia, and presence of coconut-like odor on PDA or CMD.

A pigment in some shade of yellow that diffuses through the agar is observed in the underside of PDA cultures and to a lesser extent on CMD and, at best, rarely on SNA. The formation of a yellow pigment is of limited diagnostic significance but is common in some clades. Members of the Longibrachiatum clade, especially those related to *T. citrinoviride*, *T. longibrachiatum*, and *T. reesei*, typically produce an intense yellow pigment. Yellow pigment is also often seen in members of the Viride clade, often in cultures that are sporulating poorly, and in members of the *T. harzianum* species complex (but not *T. harzianum* s. str.). Stale cultures of any *Trichoderma* species tend to develop a "dirty" yellow pigment in PDA cultures.

Odors can be detected in *Trichoderma* cultures but, for the present work, the only odor considered to be significant is the coconut-like odor of 6-pentyl-α-pyrone (6-PAP). This nontoxic aromatic compound is used as a flavoring agent but also can inhibit the germination of spores of *Phytophthora* and *Rhizoctonia* species (Cutler et al., 1986; references in Sivasithamparam and Ghisalberti, 1998). Although the compound is produced by unrelated *Trichoderma* species (Aneja et al., 2005; Sivasithamparam and Ghisalberti, 1998), the amount produced by any species varies with the strain. We have only detected the coconut odor in cultures of members of the Viride clade. The odor is detected on PDA but not SNA and its detection can depend upon the strain. **A note of caution: Because some *Trichoderma* species can infect humans and because they produce vast numbers of dry conidia, extreme care must be taken when examining a culture for coconut odor. Usually, when the odor is present, it is readily noted when the lid of a petri plate is lifted even slightly.**

Micromorphology: Characters observed using a compound microscope

When to observe micromorphology. Micromorphology described here is based on observations made from SNA or, less frequently, CMD. Unless otherwise noted, there are no significant differences between the two media. In the many publications from Bissett, micromorphology is taken from 2% malt agar (Bissett, 1984, 1991a,b,c, 1992; Bissett et al., 2003). The main difference among these media is the degree of robustness of conidial pustules. An exception to this is seen in some members of the Viride clade, in which morphology of conidiophores can be considerably different, with percurrently proliferated phialides forming moniliform chains on CMD and PDA (e.g., *T. gamsii*, Fig. 6.19G, H; *T. koningiopsis*, Fig. 6.24I; *T. viridescens*, Fig. 6.49E). As was noted above, *Trichoderma* cultures quickly become overmature, usually within 7–10 days, and are then useless for observing micromorphology. At this stage, the phialides degenerate. The best time to observe micromorphology for most species is when conidia are beginning to become green. When both white (uncolored) and green conidia are visible, the morphology of the conidiophores are in optimum condition. The measurements reported here are based on 30 measurements of individual conidia and phialides for each collection. The summary statistics represent the average of all individual measurements for each available collection of a species plus and minus the standard deviation. All measurements are made in water or 3% potassium hydroxide (KOH).

Preparing mounts for microscopy. It is important to make a flat preparation for microscopy, but it is difficult to do this with *Trichoderma* species because of the intertwined hyphae of the pustules, which must be separated. The use of lactic acid as a mounting medium makes preparing a truly flat preparation nearly impossible. Using a sharp, alcohol-sterilized scalpel tip (No. 11), place the smallest possible amount of material from a pustule into a drop of 3% (aqueous) KOH on a microscope slide. The KOH wets the conidiophores and allows them to spread. The hyphae of a pustule can be separated using two sharp scalpels or other fine implements. Sometimes, we use a pair of fine glass needles made by drawing out 0.7-mm (external diameter) capillary tubes in an alcohol flame. Place a coverslip on the preparation and, while observing under the low power (approximately 5×) of a compound microscope, lightly tap the coverslip with a sharp No. 11 scalpel. While doing this, the liquid will surge into and out of the mass of hyphae, separating them. When the hyphae have been separated as much as possible, place a small piece of tissue at the edge of the coverslip to draw out the liquid. Within a few minutes using this method, a perfectly flat preparation can be obtained that will remain useful for some time. It may be necessary to add a drop of water to the edge of the coverslip if the preparation does not flatten, again tapping with a scalpel tip while observing with the microscope, or if the preparation begins to dry.

Conidial dimensions. Conidia of most *Trichoderma* species fall within the range of 2.0 to 3.5 μm long and 2.0 to 3.0 μm wide. Although this is a narrow range, conidial dimensions of species are even more narrowly circumscribed. It is important to measure at least 30 conidia for each collection. In some cases, it is important to determine the 95% confidence interval of the range since this statistic can point to one species or another when trying to decide from among closely similar species.

Conidial morphology. Conidia of most *Trichoderma* species are ellipsoidal to nearly oblong (side walls parallel, short cylindrical) and the ratio of length to width (L/W) is 1.5 or greater. In a smaller number of unrelated species, conidia are nearly globose (subglobose) and the L/W is 1.0–1.2 or ovoidal. Members of the Harzianum and Brevicompactum clades tend to have globose or subglobose conidia.

Conidial ornamentation. Conidia of most species are smooth when viewed with a light microscope. Conidia of a few species are roughened, but this ornamentation is often very fine and difficult or, indeed, impossible to see (e.g., *T. asperelloides*; *T. asperellum*, Fig. 6.7H, I). In other species, some or all of the conidia are tuberculate or blistered (e.g., *T. ghanense*, Fig. 6.20I, J; *T. saturnisporum*, Fig. 6.40I; *T. viride*, Fig. 6.48L). In *T. ghanense*, only some conidia are blistered and only one or two blisters may be visible on some conidia.

Conidiophores. Unlike genera of hyphomycetes such as *Gliocladium* and *Verticillium*, to name just two, most *Trichoderma* species lack a discrete conidiophore. Exceptions are those described below as having Gliocladium- and Verticillium-type branching patterns. However, in most *Trichoderma* species, conidia form in what were referred to as "pustules" by Bissett (1984, 1991a,b,c) or as "shrubs" by Jaklitsch (2009, 2011). Pustules or shrubs are aggregations of interwoven, branched hyphae, and phialides arise in characteristic fashion from the ends of these hyphae, near the surface of the pustule or shrub. These fertile parts comprise what we refer to here as the conidiophores. In many species, the conidiophores elongate for a distance beyond the point where most of the phialides arise. In the present work, we refer to these extensions as hairs. The hairs can be sterile or mostly sterile (e.g., *T. polysporum*, Fig. 6.34; *T. stromaticum*, Fig. 6.44), or there can be a whorl of phialides or a few short branches bearing phialides at the tip. Typically, the phialides that arise from the tips of the hairs are longer than those produced lower down, within the pustule. The hairs are smooth, thin walled, septate, and mostly unbranched. They can be straight (e.g., *T. brevicompactum*, Fig. 6.10; *T. strigosum*, Fig. 6.43; *T. stromaticum*, Fig. 6.44), sinuous and often whiplike (e.g., *T. capillare*, Fig. 6.11; *T. citrinoviride*, Fig. 6.12; *T. tomentosum*, Fig. 6.45), or spiraled (e.g., *T. polysporum*, Fig. 6.34B; *T. spirale*, Fig. 6.41). There are three more-or-less distinct types of *Trichoderma* conidiophores. Examples of these branching types can be seen in the figures of the cited species.

Gliocladium type: The degree of formation of pustules is minimal and the conidiophores may appear to be discrete, and in these cases, the phialides are arranged in a penicillus. In *T. deliquescens*, the conidiophore is discrete and may reach 1 mm in length. Examples are *T. deliquescens* (Fig. 6.14), *T. gelatinosum*, *T. pleuroti* (Fig. 6.32), and *T. virens* (Fig. 6.47).

Pachybasium type: Pustules are more-or-less distinct, 0.25–3 mm in diameter, and often hemispherical or pulvinate. Hyphae of the pustules are densely or loosely intertwined. In *T. stromaticum*, hyphae of the pustule are short and disarticulate easily. Conidiophores arise from hyphae of the pustule. They usually consist of a strongly developed main axis with fertile branches tending to be paired, arising at right angles from the main axis, and increasing in length with distance from the tip of the conidiophores and rebranching. The primary fertile branches are often no more than three or four cells in length. The secondary fertile branches are often reduced to single cells. The phialides tend to be crowded at the tips of branches and typically are ampulliform with a L/W of 1.5–2.0. The main axis and the fertile branches are typically broad, and the length of the phialide is often not much greater than the width of the cell that produced it. The conidiophore of most species having a Pachybasium-type branching pattern is extended into a fertile or sterile hair, with the fertile branches arising at the base of the extension. The extension of conidiophores may be sterile or may produce one or a few lageniform phialides from its tip and, less frequently, along the length on short branches. The extensions are thin walled, septate, branched to a greater or lesser degree, subacute or rounded at the tip, and smooth or, infrequently, ornamented. They are usually sinuous but may be spiraled or straight and more-or-less rigid or even awl shaped (e.g., *T. stromaticum*). Examples are *T. evansii* (Fig. 6.17), *T. hamatum* (Fig. 6.21), *T. polysporum* (Fig. 6.34), *T. strigosum* (Fig. 6.43), and *T. stromaticum* (Fig. 6.44).

Pyramidal type: Pustules are usually not well formed, and loosely joined conidiophores often cover extensive areas of a colony. Conidiophores are composed of a well-developed and often long main axis from which pairs or verticils of a few fertile branches often arise at regular and closely or widely spaced intervals. Fertile branches increase in length with distance from the tip of the main axis and rebranch. The secondary branches decrease in length with distance from the branching point. Thus, the conidiophores have a pyramidal appearance. Conidiophores do not extend into hairs, and there are no sterile hairs apart from conidiophores. Phialides arise in whorls at the tips of secondary branches and from the tip of the main axis. The whorls of phialides usually are composed of fewer than five phialides and are typically divergent or cruciate, the phialides arising at right angles to the cell that produced them; although phialides may also arise singly or in pairs. Sometimes, the terminal phialide in a whorl is longer and narrower than the rest. Phialides are typically lageniform, with a L/W greater than 2, and are greater than two times the width of the cell that produced them, but there is considerable variation in the length of the phialide, depending upon the species. The Pyramidal-type branching pattern intergrades with the Pachybasium-type pattern in species such as *T. harzianum* because of the close spacing of the fertile branches. The Pyramidal-type conidiophore has a decidedly pyramidal appearance thanks to the long or at least distinct main axis and the regularly and closely spaced, frequently paired primary and secondary fertile branches that arise at right angles from the main axis and increase in length with distance from the tip of the conidiophore. Bissett (1991a) included species such as *T. harzianum* in which the internode between fertile branches is very short in section *Pachybasium*. Examples are *T. atroviride* (Fig. 6.8) and *T. harzianum* (Fig. 6.22).

Longibrachiatum type: Pustules are small, less than 0.5 mm in diameter, and formed of loosely intertwined hyphae with long, completely fertile conidiophores extending outwardly to give the pustule a dentate, or spiky, appearance. Conidiophores typically are composed of a well-developed central axis from which solitary phialides arise for a considerable distance below the tip before the first fertile branches arise at right angles. Branches are usually solitary, not paired, and produce solitary phialides. A characteristic of Longibrachiatum-type branching is the formation of branches that consist of two cells: a basal cell arising from the main axis or a fertile branch, and a single terminal phialide. A short, spurlike, phialidic opening arises laterally from the basal cell at the septum separating the terminal and basal cells. This spurlike opening is referred to here as an intercalary phialide. Intercalary phialides are also found in other groups, such as the Viride clade, but they are not common there. Conidiophores of the Longibrachiatum type often terminate in whiplike, sterile or fertile extensions that reach beyond the surface of the pustule. Phialides are typically solitary but sometimes they are disposed in cruciate whorls. Typically, they are lageniform, nearly cylindrical, often hooked or sinuous, and have a length three times the width of the supporting cell. Conidiophore branching in *T. viride* is intermediate between the Longibrachiatum and Pyramidal types, having widely spaced fertile branches; solitary and whorled, frequently sinuous or hooked phialides; and conspicuous intercalary phialides. Examples are *T. longibrachiatum* (Fig. 6.26) and *T. reesei* (Fig. 6.37).

Acremonium and Verticillium types: *Acremonium*-like conidiophores, which bear one or two phialides (Gams, 1971), or *Verticillium*-like conidiophores, which bear phialides in divergent whorls, are formed by members of the Hypocreanum clade, such as *T. citrinum* and *T. sulphureum* (Bissett, 1991a; Jaklitsch, 2011; Overton, 2006a,b) and the agaric parasite *T. avellaneum* (Canham, 1969). These anamorphs are not found apart from their teleomorphs and are not included in the present work. Condiophores of *T. scalesiae* (Jaklitsch et al., 2006) and *T. paucisporum* are more-or-less *Verticillium*-like, as are the synanamorphs of several species of other more typical *Trichoderma* species (e.g., *T. brevicompactum*). Sometimes (e.g., *T. spirale*), *Verticillium*-like conidiophores are the only conidiophores produced in culture. Examples are *T. brevicompactum* (Fig. 6.10J) and *T. spirale* (Fig. 6.41M).

Indeterminate branching: Conidiophore branching of most *Trichoderma* species can be classified into one of the preceding five types, but in many cases, a single repeating branching pattern cannot be discerned. *Trichoderma viride* is cited above as an example of branching that is intermediate between the Longibrachiatum and Pyramidal types, having widely spaced fertile branches and both solitary and whorled phialides. *Trichoderma effusum* (Longibrachiatum clade) is morphologically unusual for a *Trichoderma* species. Many *Trichoderma* species that have green ascospores (Chaverri and Samuels, 2003; Jaklitsch, 2009) fall into this category, but they are not found in nature apart from their teleomorphs and none of them are included here. Examples are *T. aureoviride* (Fig. 6.9) and *T. effusum* (Fig. 6.15).

Phialides. One of the challenges to identifying *Trichoderma* species is the fact that cultures become overmature within a few days of the appearance of phialides. This is especially true of cultures on SNA and CMD and less so on PDA. When cultures are overmature, phialides can no longer be seen. They appear to dissolve after they produce a finite number of conidia and all that remains is a flat scar denoting their position on the conidiophore. This is why it is best to examine cultures before all conidia in a pustule have become green.

The following parameters of phialides are observed.

Phialide disposition: In *Trichoderma* species, phialides typically arise from a single point at the tip of a branch; are held in whorls of three to five or six clustered around a central phialide, and are inserted in such a way that the whorls diverge, that is, an angle of approximately 90° separates the phialides from each other. This can be seen in species such as *T. harzianum* (e.g., Fig. 6.22C, D) or *T. koningii* (e.g., Fig. 6.23D–F, H). In the Viride clade, pseudowhorls of phialides are often seen (e.g., *T. koningii*, Fig. 6.23G; *T. ovalisporum*, Fig. 6.30G, H) in which phialides are clustered at the end of a branch but they do not arise from a single point; rather, their insertion points are separated by a short internode. In species such as *T. deliquescens* (Fig. 6.14) and *T. virens* (Fig. 6.47D–J), the phialides are appressed to each other and form a penicillus. With Longibrachiatum-type branching (e.g., *T. longibrachiatum*, Fig. 6.26C–G; *T. orientale*, Fig. 6.29C–H), phialides tend to be solitary, especially for some distance behind the terminal part of the fertile branch. With Pachybasium-type branching, phialides tend to be squat, short, and broad at the base and are crowded along fertile branches to form grapelike clusters (botryose) (e.g., *T. minutisporum*, Fig. 6.28D–F; *T. polysporum*, Fig. 6.34G, H). There are intergradations among these types and in the Hypocreanum clade, which is not included here; phialides are solitary and *Acremonium*-like and, in *T. avellaneum*, also not included here, phialides are *Verticillium*-like.

Phialide shape and length: Phialides are described here as lageniform or ampulliform. Lageniform phialides (e.g., *T. longibrachiatum*, Fig. 6.26; *T. virens*, Fig. 6.47; *T. viride*, Fig. 6.48) are much longer than the width of the cell that produced them. Usually, they are somewhat swollen at or below the middle, but often they are nearly cylindrical. Especially in members of the Harzianum clade, in which phialides are relatively short and broad, the tip of the phialide is drawn to a fine tip. Lageniform phialides are usually straight but, in the Longibrachiatum and Viride clades, phialides are often sinuous or hooked. Ampulliform phialides (e.g., *T. hamatum*, Fig. 6.21; *T. polysporum*, Fig. 6.34) appear short and wide. They are typically no more than two times the width of the cell that produced them. Pachybasium-type phialides are typically ampulliform, while Longibrachiatum-type phialides are lageniform. However, there are intergradations, such as *T. harzianum* (Fig. 6.22). In the same culture, phialides can be either short and wide or long and narrow. The more the phialides are crowded on a conidiophore, the more they tend to be short and broad. This is seen in members of the *T. harzianum* complex, in which distance between fertile branches can be short, or in the Pachybasium type. Where there is a greater distance between fertile branches, the phialides tend to be longer (e.g., *T. asperellum*, Fig. 6.7E, F; *T. gamsii*, Fig. 6.19D–F). Often, when phialides are held in a cruciate whorl, the terminal phialide is longer than those formed below. Phialides that terminate hairs tend to be long and narrow, lageniform. Phialide length is measured from the tip to the septum that separates the phialide from the cell from which it arose. Phialide length can vary greatly within a single culture. In reporting phialide length for the present work, we averaged the length of the phialides taken from all levels of a single culture of all collections of a species.

Phialide width: Phialide width is the widest point of the phialide.

Ratio of phialide length to its width: In species having Pachybasium-type branching, the L/W is less than 2. The L/W in the other groups included here (Gliocladium, Longibrachiatum, Pyramidal, and Indeterminate types) is greater than 2, but within each group, the L/W varies considerably among the species.

Width of the cell that supports the phialide: This is the width of the hyphal cell from which the phialide arises. With Pachybasium-type branching especially, the cell that produces the phialides appears to be quite broad, at least as broad as the length of the phialide, whereas with Longibrachiatum-type branching, the supporting cell is nearly the same width as the widest part of the phialide. With Gliocladium-type branching, cells that produce the phialides, metulae, arise at up to three or four levels. As the distance from the stipe of the conidiophore increases, the length and width of these metulae decreases. The width of the metulae reported here is an average of measurements taken at all levels.

Ratio of phialide length and width to width of the supporting cell: These ratios serve to divide species into broad groups that tend to reflect the branching type of the conidiophore. Those with Pachybasium- and Gliocladium-type branching tend to have ratios below 2.7. The morphology of the phialide with Pyramidal-type branching is variable in shape and length, and thus, there is great variation in the relationship of the length and width of the phialide to the width of the cell that produced it. Members of the Viride clade, for example, have larger ratios and members of the Harzianum clade have smaller ratios approaching 1.

Intercalary phialides: The intercalary phialide is a short, lateral, phialidic spur that forms at the septum below the phialide and on the side of the cell that separates the phialide from the cell from which it arose. These are characteristic of Longibrachiatum-type branching (e.g., *T. citrinoviride*, Fig. 6.12J; *T. longibrachiatum*, Fig. 6.26E, H; *T. reesei*, Fig. 6.37E, F, I, J) but are also seen in the Viride clade (e.g., *T. gamsii*, Fig. 6.19E, F; *T. koningiopsis*, Fig. 6.24J). In *T. effusum* (Fig. 6.15B, C, I), short phialidic openings occur at the base of the phialides or along hyphae; these are termed aphanophialides (Bissett et al., 2003; Gams, 1971). Aphanophialides and intercalary phialides appear to be developmentally distinct. The former appear to develop at random along hyphae, whereas development of the intercalary phialides seems to be highly specific, always forming at the septum that delimits a terminal phialide.

Phialide proliferation: Within a few days of the onset of conidial production, phialides of most species stop producing conidia and disintegrate. Phialides of some members of the Viride clade are unusual in that they proliferate through the opening of the phialide to produce a second phialide or, in some cases, a chain of several phialides. Phialides in these cases tend to be elongated and narrow and often have an abrupt subglobose swelling toward the middle. The older phialides sometimes become greatly swollen and distorted and conidiophores may lose all sense of pattern. *Eidamia viridescens* (= *T. viridescens*), the type of a monotypic genus, was isolated from moldy apples in the United Kingdom and was characterized by these extreme characters (Jaklitsch et al., 2006). In other species, no greatly swollen cells or chains of proliferated phialides are seen but there are occasional elongated, sometimes once-proliferated phialides. Development of the *Eidamia* morphology occurs on richer media, such as PDA or CMD, but not on the carbohydrate-poor medium SNA. We do not know the biological significance of this phenomenon, but from a taxonomic point of view, the percurrent proliferation of phialides is diagnostic of the Viride clade (e.g., Jaklitsch et al., 2006). Examples are *T. gamsii* (Fig. 6.19G, H), *T. koningiopsis* (Fig. 6.24I), and *T. viridescens* (Fig. 6.49E, F).

Chlamydospores. These thick-walled cells form within hyphal cells or they terminate short branches of hyphae. They are subglobose to ovoidal or pyriform, or they conform to the shape of the cell in which they form. They are usually solitary and 20 µm or less in diameter, but they may form large aggregates or may form in great numbers along an individual hypha. For the present work, the presence or absence of chlamydospores is observed within 10 days on CMD or SNA. Most *Trichoderma* species produce chlamydospores but not all produce them within 10 days. They are most easily observed by inverting a petri plate on the stage of a compound microscope and searching from one edge of the plate to the other using a 40× objective. Chlamydospores are particularly common in cultures of *T. virens*. *Trichoderma peltatum* is known in nature only from its distinctive stromata; in culture, it does not produce a recognizable *Trichoderma* morphology. Rather, in culture, it produces great numbers of chlamydospore-like structures at the ends of hyphae (Samuels and Ismaiel, 2011). Examples are *T. reesei* (Fig. 6.37L), *T. stromaticum* (Fig. 6.44M), and *T. virens* (Fig. 6.47L).

CHAPTER 5

Synpotic Key to Forty-Five *Trichoderma* Species

Using the Key

This key includes the characters that are quantifiable or otherwise easily seen using a light microscope or stereomicroscope. Use of the key will most likely not lead to an unequivocal determination but rather, in most cases, lead to two or more phenotypically similar species, which may be distinguished by comparing illustrations and reading descriptions. It must be borne in mind that we have included the most common species along with some less common and other morphologically atypical species, but many species are not included.

The descriptions are based on our own observations and are a composite of type and other reliably identified strains. The measurements of conidia and phialides included in the key are based on the mean of all measurements of each parameter, a minimum of 30 individual measurements for each available strain of a species. In most cases, measurements and descriptions of conidiation are taken from Spezieller Nährstoffarmer agar (SNA), less frequently from cornmeal dextrose agar (CMD), after approximately 1 week at 25°C under a regimen of alternating 12 h of cool white fluorescent light and 12 h of darkness. The unit of measurement for growth is millimeters (mm). Colony characters are taken from potato dextrose agar (PDA) and SNA cultures incubated in 9-cm-diameter petri plates for 72 h at 25 and 35°C under light, as described above. A **bold** species number indicates that the number is found also in another of the options for that character and that the average measurement straddled the boundary between the two options, except in the case of conidium color and ornamentation, in which the alternatives are green and smooth, respectively.

The descriptions of species provided here are brief since the emphasis in this book is on illustrations. More complete descriptions, including descriptions of teleomorphs, are available in the references that we provide. We provide GenBank numbers for reliably identified exemplars of the species.

Synoptic Key

These are morphological characters taken from young pustules that have newly formed conidia in cultures grown on SNA or CMD at 25°C under alternating cool white fluorescent light and darkness. The ratio of length to width is abbreviated as L/W.

Trichoderma Species Included in the Synoptic Key

1. *T. aggressivum*
2. *T. amazonicum*
3. *T. asperellum*
4. *T. atroviride*
5. *T. aureoviride*
6. *T. brevicompactum*
7. *T. capillare*
8. *T. citrinoviride*
9. *T. crassum*
10. *T. deliquescens*
11. *T. effusum*
12. *T. erinaceus*
13. *T. evansii*
14. *T. flagellatum*
15. *T. gamsii*
16. *T. ghanense*
17. *T. hamatum*
18. *T. harzianum*
19. *T. koningii*
20. *T. koningiopsis*
21. *T. lieckfeldtiae*
22. *T. longibrachiatum*
23. *T. longipile*
24. *T. minutisporum*
25. *T. orientale*
26. *T. ovalisporum*
27. *T. paucisporum*
28. *T. pleuroti*
29. *T. pleuroticola*
30. *T. polysporum*
31. *T. pseudokoningii*
32. *T. pubescens*
33. *T. reesei*
34. *T. rossicum*
35. *T. saturnisporopsis*
36. *T. saturnisporum*
37. *T. spirale*
38. *T. stilbohypoxyli*
39. *T. strigosum*
40. *T. stromaticum*
41. *T. tomentosum*
42. *T. velutinum*
43. *T. virens*
44. *T. viride*
45. *T. viridescens*

CONIDIA

Conidium Average Length

≤3.7 µm: 1, 2, **3**, 4, **5**, 6, **7**, **8**, 10, **12**, 13, **14**, 18, 20, **21**, **22**, **24**, **26**, 27, **28**, **29**, 30, **31**, **33**, **38**, 41, **42**, **45**

>3.7 to ≤4.5 µm: **3**, **5**, **7**, **8**, **9**, **12**, **14**, **15**, 17, 19, **20**, **21**, **22**, **24**, **25**, **26**, **28**, 29, **31**, **32**, **33**, **34**, **35**, 36, 37, **39**, 40, **42**, **43**, 44, **45**

>4.5 µm: **5**, **8**, **9**, 11, **14**, **15**, 16, 23, **25**, **27**, **31**, **32**, **34**, **35**, **38**, **39**, **43**

Conidium Average Width

≤2.7 µm: **1**, 2, **5**, 6, **7**, 8, **9**, 10, **12**, 14, **15**, **17**, **18**, **19**, **20**, 21, **22**, **23**, 24, **25**, **27**, 28, **29**, 30, **31**, 32, **33**, 37, **38**, 39, 41, 42

>2.7 µm: **1**, 3, 4, **5**, **7**, **9**, 11, **12**, 13, **15**, 16, **17**, **18**, **19**, **20**, **22**, **23**, **25**, 26, **27**, **29**, **31**, **33**, 34, 35, 36, **38**, 40, 43, 44, 45

Conidium Average L/W

≤1.3: 1, **2**, 3, 4, **5**, 6, **7**, **9**, **10**, **12**, 13, **17**, 18, **24**, 26, **27**, **29**, **33**, **35**, **36**, **38**, 43, 44, 45

>1.3 to ≤1.5: **2**, **5**, 7, **9**, **10**, **12**, **14**, **15**, **17**, **20**, **21**, **22**, **24**, 25, 27, **28**, **30**, **31**, **33**, **34**, **35**, **36**, **38**, **40**, **41**

>1.5: **5**, 8, **10**, 11, **14**, **15**, 16, **17**, 19, **20**, **21**, **22**, 23, **25**, **27**, **28**, **29**, **30**, **31**, 32, **33**, **34**, **36**, 37, 39, **40**, **41**, 42

Conidium Ornamentation

(All appear smooth in the light microscope, except the following.)

Roughened or tuberculate: 3, 7, **16**, 35, 36, 44, **45**

Conidium Color

(All are some shade of green, except the following. Bold numbers indicate the possibility of green or another color of conidia in the same petri dish.)

White in mass: **7**, 30

Yellow in mass: **27**, **40**, **45**

PHIALIDES

Phialide Average Length

≤6.5 µm: **1**, 2, **3**, **6**, **8**, **12**, 13, **14**, **16**, **17**, **18**, **20**, 21, **22**, **23**, **24**, **26**, **28**, **29**, 30, **31**, 32, **34**, **35**, **37**, **38**, **39**, **40**, 41, **42**

>6.5 to ≤8.5 µm: **1**, **3**, **4**, **6**, 7, **8**, **9**, **10**, **12**, **14**, **16**, **17**, **18**, 19, **20**, **22**, **23**, **24**, **25**, **26**, **27**, **28**, **29**, **31**, 33, **34**, **35**, **36**, **37**, **38**, **39**, **40**, **42**, **44**, **45**

>8.5 µm: **4**, 5, 7, **8**, **9**, **10**, 11, **12**, 15, **16**, **20**, **22**, **25**, **26**, **27**, **28**, **31**, **35**, **36**, **38**, **39**, 43, **44**, **45**

Phialide Average Width

≤2.0 µm: **10**, **14**, **44**

>2.0 to ≤3.2 µm: 1, **2**, **3**, **4**, 5, **6**, 7, **8**, **9**, **10**, 11, **12**, **14**, 15, **16**, 19, **20**, **21**, 22, **24**, **25**, **26**, **27**, **28**, **29**, **30**, **31**, 33, **35**, **36**, **38**, **39**, **40**, **41**, **42**, **44**, 45

>3.2 µm: **2**, **3**, **4**, **6**, 7, **8**, **9**, **12**, 13, **16**, 17, 18, **20**, **21**, 23, **24**, **25**, **26**, **27**, **28**, **29**, **30**, **31**, 32, 34, **35**, **36**, 37, **38**, **39**, **40**, **41**, **42**, 43

Phialide Average L/W

≤2.0: **1**, **2**, **3**, **4**, **6**, 7, **8**, **12**, 13, **16**, **17**, **18**, 19, **20**, 21, **22**, 23, **24**, **25**, **26**, 27, **28**, **29**, 30, **31**, 32, 34, **35**, **36**, 37, **38**, **39**, 40, 41, 42

>2.0 to ≤2.7: **1**, **2**, **3**, **4**, **6**, 7, **8**, **9**, **12**, 14, **15**, **16**, **17**, **18**, **20**, **22**, **24**, **25**, **26**, **28**, **29**, **31**, 33, **35**, **36**, **38**, **39**, 43, **44**, **45**

>2.7: **3**, **4**, 5, 7, **8**, **9**, 10, 11, **12**, **15**, **20**, **22**, **25**, **26**, **28**, **29**, **31**, **35**, **36**, **38**, **39**, **44**, **45**

Average Width of Cell Supporting the Phialide

≤2.5 µm: **4**, **5**, 7, **8**, **9**, **10**, **12**, 15, 18, 19, **20**, **21**, **22**, **25**, 27, **35**, **38**, **45**

>2.5 to ≤3.2 µm: 1, **2**, 3, **4**, **5**, **6**, 7, **8**, **9**, **10**, 11, **12**, 14, 16, **17**, **20**, **21**, **22**, **23**, **25**, **26**, **27**, **28**, 29, **30**, 31, 33, **35**, **36**, **37**, **38**, **39**, **41**, 42, 44, **45**

>3.2 µm: **2**, **6**, 7, **9**, **10**, **12**, 13, **17**, **21**, **23**, 24, **25**, **26**, **28**, **30**, 32, 34, **36**, **37**, **39**, 40, **41**, 43

Ratio of Phialide Length to Width of Its Supporting Cell

≤1.7:	**2**, **6**, **8**, 10, **13**, **14**, **17**, **21**, **23**, **24**, **29**, 30, 32, 34, **37**, **39**, **40**, **41**, 42
>1.7 to ≤2.7:	**1**, **2**, **3**, **4**, **6**, **7**, **8**, **13**, **14**, **15**, **16**, **17**, **18**, **20**, **21**, **22**, **23**, **24**, **25**, **26**, 28, **29**, **31**, **35**, **36**, **37**, **38**, **39**, **40**, **41**, 43, **45**
>2.7 to ≤3.5:	**1**, **3**, **4**, **7**, **8**, **9**, 12, **14**, **15**, **16**, **18**, 19, **20**, **22**, **25**, **26**, **27**, **31**, 33, **35**, **36**, **38**, **39**, **44**, **45**
>3.5:	5, 7, **8**, **9**, 11, **15**, **22**, **25**, **26**, **27**, **31**, **35**, **36**, **38**, **44**, **45**

Ratio of Phialide Width to Width of Its Supporting Cell

≤0.9:	**4**, **5**, **6**, **8**, **9**, 10, **13**, **17**, **20**, **21**, **22**, **23**, **24**, **25**, **26**, **27**, **28**, **29**, **30**, **31**, **34**, **36**, **37**, **40**, 42, **43**
>0.9 to ≤1.3:	1, 2, 3, **4**, **5**, **6**, 7, **8**, **9**, 11, **12**, **13**, 14, 15, **16**, **17**, 18, 19, **20**, **21**, **22**, **23**, **24**, **25**, **26**, **27**, **28**, **29**, **30**, **31**, 32, 33, **34**, **35**, **36**, **37**, **38**, 39, **40**, 41, **43**, 44, **45**
>1.3:	**7**, **8**, **9**, **12**, **16**, **20**, **21**, **26**, **31**, **35**, **36**, **38**, **45**

BRANCHING TYPE OF CONIDIOPHORES FROM SNA

(See description of branching types on pages 64–66.)

Pyramidal:	1, 3, 4, 12, 15, 18, 19, 20, 26, 38, 39, **44**, **45**
Pachybasium:	2, 6, **9**, 13, 17, 21, 23, 24, 29, 30, 32, 34, **37**, 40, 41, 42
Longibrachiatum:	7, 8, 14, 16, 22, 25, 31, 33, 36
Gliocladium:	**9**, 10, 28, 43
Acremonium and Verticillium:	**27**, **37**
Indeterminate:	5, 11, **27**, 35, **44**, **45**

PROTRUDING HAIRS, STERILE OR FERTILE AT THE TIP ON SNA

(Otherwise hairs are absent.)

Present:	6, 7, 12, 13, 17, 21, 23, 30, 32, 34, 35, 37, 39, 40, 41, 42

GROWTH RATE

(Colonies are inoculated with a 5-mm plug taken from the edge of an actively growing colony on CMD and grown in darkness in 9-cm-diameter petri dishes containing freshly prepared medium.)

On SNA, Colony Radius (Average) After 72 h at 35°C

No growth:	**2**, 5, 9, 13, **15**, 17, **20**, 21, **26**, **27**, 32, 34, **40**, **41**, 44, 45
≤20 mm:	**1**, **3**, 4, 6, 11, **12**, **15**, 19, **20**, 23, 24, **26**, **27**, **28**, 29, 30, **35**, 36, 37, 38, 39, **40**, **41**, **42**
>20 to ≤50 mm:	**1**, **2**, **3**, 10, **12**, 14, 18, **22**, **28**, **31**, **33**, **35**, **42**, 43
>50 mm:	7, 8, 16, **22**, 25, **31**, **33**

On SNA, Colony Radius (Average) After 72 h at 25°C

≤29 mm:	**3**, 5, **10**, 11, **23**, **24**, 27, 30, **32**, 34, **35**, 36, **38**, **39**, 40, **44**, 45
>29 to ≤50 mm:	2, **3**, 4, 6, 7, **8**, **9**, **10**, **12**, 13, 14, 17, **18**, 19, 20, 21, **22**, **23**, **24**, **25**, 26, 28, **31**, **32**, **33**, **35**, **37**, **38**, **39**, 41, 42, **43**, **44**
>50 mm:	1, **8**, **9**, **12**, 15, 16, **18**, **22**, **25**, 29, **31**, **33**, **37**, **43**

On PDA (Difco), Colony Radius (Average) After 72 h at 35°C

No growth:	**2**, 5, 9, 13, 17, 21, **26**, **27**, 32, 34, **40**, **41**, 44, 45
≤15 mm:	1, 4, 6, **12**, 19, 20, 24, **26**, **27**, 29, 30, 37, 38, **39**, **40**, **41**, **42**
>15 to ≤50 mm:	**2**, 3, 10, 11, **12**, 23, 28, **35**, 36, **39**, **42**, 43
>50 mm:	**2**, 7, 8, 14, 15, 16, 18, 22, 25, 31, 33, **35**

On PDA (Difco), Colony Radius (Average) After 72 h at 25°C

≤40 mm: 5, **6**, **10**, 11, **23**, 24, 27, **28**, 30, 32, 36, **37**, **38**, **40**, **43**, 44, 45

>40 to ≤60 mm: 1, 2, 3, **4**, **6**, **8**, 9, **10**, 13, 17, 18, 19, **20**, **23**, **28**, 29, **31**, **33**, 34, **35**, **37**, **38**, **39**, **40**, 41, **42**, **43**

>60 mm: **4**, 7, **8**, 12, 14, 15, 16, **20**, 21, 22, 25, 26, **31**, **33**, **35**, **39**, **42**

CHAPTER 6

Trichoderma Species Descriptions

In the following descriptions, growth rates are reported as the colony diameter after incubation in darkness (with intermittent light when cultures were measured). The radius is measured from the edge of the 3- to 5-mm plug of inoculum to the most distant point of the colony. The colony radius is given in millimeters (mm). The description of conidiation and the observations of conidia and phialides are taken from colonies grown on Spezieller Nährstoffarmer agar (SNA) for 1 week at 25°C under a 12-h darkness/12-h cool white fluorescent light regime. Chlamydospores are observed at 20× by scanning a colony in an inverted 9-cm-diameter petri plate using bright-field microscopy. Illustrations of colonies grown in petri plates are shown in Figures 6.1–6.4.

(*figure appears on page 74*) ➤

Figure 6.1. Cultures of *Trichoderma* species. Except where noted, cultures were grown for 1 week at 25°C under cool white fluorescent light; potato dextrose agar (PDA) on the left, Spezieller Nährstoffarmer agar (SNA) on the right. **A,** *T. aggressivum.* **B,** *T. amazonicum.* **C,** *T. asperellum.* **D,** *T. atroviride.* **E,** *T. aureoviride*, crystals in malt extract agar. **F,** *T. brevicompactum.* **G,** *T. capillare.* **H,** *T. citrinoviride.* **I,** *T. crassum.* **J,** *T. deliquescens* on PDA. **K,** *T. effusum.* **L,** *T. erinaceus* was grown for 1 week at 25°C in the dark.

(*figure appears on page 75*) ➤

Figure 6.2. Cultures of *Trichoderma* species. Cultures were grown at 25°C under cool white fluorescent light; potato dextrose agar on the left, Spezieller Nährstoffarmer agar (SNA) on the right. **A,** *T. evansii.* **B,** *T. flagellatum.* **C,** *T. gamsii.* **D,** *T. ghanense.* **E,** *T. hamatum.* **F,** *T. harzianum.* **G,** *T. koningii.* **H,** *T. koningiopsis.* **I,** *T. lieckfeldtiae.* **J,** *T. longibrachiatum.* **K,** *T. longipile.* **L,** *T. minutisporum.* **A–C and J–L,** Cultures were grown for 1 week. **D and F–H,** Cultures were grown for 96 h. **E and I,** Cultures were grown for 10 days.

(*figure appears on page 76*) ➤

Figure 6.3. Cultures of *Trichoderma* species. Cultures were grown at 25°C under cool white fluorescent light; potato dextrose agar (PDA) on the left, Spezieller Nährstoffarmer agar (SNA) on the right. **A,** *T. orientale* (right is PDA reverse to show diffusing yellow pigment). **B,** *T. ovalisporum.* **C,** *T. paucisporum*, a conidial pustule on PDA is shown on the right. **D,** *T. pleuroti.* **E,** *T. pleuroticola.* **F,** *T. polysporum.* **G,** *T. pseudokoningii.* **H,** *T. pubescens.* **I,** *T. reesei.* **J,** *T. rossicum.* **K,** *T. saturnisporopsis.* **L,** *T. saturnisporum.* **A, G, I, and K,** Cultures were grown for 1 week. **B, C, and L,** Cultures were grown for 96 h. **D, E, and J,** Cultures were grown for 10 days. **F and H,** Cultures were grown for 2 weeks.

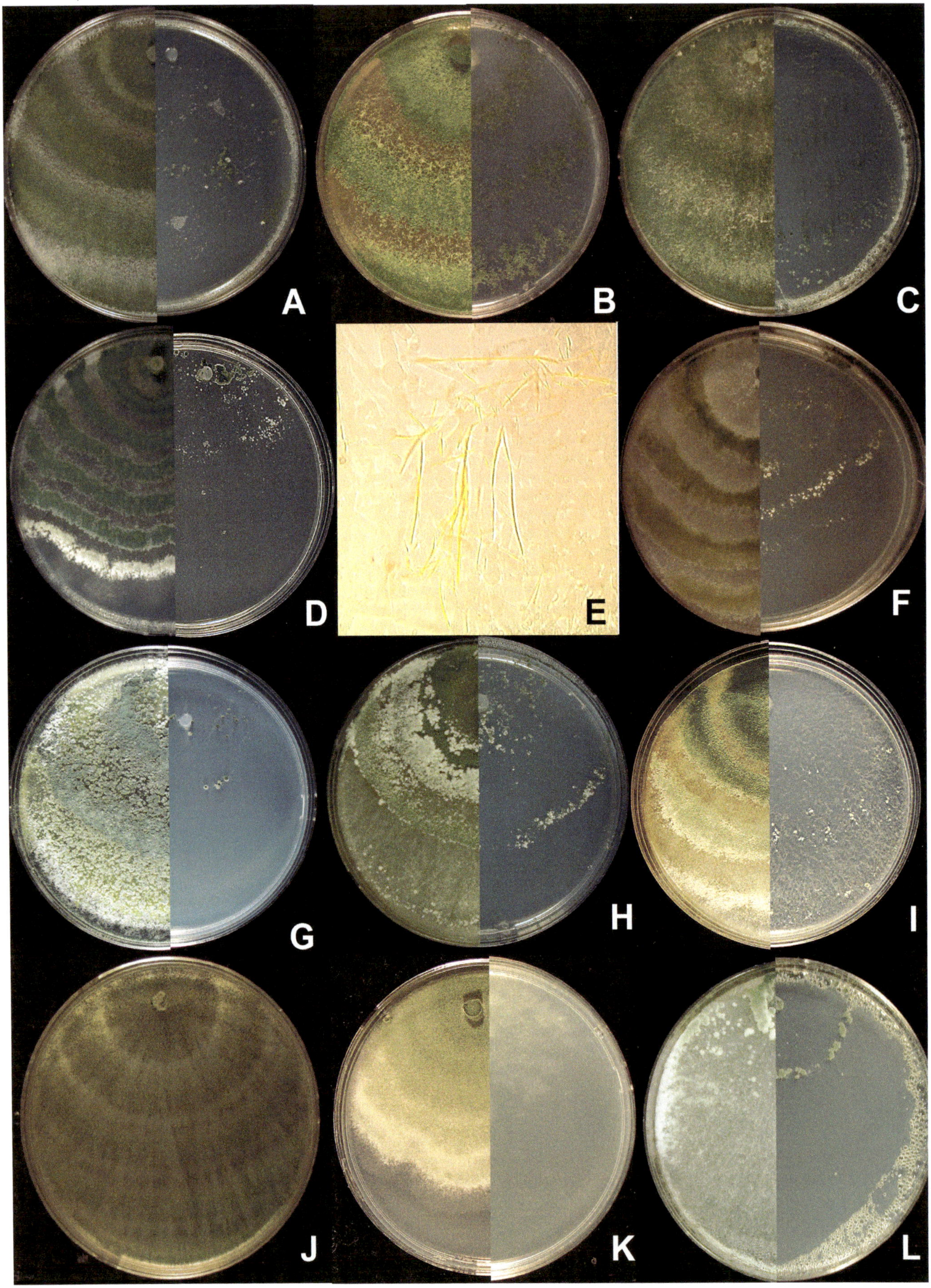

Figure 6.1

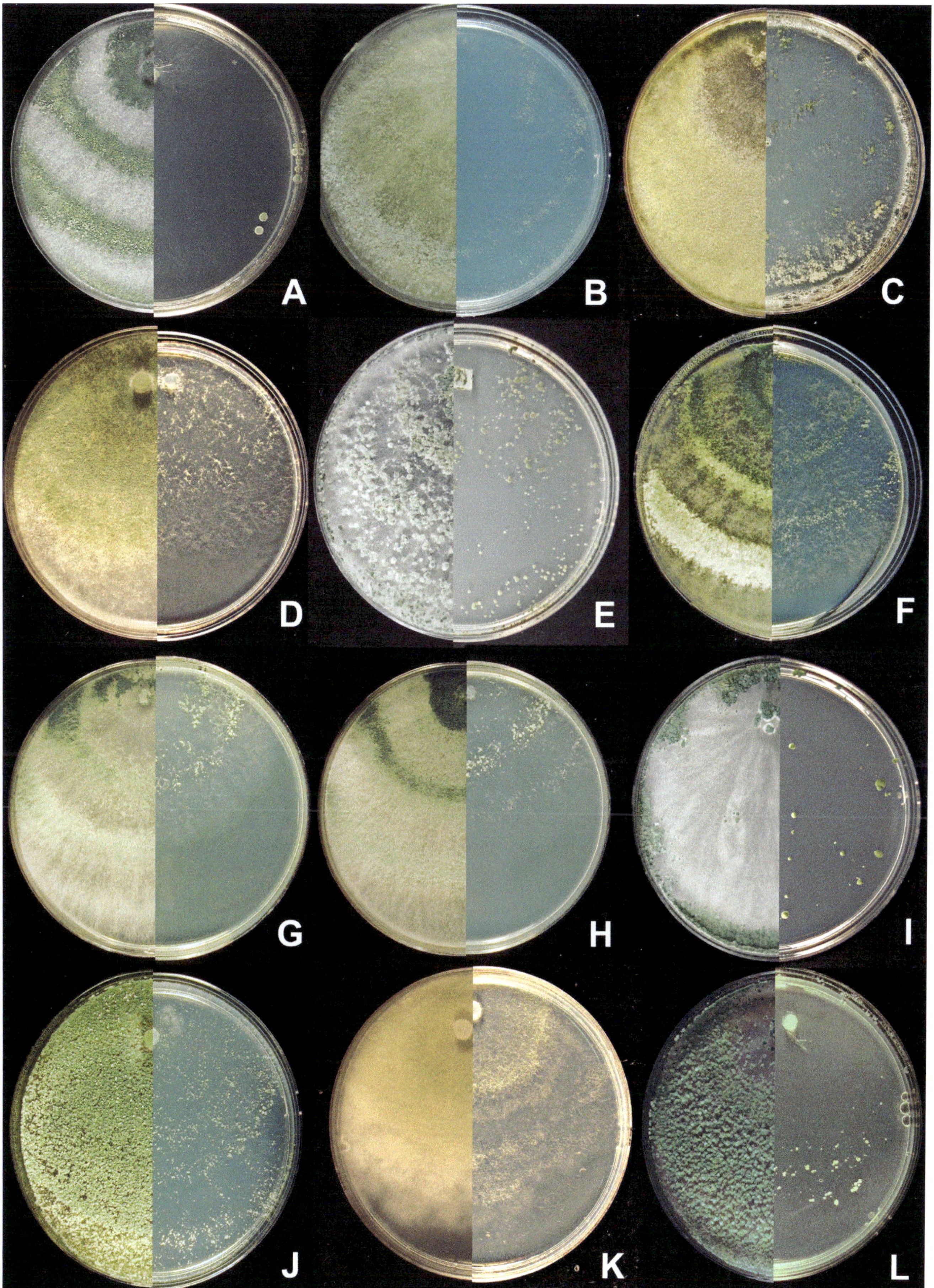

Figure 6.2

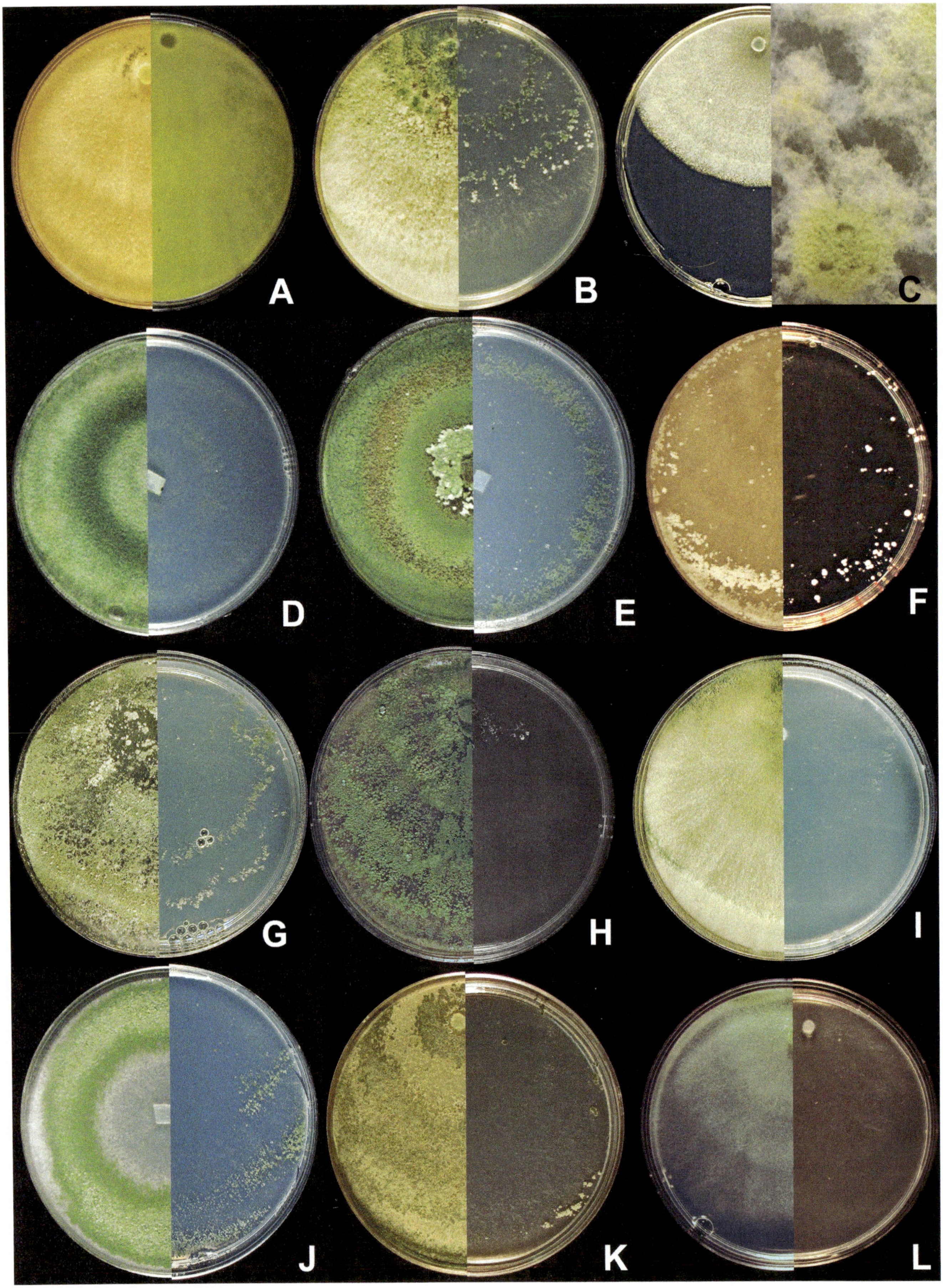

Figure 6.3

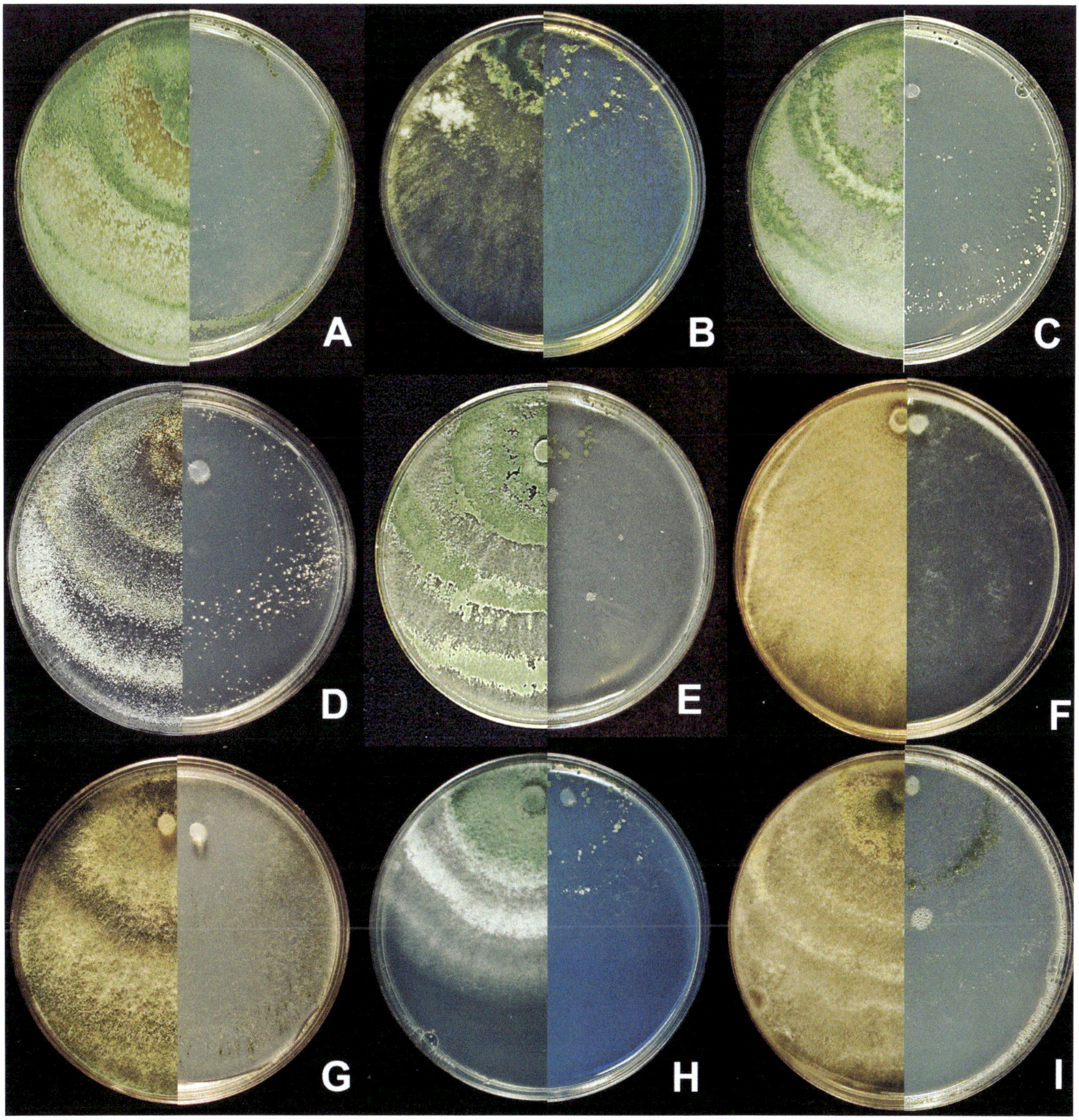

Figure 6.4. Cultures of *Trichoderma* species. Cultures were grown at 25°C under cool white fluorescent light (except where noted); potato dextrose agar on the left, Spezieller Nährstoffarmer agar (SNA) on the right. **A,** *T. spirale*. **B,** *T. stilbohypoxyli*. **C,** *T. strigosum*. **D,** *T. stromaticum*. **E,** *T. tomentosum*. **F,** *T. velutinum*, grown in darkness. **G,** *T. virens*. **H,** *T. viride*. **I,** *T. viridescens*. **A, C–E, G, and I,** Cultures were grown for 1 week. **B, F, H,** Cultures were grown for 96 h.

Trichoderma aggressivum

Samuels & W. Gams *in* Dodd et al., Mycologia 94:167. 2002. Figs. 6.1A, 6.5.

Ex-type culture: DAOM 222156 = IMI 393971
GenBank: *rpb2*: AF545541; *tef1*: AF348098

Clade: Harzianum

Colony radius at 25°C: PDA (33–)45–60. **SNA** (42–)55–65(–68). No diffusing pigment or distinctive odor on PDA.

Colony radius at 35°C: PDA (1–)3–10(–12). **SNA** (4–)8–25(–32).

Conidiation: Pustules scattered throughout the colony, 1–2 mm in diameter, loosely cottony, grayish green; conidia appearing dry.

Microscopic characteristics: Branching pattern typically Harzianum type with phialides held in divergent whorls on frequently paired or verticillately whorled branches; solitary phialides and unilateral branching and extended conidiophores common. **Sterile hairs** not formed. **Phialides** ampulliform, sharply constricted below the tip to form a narrow neck and slightly constricted at the base, (4.0–)5.7–7.7(–21.0) µm long, (1.3–)2.7–3.5(–4.3) µm wide at the widest point, L/W (1.2–)1.7–2.3(–8.0). **Supporting cell** (1.5–)2.5–3.7(–5.2) µm wide; ratio of length of phialide to width of supporting cell (1.0–)1.7–3.2(–8.0); ratio of width of phialide to width of supporting cell 1. **Conidia** subglobose to ovoidal, (2.0–)2.5–3.2(–3.7) × (2.0–)3.0–3.5(–5.5) µm, L/W (0.7–)1.0–1.3(–1.7), smooth, green. **Chlamydospores** few.

Distinguishing characters: *Trichoderma aggressivum* is most similar to *T. harzianum* from which it differs in its frequently inequilaterally branched conidiophores and generally very slow rate of growth at 35°C.

Habitat: Mushroom compost, causing green mold disease of commercial mushrooms (*Agaricus bisporus*). Not found apart from mushroom compost. On *Pleurotus ostreatus*, compare with *T. pleuroti* and *T. pleuroticola.*

Distribution: North America, Europe.

Literature: Samuels et al. (2002), Seaby (1998; as *T. harzianum* Group IV).

Figure 6.5. *Trichoderma aggressivum.* **A–H,** Conidiophores. **I,** Conidia. **A, B, E, F, H, and I,** From cornmeal dextrose agar. **C, D, and G,** From Spezieller Nährstoffarmer agar (SNA). Scale bars: **A–G** = 20 µm; **H, I** = 10 µm.

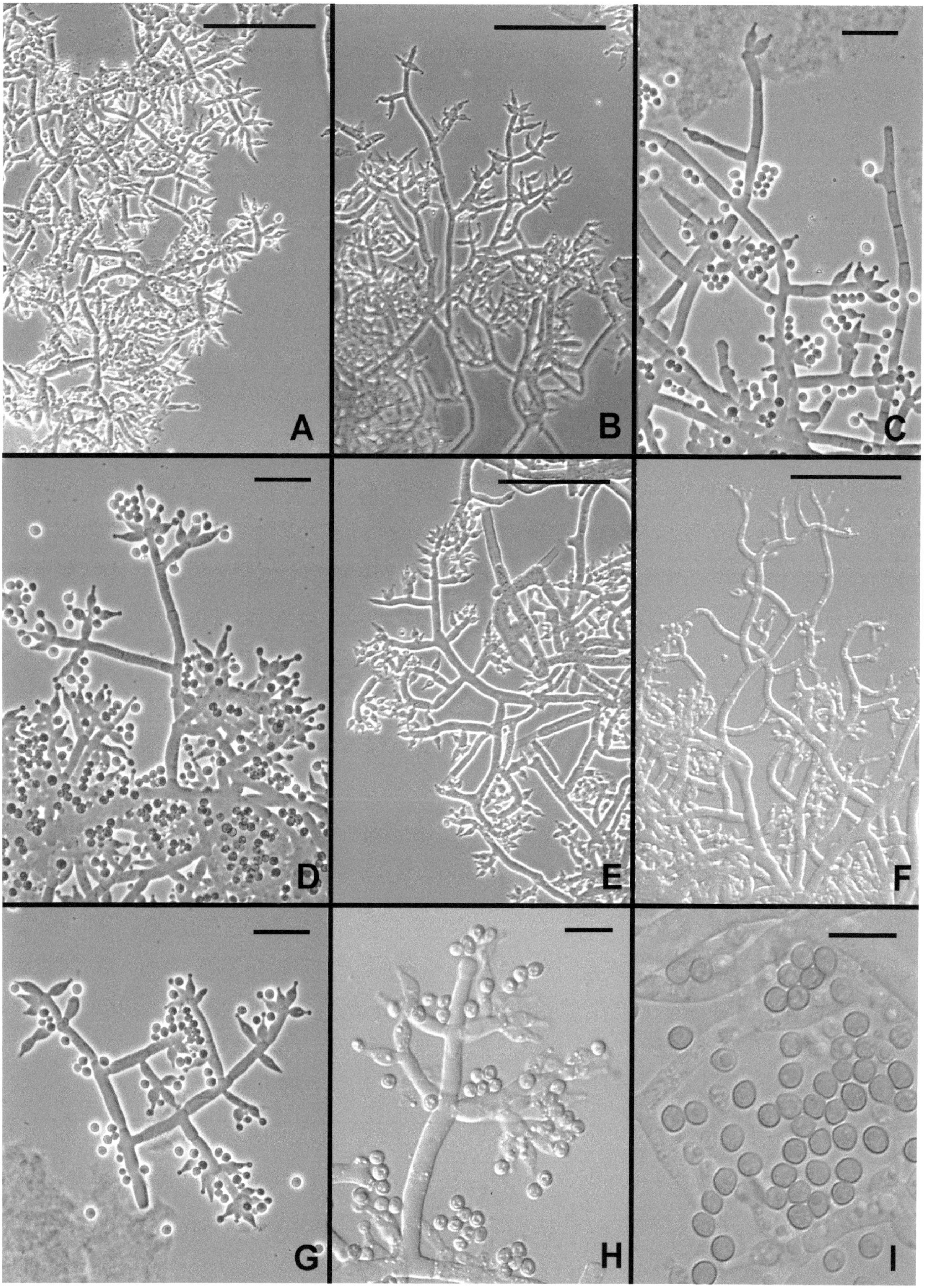
A
B
C
D
E
F
G
H
I

Trichoderma amazonicum

P. Chaverri & Gazis *in* Chaverri et al., Mycologia 103:146. 2011. Figs. 6.1B, 6.6.

Ex-type culture: CBS 126898
GenBank: *rpb2*: HM142368; *tef1*: HM142376

Clade: Harzianum

Colony radius at 25°C: PDA 55–60. **SNA** 45–50. No diffusing pigment or distinctive odor on PDA or a slightly fruity odor and/or brownish diffusing pigment observed in some cultures.

Colony radius at 35°C: PDA 0 or 40–55. **SNA** 0 or 35–45.

Conidiation: Discrete pustules not forming or, at most, conidiophores loosely aggregated into ill-defined pustules in the colony center, typically conidia forming in broad, more-or-less continuous, cottony concentric rings, green; conidia appearing dry.

Microscopic characteristics: Branching pattern Pachybasium type with a well-defined, conspicuously broad (3–5 µm wide) main axis and broad primary branches arising along the length in whorls of two or three, increasing in length with distance from the tip, rebranching or producing whorls of phialides directly; secondary branches one or a few cells in length, terminating in three to several phialides. **Phialides** ampulliform, typically sharply constricted to a narrow neck, 4.5–6.5 µm long, 2.7–3.7 µm at the widest point, L/W 1.5–2.2. **Supporting cell** 3–4 µm wide; ratio of length of phialide to width of supporting cell 1.4–2.0; ratio of width of phialide to width of supporting cell approximately 1. **Conidia** broadly ellipsoidal to subglobose, (1.7–)2.2–2.7(–3.0) × (2.0–)2.5–3.2(–3.5) µm, L/W (0.8–)1.0–1.4(–1.6), smooth, green. Discrete **chlamydospores** not observed but tight clusters of globose to subglobose chlamydospore-like structures observed after 3 weeks on PDA; these structures are capable of germinating.

Distinguishing characters: The conidiophores with a strongly developed and conspicuously broad main axis and uniformly spaced whorls of branches are characteristic. The closest morphological comparison is with *T. pleuroticola*, but conidia of *T. pleuroticola* are formed in more-or-less discrete pustules. Chaverri et al. (2011) reported no growth for endophytic strains of *T. amazonicum* at 35°C and somewhat larger conidia; the description given here is based on strains isolated from soil in Peru, which did grow at 35°C.

Habitat: Endophytic in sapwood and leaves of *Hevea* species and in soil.

Distribution: Common in Amazonian Peru, not known from elsewhere.

Literature: Chaverri et al. (2011).

Figure 6.6. *Trichoderma amazonicum.* **A and B,** Pustules of conidia. **C–H,** Conidiophores. **I and J,** Phialides. **K,** Conidia. **A–K,** From Spezieller Nährstoffarmer agar (SNA). Scale bars: **A** = 1 mm; **B** = 0.5 mm; **C–E, G, H** = 20 µm; **F, I–K** = 10 µm.

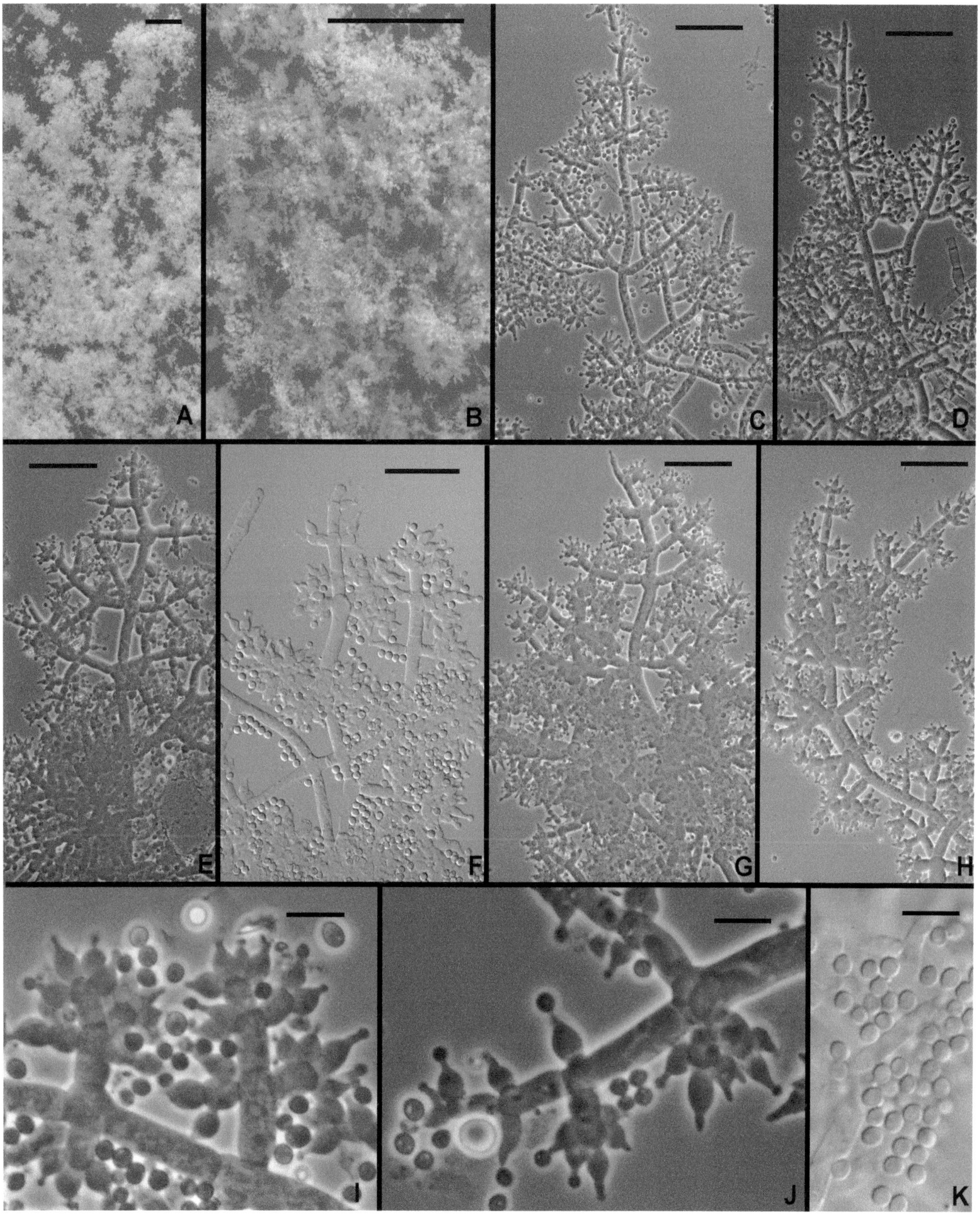
A
B
C
D
E
F
G
H
I
J
K

Trichoderma asperellum

Samuels, Lieckf. & Nirenberg *in* Samuels et al., Sydowia 51:81. 1999. Figs. 6.1C, 6.7.

Ex-type culture: CBS 433.97
GenBank: *rpb2*: EU248617; *tef1*: AF456907, AY376058

Clade: Viride

Colony radius at 25°C: PDA (10–)37–57(–60). **SNA** (10–)28–46(–48). No pigment or distinctive odor on PDA.

Colony radius at 35°C: PDA (2–)17–37(–45). **SNA** (1–)13–37(–43).

Conidiation: Pustules forming abundantly in two or more often conspicuous concentric rings, 1–2 mm in diameter, hemispherical, loosely organized with conidiophores visible with stereomicroscope, dark green; conidia appearing dry.

Microscopic characteristics: Branching pattern Pyramidal type with verticillate or frequently conspicuously paired lateral branches arising from a well-developed main axis; lateral branches rebranching; all branches arising at approximately 90° with respect to the subtending branch. **Sterile hairs** not formed. **Phialides** terminating branches, held in divergent whorls of two to four, lageniform, (4.0–)6.2–9.7(–15.2) µm long, (2.5–)3.0–3.5(–5.0) µm at the widest point, L/W (1.3–)1.7–3.1(–5.4). **Supporting cell** (1.7–)2.5–3.5(–5.7) µm wide; ratio of length of phialide to width of supporting cell (1.7–)2.0–3.4(–4.8); ratio of width of phialide to width of supporting cell (0.6–)0.9–1.3(–1.9). **Conidia** subglobose to ovoidal, (2.2–)3.0–3.5(–4.7) × (3.0–)3.5–4.0(–4.7) µm, L/W (0.9–)1.0–1.3(–1.7), green, finely warted (often appearing smooth). **Chlamydospores** not observed in all cultures, subglobose, terminal.

Distinguishing characters: Rather large, subglobose to ovoidal, finely ornamented conidia are produced from slender phialides on long, very uniformly branched conidiophores in dark green pustules; fast rate of growth at 35°C. The conidia often appear smooth. *Trichoderma asperellum* is morphologically indistinguishable from *T. asperelloides* (Samuels et al., 2010).

Habitat: Soil.

Distribution: Cosmopolitan.

Literature: Chutrakul et al. (2008; trichotoxin, peptaibols), Lieckfeldt et al. (1999), Samuels et al. (1999, 2010).

Figure 6.7. *Trichoderma asperellum.* **A–C,** Conidial pustules. Note the projecting conidiophores in **B** and **C**. **D–G,** Conidiophores. **H and I,** Conidia (optical section in **H**; surface view in **I**). **A and H–I,** From cornmeal agar. **B and D–G,** From cornmeal dextrose agar. **C,** From Spezieller Nährstoffarmer agar (SNA). Scale bars: **A, B** = 1 mm; **C** = 0.5 mm; **D–G** = 20 µm; **H, I** = 10 µm. ➤

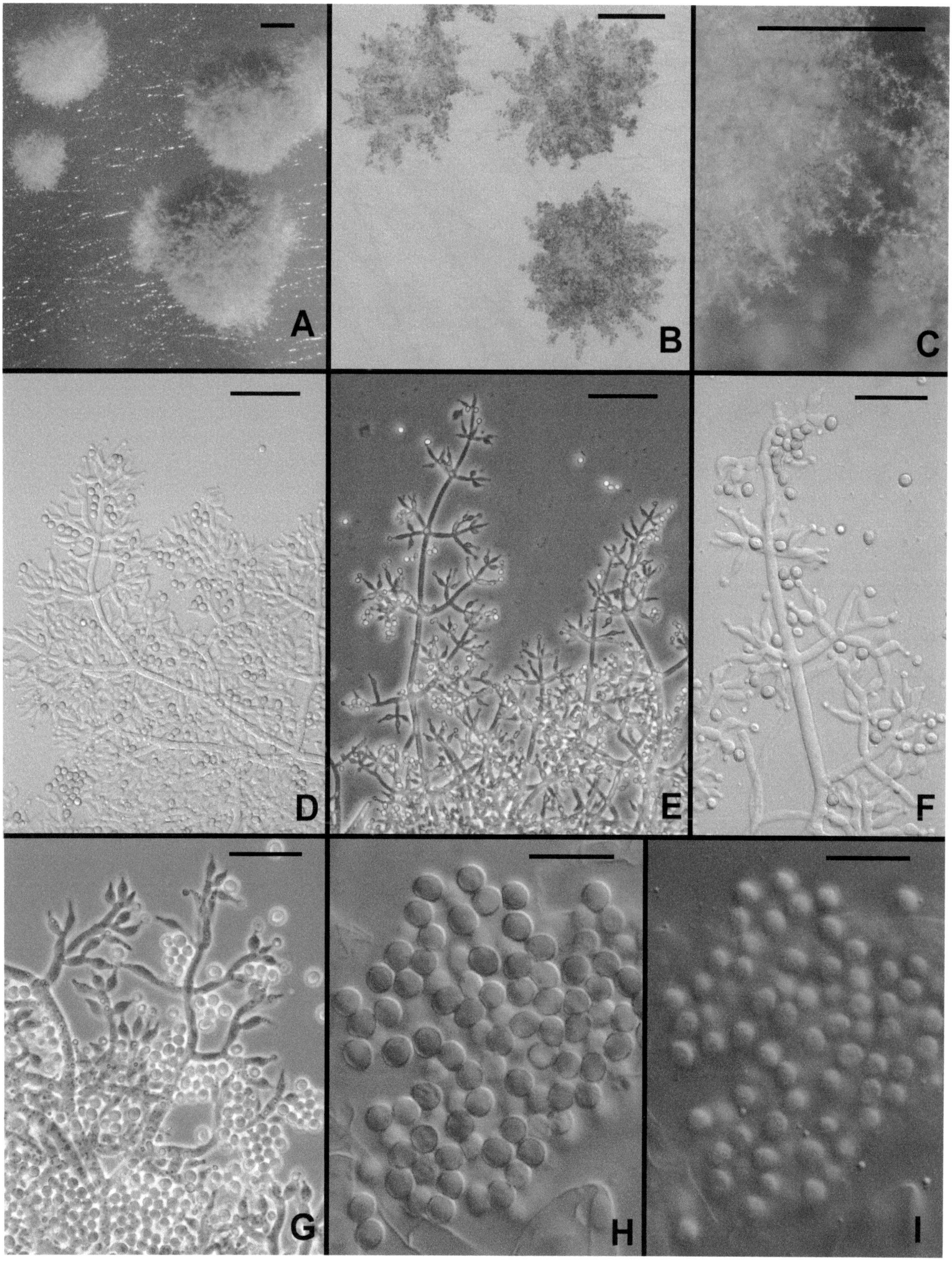
A
B
C
D
E
F
G
H
I

Trichoderma atroviride

P. Karst. *in* Karsten, Finlands Mögelsvampar p. 21. 1892. Figs. 6.1D, 6.8.
= *Hypocrea atroviridis* Dodd, Lieckf. & Samuels *in* Dodd et al., Mycologia 95:36. 2003.

Ex-epitype culture: CBS 142.95 (Gams and Meyer, 1998)
GenBank: *rpb2*: EU341801; *tef1*: AF456891, AY376051

Clade: Viride

Colony radius at 25°C: PDA (42–)50–70. **SNA** (25–)30–40(–50). Colonies on PDA typically have a distinctive, sweet coconut odor.

Colony radius at 35°C after 72 h: PDA (0–)1–3(–8). **SNA** (0–)2–7.

Conidiation: Pustules forming around the original inoculum or in one or two concentric rings, 1–2 mm in diameter, cottony to compact, dark green; conidia appearing dry.

Microscopic characteristics: Branching pattern Pyramidal type with phialides held in divergent whorls of three or four or arising singly near the tip of the main axis but unilateral branches are common as are etiolated conidiophores with widely spaced branches; extended conidiophores or **hairs** not seen. **Phialides** ampulliform when closely spaced, lageniform when more distantly spaced, straight or sinuous, sharply constricted at the neck when ampulliform, (4.2–)6.0–9.7(–15.0) μm long, (1.8–)2.5–3.5(–4.8) μm wide at the widest point, L/W (1.1–)1.5–3.5(–6.3). **Supporting cell** (1.7–)2.3–3.3(–4.2) μm wide; ratio of length of phialide to width of supporting cell (1.2–)2.0–3.6(–6.0); ratio of width of phialide to width of supporting cell (0.4–)0.8–1.2(–1.9). **Conidia** subglobose to ovoidal, (2.2–)2.7–3.5(–4.0) × (2.7–)3.2–3.7(–4.7) μm, L/W (0.8–)1.0–1.3(–1.6), green, smooth. **Chlamydospores** abundant, globose to subglobose, terminal or intercalary.

Distinguishing characters: *Trichoderma atroviride* is similar to the unrelated *T. harzianum*, from which it differs in the dark green conidia, coconut odor, and near constant production of chlamydospores. Some strains of *T. harzianum* produce chlamydospores, while others do not. *Trichoderma paratroviride* is morphologically indistinguishable from *T. atroviride*; the two differ in cultural characteristics (Jaklitsch and Voglmayr, 2015).

Habitat: Soil; more common in temperate regions than in tropical regions. Common in commercial mushroom production. Teleomorph occurring on decaying wood. This species was among the most effective species in degrading cells of grass *Miscanthus* × *giganteus* in the United States (Shrestha et al., 2011).

Distribution: Cosmopolitan but more common in north temperate regions.

Literature: Bissett (1992), Dodd et al. (2003), Gams and Meyer (1998), Samuels et al. (2002), http://genome.jgi-psf.org/Triat2/Triat2.home.html (genome).

Figure 6.8. *Trichoderma atroviride.* **A and B,** Conidial pustules. **C–H,** Conidiophores. **I,** Conidia. **J,** Chlamydospores. **A, B, and H,** From Spezieller Nährstoffarmer agar (SNA). **C–G, I, and J,** From cornmeal dextrose agar. Scale bars: **A** = 1 mm; **B** = 0.5 mm; **C–G, J** = 20 μm; **H, I** = 10 μm.

➤

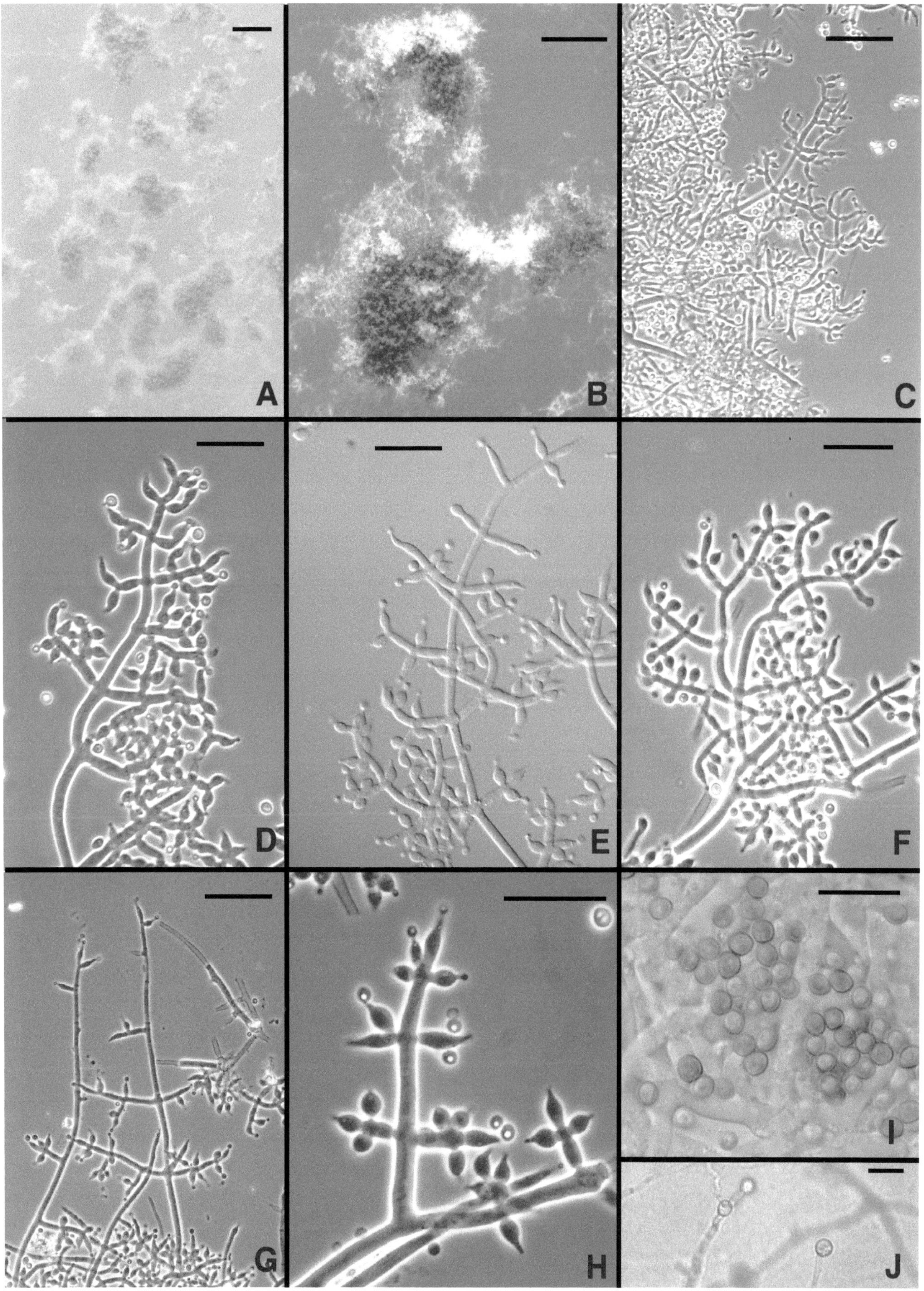
A
B
C
D
E
F
G
H
I
J

Trichoderma aureoviride

Rifai, Mycol. Pap. 116:34. 1969. nom cons. prop. Figs. 6.1E, 6.9.

= *Hypocrea aureoviridis* Plowr. & Cooke *in* Phillips and Plowright, Grevillea 8(47):104. 1879 [1880].

Ex-neotype culture: CBS 245.63 (*Hypocrea aureoviridis* epitype, Jaklitsch, 2009)
GenBank: *rpb2*: FJ179602; *tef1*: Z956322

Clade: Aureoviride

Colony radius at 25°C: PDA and **SNA** <10 mm, slower on PDA than SNA; temperature optimum 20°C and only barely growing at 30°C. On PDA and cornmeal dextrose agar (CMD), a yellowish pigment diffusing through the agar; dendroidal, yellow crystals observed on PDA, SNA, and CMD at 15°C; no distinctive odor.

Colony radius at 35°C after 72 h: No growth at 35°C.

Conidiation: Conidiation dispersed throughout the colony, often occurring in concentric rings; conidiophores arising in the aerial mycelium and in loose, green, 0.5- to 1.5-mm-diameter pustules.

Microscopic characteristics: Branching pattern typically no repeating pattern of branching is apparent, with branches arising from a more-or-less discernible main axis at irregularly spaced intervals; branches not paired, variable in length; sometimes arranged in a more-or-less verticillate pattern. **Phialides** solitary or in verticillate or cruciate whorls of up to six, lageniform to ampulliform, (6.0–)11.5–17.5(–28.5) µm long, (1.7–)2.2–3.2(–5.0) µm at the widest point, L/W (2.0–)3.5–6.7(–9.3). **Supporting cell** (1.7–)2.2–3.2(–5.0) µm wide; ratio of length of phialide to width of supporting cell (2.0–)3.5–6.7(–12.5); ratio of widest part of phialide to width of supporting cell (0.5–)0.8–1.2(–2.0). **Conidia** clavate, ovoid to ellipsoidal or subglobose, often with a flat base, (1.2–)2.0–3.0(–4.0) × (2.0–)3.5–5.2(–7.2) µm, L/W (0.9–)1.1–2.3(–4.6), green, smooth. **Chlamydospores** not observed.

Distinguishing characters: Outstanding characteristics of *T. aureoviride* include its slow rate of growth, formation of dendroidal crystals at 15°C, irregularly branched conidiophores, and often clavate conidia. Despite being commonly cited in the literature, it has not been isolated directly from substrates; we only know it from cultures derived from ascospores. Literature reports of this species should be viewed with suspicion. Jaklitsch (2009) provided the only modern description of this species based on new collections made in northern Europe.

Habitat: On rotting wood.

Distribution: United Kingdom, northern Europe, Atlantic coast of France; known only as its teleomorph. Commonly misidentified in the literature.

Literature: Chaverri and Samuels (2003), Gams and Bissett (1998), Jaklitsch (2009), Lieckfeldt et al. (2001).

Figure 6.9. *Trichoderma aureoviride*. **A,** Dispersed conidiation. **B–H,** Conidiophores. **I,** Conidia. **A,** From Spezieller Nährstoffarmer agar (SNA). **B,** From cornmeal dextrose agar. **C–I,** From malt extract agar. Scale bars: **A** = 0.5 mm; **B, C, E, F** = 20 µm; **D, G–I** = 10 µm.

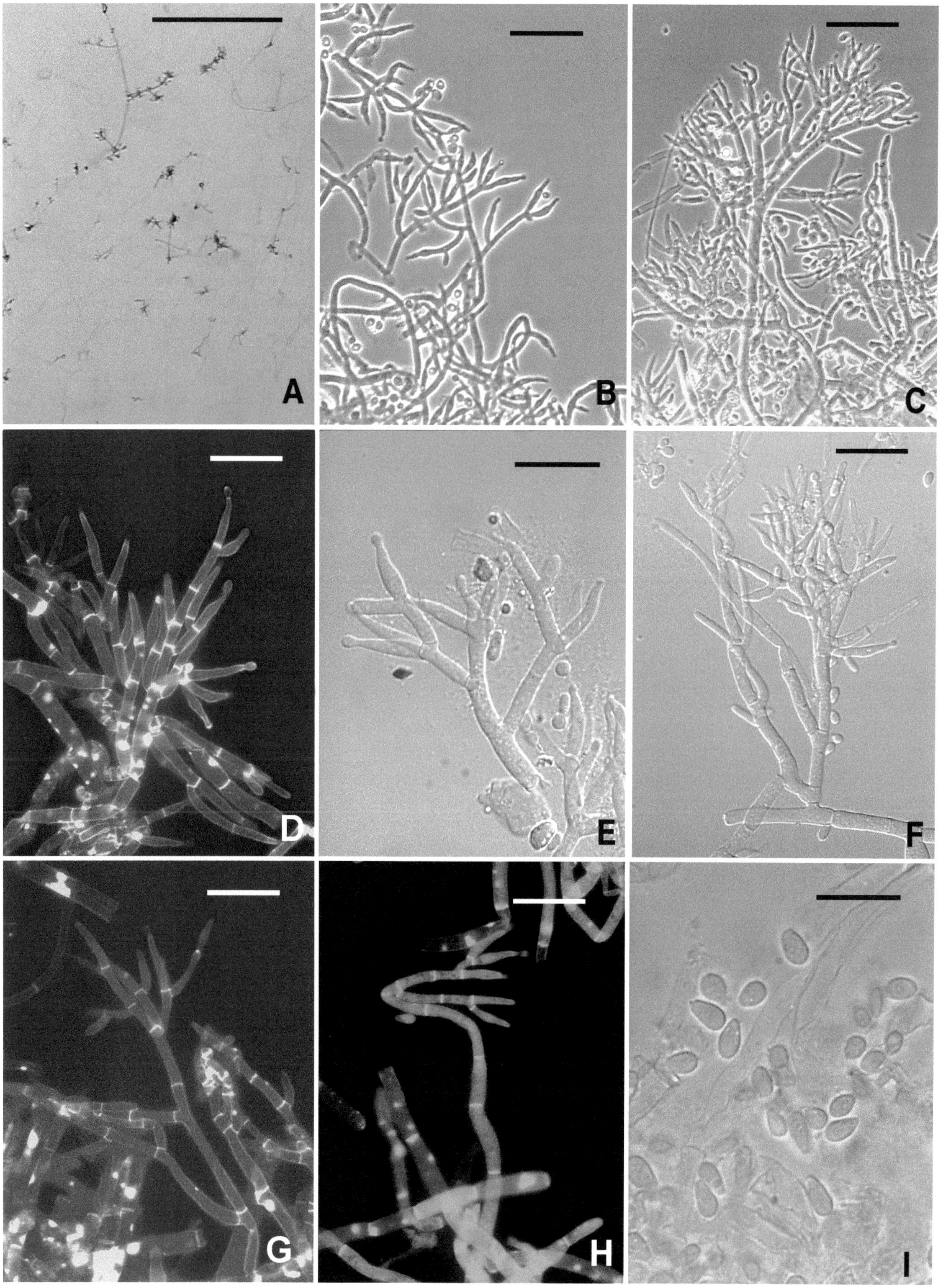
A
B
C
D
E
F
G
H
I

Trichoderma brevicompactum

G. F. Kraus, C. P. Kubicek & W. Gams *in* Kraus et al., Mycologia 96:1063. 2004. Figs. 6.1F, 6.10.

Ex-type culture: CBS 109720
GenBank: *rpb2*: EU338317; *tef1*: AY937453, EU338299

Clade: Brevicompactum

Colony radius at 25°C: PDA (30–)37–47(–50). **SNA** (19–)30–40(–48). No pigment or distinctive odor on PDA.

Colony radius at 35°C: PDA (0–)2–10(–18). **SNA** (0–)2–11(–22).

Conidiation: After 1 week, pustules scattered throughout the colony but often more concentrated at the periphery of the colony, to 2 mm in diameter, flat, compact, grayish green; extended conidiophores reaching beyond the surface of the pustule; conidia appearing moist.

Microscopic characteristics: Branching pattern Pachybasium type; phialides clustered on short broad branches at the base of extended hairlike conidiophores; hairlike conidiophores conspicuously projecting beyond the surface of the pustule, producing phialides in divergent whorls from near the tip or branched variously along the length, and branches often sterile for a considerable length or conidiophores shorter and extensively rebranched in a Harzianum pattern. **Phialides** ampulliform when crowded, lageniform when formed at the tip of extended conidiophores, (3.7–)4.7–6.7(–15) µm long, (2.5–)3.0–3.7(–4.5) µm at the widest point, L/W (1.1–)1.3–2.2(–5.1). **Supporting cell** (1.7–)2.7–4.0(–5.7) µm wide; ratio of length of phialide to width of supporting cell (1.0–)1.3–2.3(–4.8); ratio of width of phialide to width of supporting cell (0.6–)0.8–1.2(–1.7). **Conidia** subglobose to ovoidal, (2.0–)2.2–2.7(–3.0) × (2.2–)2.7–3.0(–3.7) µm, L/W (0.9–)1.1–1.3(–1.7), green, smooth. **Chlamydospores** not produced in all cultures, terminal, subglobose. **Synanamorph** *Verticillium*/*Gliocladium*-like, forming abundantly on cornmeal dextrose agar (CMD).

Distinguishing characters: Minute, subglobose, green conidia produced on Pachybasium-like conidiophores in moist masses at the base of extended conidiophores in dense, flat pustules. Morphologically nearly indistinguishable from the closely related species *T. arundinaceum*; *T. protrudens*; *T. rodmanii*; and *T. turrialbense*. *Trichoderma brevicompactum* produces the mycotoxin trichodermin; *T. arundinaceum*, *T. protrudens*, and *T. turrialbense* produce harzianum A; no mycotoxins were detected in *T. rodmanii* (Degenkolb et al., 2008a,b).

Habitat: Soil.

Distribution: Cosmopolitan, temperate, tropical.

Literature: Degenkolb et al. (2008a), Kraus et al. (2004), Nielsen et al. (2005; trichothecene production).

Figure 6.10. *Trichoderma brevicompactum*. **A and B,** Conidial pustules. **C–H,** Conidiophores. **I,** Conidia. **J,** Synanamorph conidiation. Note the wet heads. **A, C, and J,** From cornmeal dextrose agar. **B and D–I,** From Spezieller Nährstoffarmer agar (SNA). Scale bars: **A, J** = 0.5 mm; **B** = 1 mm; **C–F** = 20 µm; **G–I** = 10 µm. (C, G, and J reproduced from Degenkolb et al., 2008a, with kind permission of Springer Science+Business Media)

➤

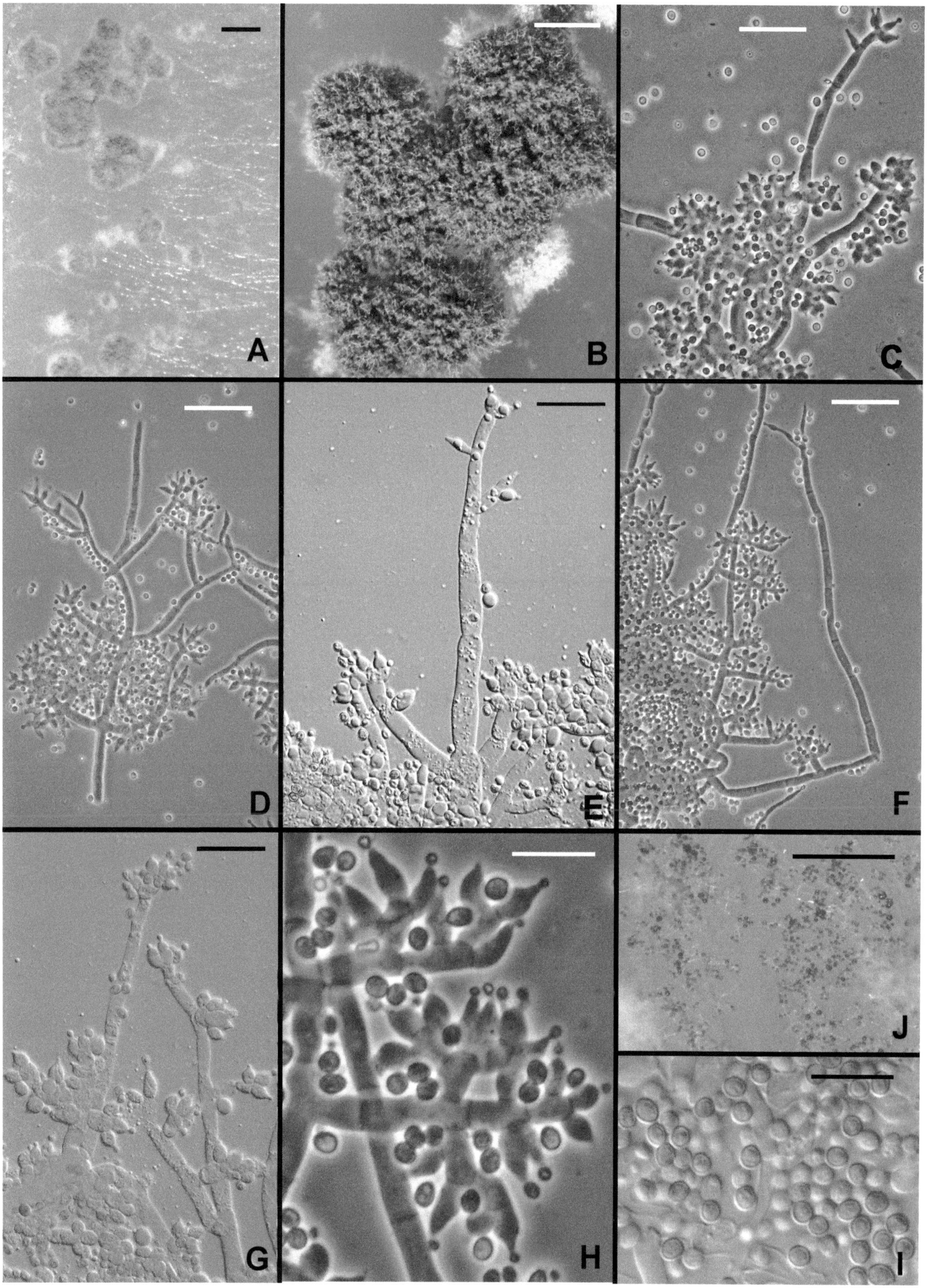
A
B
C
D
E
F
G
H
J
I

Trichoderma capillare

Samuels & C. P. Kubicek *in* Samuels et al., Fungal Diversity 55:83. 2012. Figs. 6.1G, 6.11.

Ex-type culture: CBS 130629
GenBank: *rpb2*: JN175530; *tef1*: JN182283

Clade: Longibrachiatum

Colony radius at 25°C: PDA 60–70. **SNA** 40–50. On PDA a pale diffusing pigment sometimes seen; no distinctive odor detected.

Colony radius at 35°C after 72 h: PDA and **SNA** 65–70.

Conidiation: Pustules scattered pulvinate, 0.5–1.5 mm in diameter, individual conidiophores visible within pustules; hairs arising from pustules, conspicuous or not, flexuous, long, sterile; conidia dark green to gray-green, in one strain conidia white, only very slowly becoming green.

Microscopic characteristics: Branching pattern conidiophores arising from hyphae within pustules, highly variable in form; commonly fertile branches Longibrachiatum type, producing solitary phialides, **intercalary phialides** infrequent; often conidiophores producing fertile branches laterally with branches terminating in whorls of a few phialides; sometimes fertile branches lacking any obvious pattern, cells of fertile branches sometimes nodose and producing numerous phialides. **Hairs** arising as outgrowths of the hyphae of the pustule. **Phialides** lageniform, nearly cylindrical or conspicuously swollen below the middle, straight or less frequently asymmetric or hooked, (4.0–)6.5–10.5(–14.0) μm long, (2.2–)3.0–3.5(–4.5) μm at the widest point, L/W (1.5–)1.6–3.2(–5.5). **Supporting cell** (1.7–)2.2–3.5(–4.5) μm wide; ratio of length of phialide to width of supporting cell (1.3–)2.0–4.2(–7.3); ratio of width of phialide to width of supporting cell (0.8–)1.0–1.6(–2.0). **Conidia** subglobose to broadly ellipsoidal, (1.7–)2.5–3.5(–4.0) × (2.2–)2.7–4.0(–4.5) μm, L/W (0.9–)1.0–1.4(–1.6), green, roughened, less frequently smooth. **Chlamydospores** not observed.

Distinguishing characters: *Trichoderma capillare* is unusual in the Longibrachiatum clade because its conidiophores often lack a discernible branching pattern and are often nodose, the phialides are frequently whorled, not solitary, and the conidia are roughened. Pustules formed by one strain from New York remained white, becoming green only after several days.

Habitat: Soil; isolated once from an *Agaricus* farm (Hungary).

Distribution: Cosmopolitan, known from North America, Europe, Vietnam; not common.

Literature: Samuels et al. (2012).

Figure 6.11. *Trichoderma capillare*. **A and B,** Conidial pustules. **C–L,** Conidiophores. **M,** Conidia. **A–M,** From Spezieller Nährstoffarmer agar (SNA). Scale bars: **A** = 1 mm; **B** = 0.5 mm; **C–E, I** = 20 μm; **F–H, J–M** = 10 μm. (A, B, D, G–I, and M reproduced, by permission, from Samuels et al., 2012)

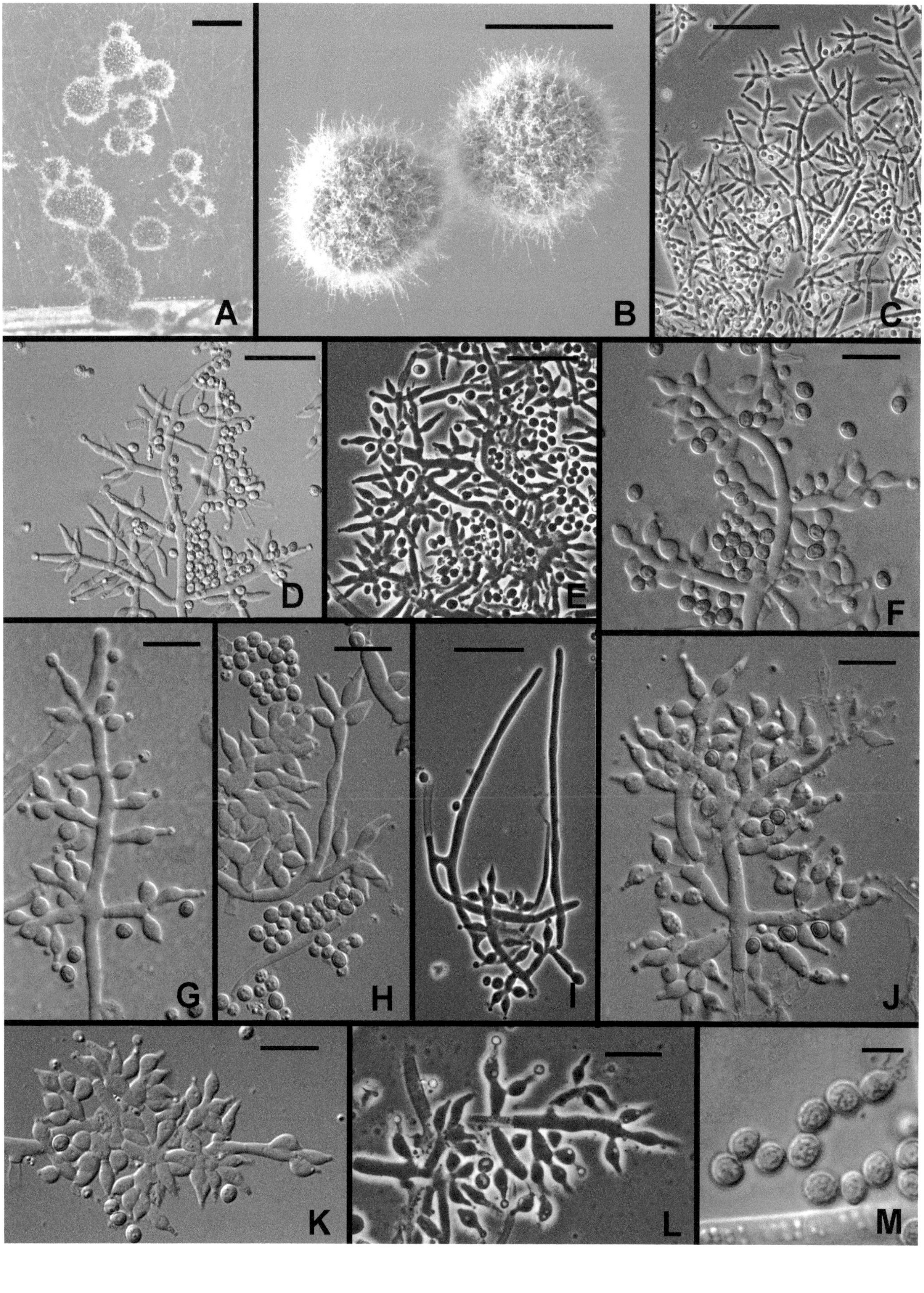
A
B
C
D
E
F
G
H
I
J
K
L
M

Trichoderma citrinoviride

Bissett, Can. J. Bot. 62:926. 1984. nom cons. prop. Figs. 6.1H, 6.12.
= *Hypocrea schweinitzii* (Fr.) Sacc. *in* Saccardo, Syll. Fung. 2:522. 1883.

Ex-type culture: DAOM 172792 = CBS 259.85
GenBank: *rpb*2: JN175544; *tef*1: EU280036

Clade: Longibrachiatum

Colony radius at 25°C: PDA (36–)50–70. **SNA** (37–)42–55(–59). A diffusing yellow pigment on PDA; no distinctive odor.

Colony radius at 35°C: PDA (35–)57–70. **SNA** (40–)60–65(–70). **At 40°C after 72 h: PDA** (13–)28–58(–63).

Conidiation: Pustules abundant, uniformly dispersed, sometimes in concentric rings, minute, <0.25 mm in diameter, becoming confluent and often forming a continuous lawn; cottony or wooly, completely fertile conidiophores visible, long sinuous sterile white hairs visible especially in young pustules; conidia dark green, dry.

Microscopic characteristics: Branching pattern Longibrachiatum type; fertile branches typically short, composed of three to six lateral branches, the branches near the tip monophialidic, branches distal to the tip progressively longer, rebranched in the same pattern as the main fertile branch; secondary branches often crowded on the main axis; main axis often nodose; fertile branches often arising at the base of sinuous, **sterile hairs**. **Phialides** lageniform to ampulliform, swollen to a greater or lesser extent in the middle, solitary but often in whorls, often terminating in a single cell, (3.0–)4.7–8.7(–14.5) µm long, (1.2–)2.5–3.5(–4.7) µm at the widest point, L/W (1.2–)1.5–3.1(–6.6). Supporting cell (1.0–)2.0–3.0(–4.0) µm wide; ratio of length of phialide to width of supporting cell 1.4–3.4; ratio of width of phialide to width of supporting cell 0.3–0.7. **Intercalary phialides** present but not abundant. **Conidia** ellipsoidal, (1.5–)2.0–2.7(–4.2) × (2.5–)3.0–5.0(–10.7) µm, L/W 1.5–2.0, green, smooth. **Chlamydospores** absent to abundant, terminal and intercalary, subglobose.

Distinguishing characters: Yellow pigment on PDA, rapid growth at 35°C. Distinguished from *T. longibrachiatum* and *T. reesei* by producing fertile branches at the base of sterile hairs and the shorter fertile branches having often crowded phialides, often having a pyramidal aspect. Phialides often appear to be in whorls. Compared with *T. longibrachiatum* and *T. reesei*, intercalary phialides are uncommon.

Habitat: Soil. The teleomorph occurs on corticated hardwood. Isolated from immunocompromised humans.

Distribution: Mainly north temperate.

Literature: Bissett (1984), Gams and Bissett (1998), Samuels et al. (1998).

Figure 6.12. *Trichoderma citrinoviride*. **A and B,** Conidiation. **C–J,** Conidiophores. An intercalary phialide is seen in **J** (arrow). **K,** Conidia. **A, B, D, E, H, and I,** From cornmeal agar. **G and K,** From cornmeal dextrose agar. **C, F, and J,** From Spezieller Nährstoffarmer agar (SNA). Scale bars: **A** = 1 mm; **B** = 0.5 mm; **C–H** = 20 µm; **I–K** = 10 µm.

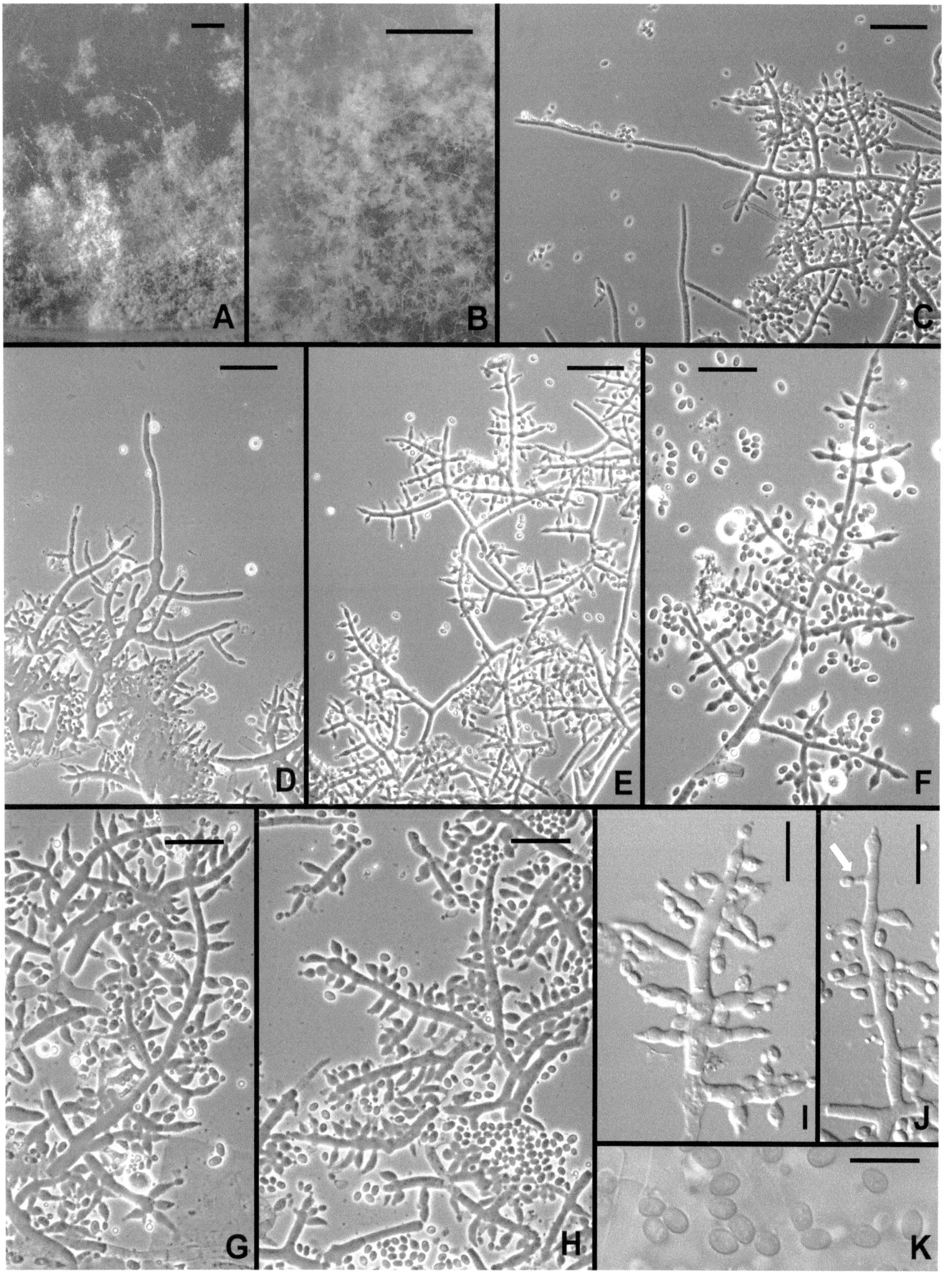
A
B
C
D
E
F
G
H
I
J
K

Trichoderma crassum

Bissett, Can. J. Bot. 69:2376. 1991 [1992]. Figs. 6.11, 6.13.

Ex-type culture: DAOM 164916
GenBank: *rpb2*: AF545542; *tef1*: AF534615

Clade: Virens

Colony radius at 25°C: PDA and **SNA** 35–55. On PDA a diffusing yellow pigment sometimes observed; no distinctive odor detected.

Colony radius at 35°C: PDA and **SNA** 0.

Conidiation: Conidiophores abundant, forming in minute, scattered, loosely formed, often confluent pustules with sterile or fertile hairs and in the aerial mycelium. Conidia held in a glistening globose drop of clear, dark green liquid at the tip of each conidiophore.

Microscopic characteristics: Branching pattern conidiophores formed in pustules intermediate between Gliocladium and Pachybasium types with phialides densely clustered at the end of short fertile branches but appressed in a vague penicillus, the branches broad, often paired and arising from a more-or-less conspicuous main axis; phialides arising from these conidiophores ampulliform, 5–8 μm long, 3–6 μm at the widest point, L/W 1.1–2.0. **Supporting cell** 3.5–6.0 μm wide; ratio of length of phialide to width of supporting cell 1–2; ratio of width of phialide to width of supporting cell 0.6–1.3, **hairs** protruding from pustules, sterile or terminating in two to four phialides arranged in a penicillus. Conidiophores arising in the aerial mycelium, 40–100 μm long, unbranched or infrequently branched, each branch terminating in *Gliocladium*- or *Verticillium*-like whorls of a few lageniform or ampulliform phialides. **Phialides** 8–15 μm long, 3–4 μm at the widest point, L/W 2.5–5.0. **Supporting cell** 2–4 μm wide; ratio of length of phialide to width of supporting cell 2.7–5.5; ratio of width of phialide to width of supporting cell 0.8–1.5. **Conidia** ellipsoidal to obovoidal, green, smooth; conidia from pustules, 2.5–4.0 × 3.7–5.2 μm, L/W 1.2–1.4; conidia from dispersed conidiophores, 4–5 × 6.0–6.5 μm, L/W approximately 1.3. **Chlamydospores** terminal and intercalary.

Distinguishing characters: *Trichoderma crassum* is closely related to *T. virens*. It differs from that species mainly in the production of pustules from which sterile hairs or long, terminally fertile conidiophores arise in addition to effused conidiophores and by the inability of *T. crassum* to grow at 35°C.

Note: The reported link to *Hypocrea crassa* P. Chaverri & Samuels is incorrect; *H. crassa* is a similar but distinct species in the *T. virens* species complex.

Habitat: Soil; teleomorph on wood with other fungi.

Distribution: North temperate (Québec, New York), possibly Belize; uncommon.

Literature: Bissett (1991b), Chaverri and Samuels (2003), Chaverri et al. (2003a), Gams and Bissett (1998).

Figure 6.13. *Trichoderma crassum*. **A and B,** Conidiation. **C–H,** Conidiophores. **I,** Conidia. **J and K,** Chlamydospores. **L,** *Verticillium*-like synanamorph conidiophores. **A,** From cornmeal dextrose agar. **B–L,** From Spezieller Nährstoffarmer agar (SNA). Scale bars: **A** = 0.5 mm; **B** = 150 μm; **C–F, J–L** = 20 μm; **G–I** = 10 μm.

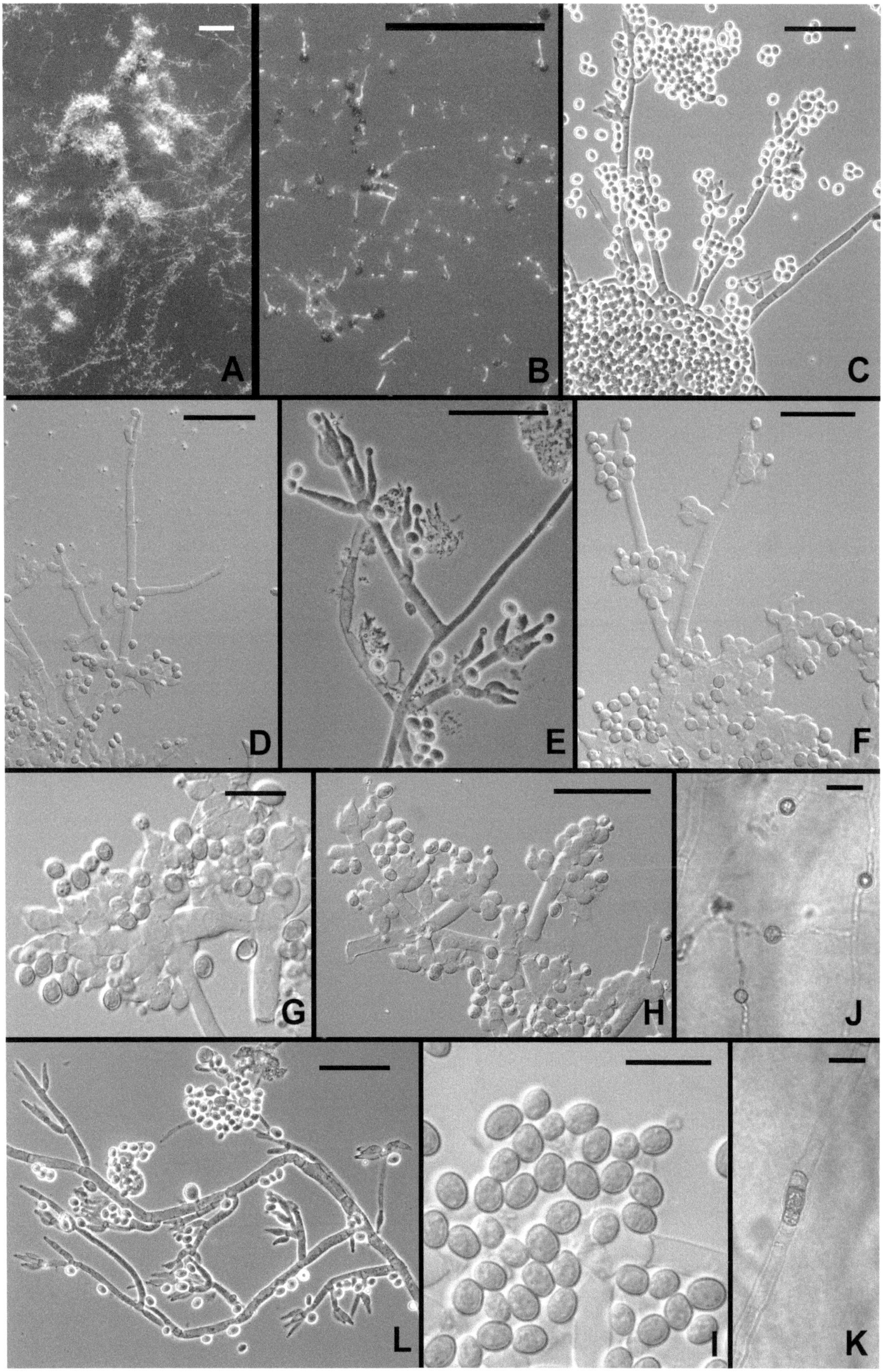
A
B
C
D
E
F
G
H
J
L
I
K

Trichoderma deliquescens

(Sopp) Jaklitsch *in* Jaklitsch, Fungal Diversity 48:176. 2011. nom cons. prop. Figs. 6.1J, 6.14.

≡ *Gliocladium deliquescens* Sopp, Skr. VidenskSelsk. Christiania, Kl. I, Math.-Natur. 11:89. 1912.

= *Gliocladium viride* Matr. *in* Matruchot, Bull. Soc. Mycol. Fr. 9:251. 1912. non *Trichoderma viride* Pers. 1794.

= *Hypocrea lutea* (Tode) Petch *in* Petch, J. Bot. Lond. 75:231. 1937.

Ex-neotype culture: CBS 121131 (Jaklitsch, 2011)
GenBank: *rpb*2: FJ179609; *tef*1: EU338287

Clade: Deliquescens

Colony radius at 25°C: PDA 30–47. **SNA** 20–40. Typically diffusing pigment or distinctive odor on PDA, sometimes a reddish pigment on PDA, the pigment resulting from red crystals.

Colony radius at 35°C: PDA and **SNA** (0–)20–45(–65).

Conidiation: Discrete pustules not forming; conidiophores arising throughout the colony, solitary or arising in fascicles of up to 10 or more; conidia dark green, held in glistening drops of clear green liquid.

Microscopic characteristics: Branching pattern Gliocladium type; conidiophores composed of a long, robust, smooth stipe terminating in a penicillus of phialides 100–200 µm long, 5–10 µm wide at the base, tapering somewhat to the penicillus; buttressing hyphae arising from the lower half of the conidiophores and anchoring the conidiophore to the substrate; penicillus up to 100 µm long, 70–120 µm wide at the apex, composed of up to four levels of branches, the ultimate level producing phialides; branches cylindrical, arising in appressed whorls of two to five; primary branches (15–)20–30(–35) µm long, 2.5–10 µm wide; secondary branches 10–20(–30) µm long, (2.5–)3.2–6.0 µm wide; tertiary and quaternary branches 8–13 µm long, 2–5 µm wide. **Phialides** cylindrical, (5–)8–12 µm long, 1.5–2.5(–3.2) µm wide, L/W (3.5–)3.8–6.1(–8.5); sometimes with a flared collaret and green, held in *Gliocladium*-like heads of several. **Supporting cell** 2–4 µm wide; ratio of length of phialide to width of supporting cell (1.0–)1.6–3.8(–10.1); ratio of width of phialide to width of supporting cell 0.6–0.7. **Conidia** subglobose to ellipsoidal or ovoidal, (1.0–)1.7–2.7(–3.5) × (1.7–)2.2–3.7(–4.7) µm, L/W (0.8–)1.1–1.7(–3.0), green, smooth. **Chlamydospores** few to abundant, terminal or intercalary in hyphae.

Distinguishing characters: Several *Trichoderma* species produce conidia in glistening drops of liquid. The long, infrequently branched conidiophores that have buttressing hyphae arising from the lower half distinguish *T. deliquescens* from all other species. The concept of *T. deliquescens* presented here includes strains from a wide geographic sampling; the greatest variation was seen in growth at 35°C. The species has not been subjected to comprehensive phylogenetic analysis, which may reveal a multiplicity of taxa. The description presented here includes some elements taken from a description of the species based on European teleomorph collections (Jaklitsch, 2011).

Habitat: Soil, litter, fungi; teleomorph on fallen wood.

Distribution: Cosmopolitan but not common, possibly more common in temperate regions.

Literature: Domsch et al. (2007), Jaklitsch (2011).

Figure 6.14. *Trichoderma deliquescens.* **A and B,** Conidiation. Note the wet heads. **C–I,** Conidiophores. Buttressing hyphae are seen in **D, E,** and **G** (arrows). **J,** Conidia. **K,** Chlamydospores. **A, B, E, I, J, and K** (bottom portion), From cornmeal dextrose agar. **C, D, F–H, and K** (top portion), From Spezieller Nährstoffarmer agar (SNA). Scale bars: **A, B** = 0.5 mm; **C, D** = 100 µm; **E–G, K** = 20 µm; **H–J** = 10 µm.

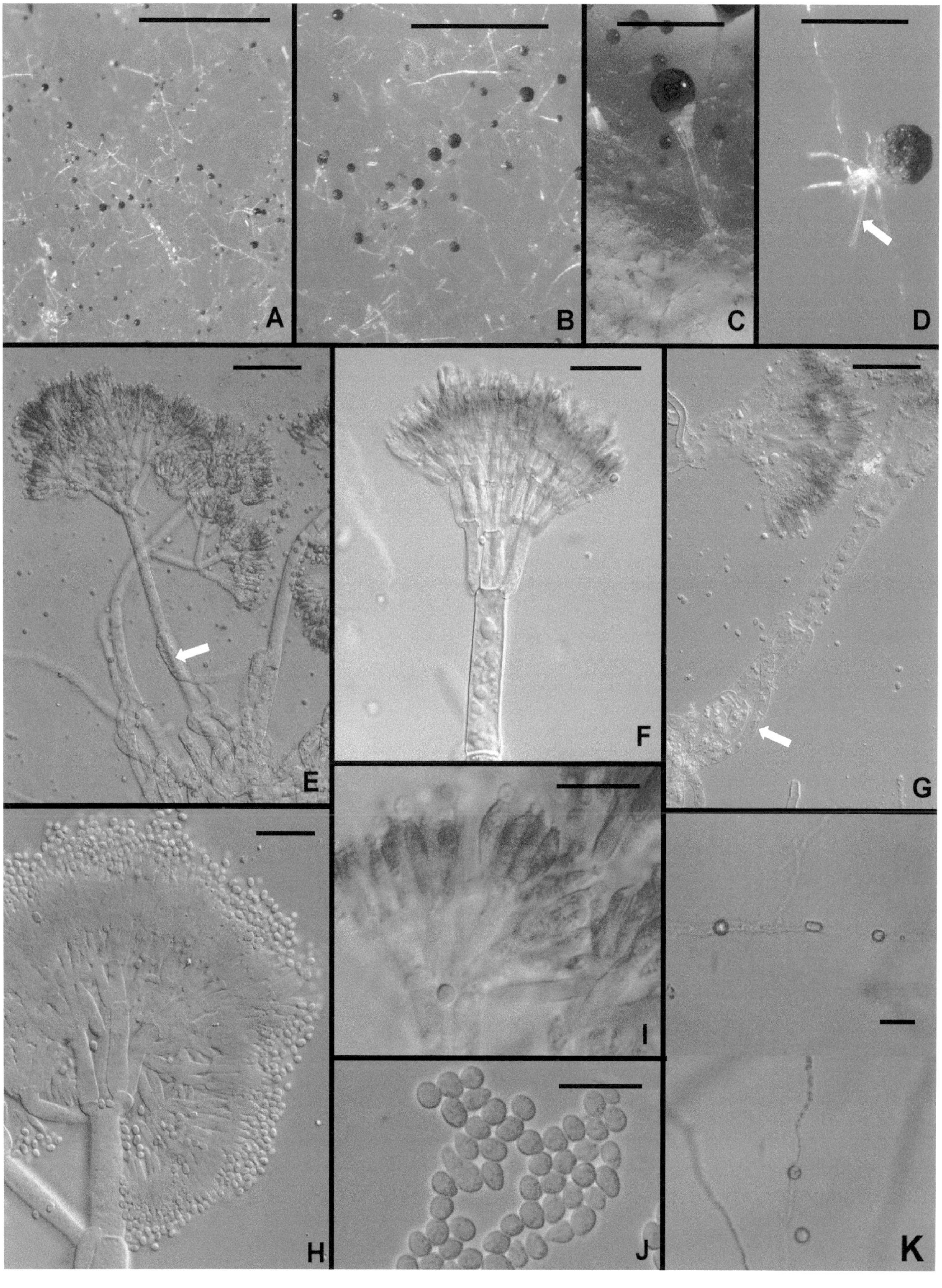
A
B
C
D
E
F
G
H
I
J
K

Trichoderma effusum

Bissett, C. P. Kubicek & Szakács *in* Bissett et al., Can. J. Bot. 81:575. 2003. Figs. 6.1K, 6.15.

Ex-type culture: DAOM 230007
GenBank: *rpb2*: JX238484; *tef1*: AF510432

Clade: Longibrachiatum

Colony radius at 25°C: PDA approximately 30. **SNA** approximately 15. Colony reverse on PDA olivaceous yellow; no distinctive odor.

Colony radius at 35°C: PDA approximately 25. **SNA** approximately 10.

Conidiation: Colony margin deeply dissected, conidia sparingly produced within 72–96 h, conidiophores arising directly from the surface of the agar and conidia also formed from phialides formed along hyphae submerged in the agar. Conidia slowly turning pale green, held in wet heads.

Microscopic characteristics: Branching pattern *Acremonium*-like or Longibrachiatum type. Differentiated conidiophores not observed on SNA; conidia produced from solitary phialides arising from erect or immersed hyphae; sometimes arising in clusters from knots of hyphae; phialides closely or distantly spaced, sometimes forming in knots. **Phialides** lageniform to ampulliform, sometimes strongly curved, at most only slightly swollen in the middle, straight or hooked, often reduced to short pegs (aphanophialides) along hyphae, sometimes aphanophialides forming at the base of a phialide or along hyphae, (4.5–)6.5–13.0(–24.0) μm long, (2.2–)2.5–3.5(–4.0) μm at the widest point, L/W (1.7–)3.6–5.0(–7.8). **Supporting cell** (2.0–)2.5–3.2(–3.7) μm wide; ratio of length of phialide to width of supporting cell 2.7–6.7; ratio of width of phialide to width of supporting cell 0.8–1.4. **Conidia** variable in shape and size, ellipsoidal to nearly oblong, (2.5–)2.7–3.5(–4.0) × (4.5–)5.0–7.7(–9.7) μm, L/W (1.2–)1.5–2.7(–3.5), green, smooth. **Chlamydospores** not observed but knots of hyphae can be observed.

Distinguishing characters: The morphology of *T. effusum* is highly atypical in *Trichoderma* species. Phylogenetically it belongs to the Longibrachiatum clade and the frequent occurrence of solitary phialides and intercalary phialides and aphanophialides suggest other members of that clade. It is somewhat similar to other unrelated species having a reduced morphology, including *T. scalesiae* and *T. paucisporum*.

Habitat: Soil.

Distribution: India (Himalaya Mountains, elevation 2,800 m), known only from the original collection.

Literature: Bissett et al. (2003), Samuels et al. (2012).

Figure 6.15. *Trichoderma effusum*. **A**, Conidiation. **B–I**, Phialides arising from hyphae. Aphanophialides are indicated by arrows in **B**, **C**, and **I**. Phialides arise from a knot of hyphae in **H**. **J**, Conidia. **A–J**, From Spezieller Nährstoffarmer agar (SNA). Scale bars: **A** = 0.5 mm; **B, C, I** = 20 μm; **D–H, J** =10 μm. (A–F, H, and I reproduced, by permission, from Samuels et al., 2012)

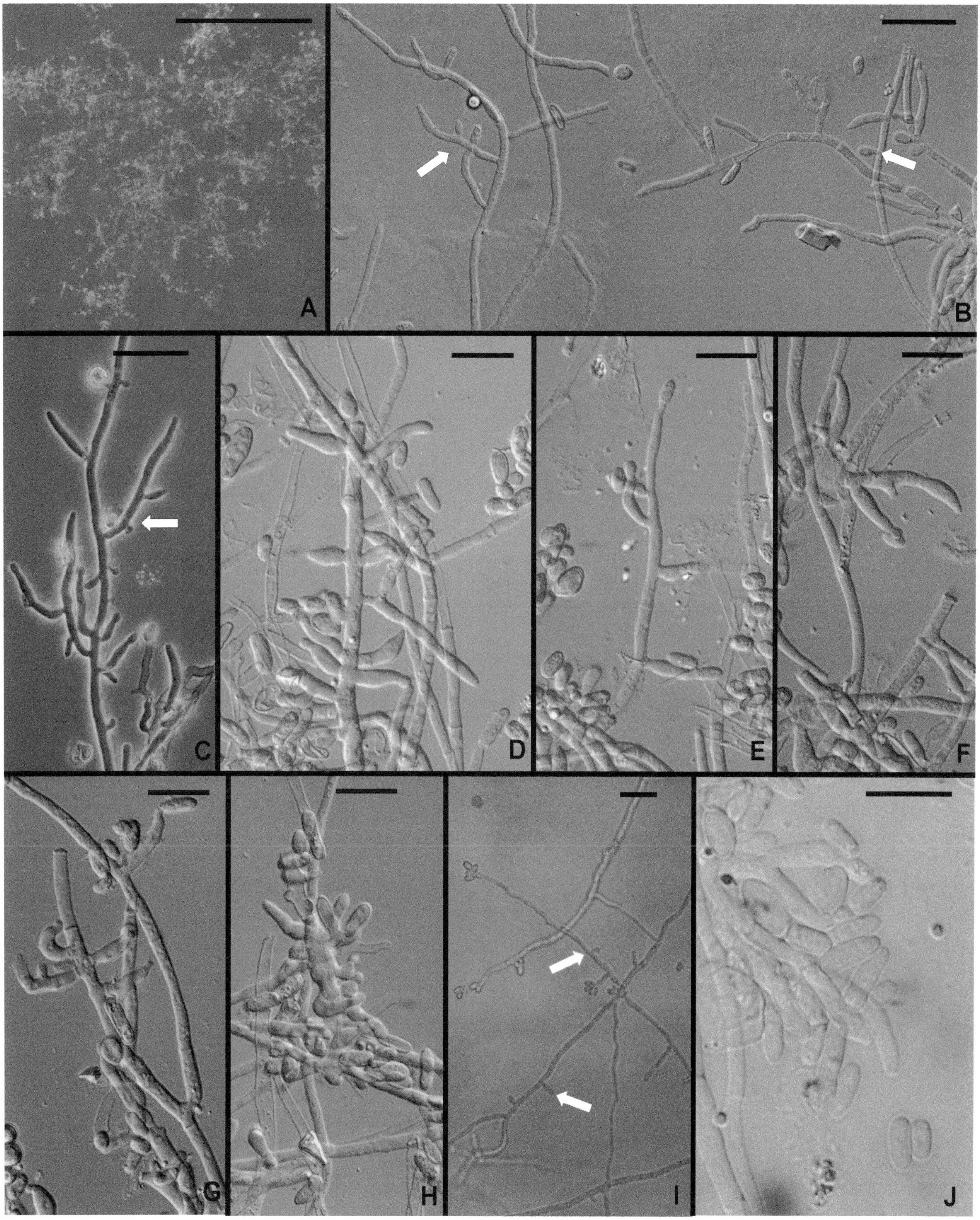
A
B
C
D
E
F
G
H
I
J

Trichoderma erinaceus

Bissett, C. P. Kubicek & Szakács *in* Bissett et al., Can. J. Bot. 81:583. 2003 [as *erinaceum*]. Figs. 6.1L, 6.16.

Ex-type culture: DAOM 230019
GenBank: *rpb2*: EU248604; *tef1*: AY750880

Clade: Viride

Colony radius at 25°C: PDA (51–)62–70. **SNA** (33–)42–60(–70). On PDA no diffusing pigment; frequently a more-or-less strong coconut odor, sometimes yellow pigment in conidia.

Colony radius at 35°C: PDA (9–)13–44(–64). **SNA** (15–)18–34(–38).

Conidiation: Pustules forming around the original inoculum and in concentric rings, flat to hemispherical, 2–3 mm in diameter, compact, grayish green, tending to become confluent; velvety from terminally fertile, hairlike conidiophores extending beyond the surface of the pustule; conidia appearing dry.

Microscopic characteristics: Branching pattern Pyramidal type with phialides held in more-or-less widely spaced, divergent whorls of two to four or arising singly near the main axis; many fertile branches producing a long, awl-like conidiophore; awl-like conidiophore typically terminating in a single phialide or in a divergent whorl of two phialides and projecting beyond the surface of the pustule. **Phialides** lageniform or ampulliform, straight, typically somewhat swollen in the middle, not sharply swollen at the neck, (4.0–)5.7–9.0(–14.5) μm long, (1.7–)2.5–3.5(–4.2) μm at the widest point, L/W (1.4–)1.5–3.5(–8.3). **Supporting cell** (1.5–)2.0–3.0(–4.0) μm wide; ratio of length of phialide to width of supporting cell (1.7–)1.9–4.2(–5.2); ratio of width of phialide to width of supporting cell (0.8–)1.1–1.5(–2.0). **Conidia** ellipsoidal to broadly ellipsoidal, (2.2–)2.7–3.2(–3.7) × (3.0–)3.5–4.0(–5.0) μm, L/W (1.1–)1.2–1.4(–1.8), green, smooth. **Chlamydospores** terminal to intercalary, globose to subglobose. *Verticillium*/*Gliocladium*-like **synanamorph** with wet, green conidia forming in aerial mycelium on cornmeal dextrose agar (CMD).

Distinguishing characters: Ellipsoidal conidia arising from lageniform phialides in pustules having abundant protruding hairlike conidiophores; typically a coconut odor.

Note: Unidentified *Hypocrea rufa*-like stromata collected once in Sri Lanka (BPI 871397, culture G.J.S. 02-103 = CBS 126393).

Habitat: Soil, found also on pseudostroma of *Moniliophthora roreri* on *Theobroma cacao.*

Distribution: Tropical; known from Peru, Sri Lanka (teleomorph), Thailand, Vietnam.

Literature: Bissett et al. (2003).

Figure 6.16. *Trichoderma erinaceus.* **A–C,** Conidial pustules. **D–G,** Conidiophores. **H,** Conidia. **I,** Chlamydospores. **J,** *Verticillium*-like conidiophores of the synanamorph. **A–C,** From Spezieller Nährstoffarmer agar (SNA). **D–J,** From cornmeal dextrose agar. Scale bars: **A** = 1 mm; **B** = 0.5 mm; **C** = 150 μm; **D, E, G, I, J** = 20 μm; **F, H** = 10 μm.

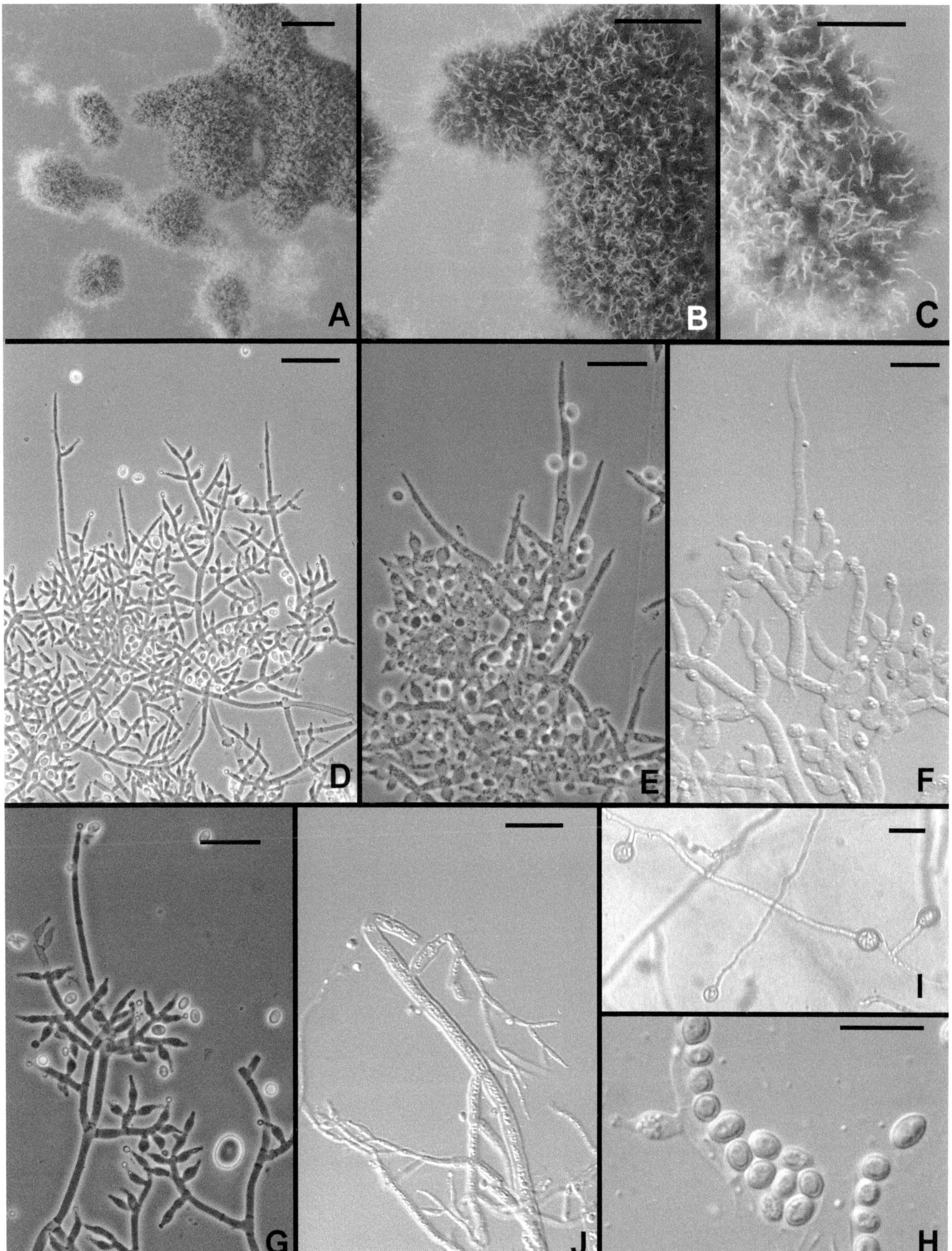
A
B
C
D
E
F
G
J
I
H

Trichoderma evansii

Samuels *in* Samuels and Ismaiel, Mycologia 101:149. 2009. Figs. 6.2A, 6.17.

Ex-type culture: CBS 123079
GenBank: *rpb2*: EU883558; *tef1*: EU883566

Clade: Viride

Colony radius at 25°C: PDA 45–60. **SNA** 30–40. No diffusing pigment or distinctive odor on PDA.

Colony radius at 35°C: PDA and **SNA** 0.

Conidiation: Pustules abundant, discrete, flat, 1–2 mm in diameter, disposed in two incomplete concentric rings, conidia gray-green; sterile hairs arising from the entire pustule, abundant, conspicuous, flexuous, white. No synanamorph observed.

Microscopic characteristics: Branching pattern Pachybasium type; fertile branches arising at the base of a sinuous, branched, septate, thin-walled, acute sterile extension; fertile branches composed of one or a few broad cells, often rebranching; secondary branches typically unicellular. **Phialides** solitary or clustered at branch tips, ampulliform to lageniform, 5–6 μm long, 3.5–4.0 μm at the widest point, L/W 1.5–1.6. **Supporting cell** 3.5–4.0 μm wide; ratio of length of phialide to width of supporting cell 1.2–1.8; ratio of width of phialide to width of supporting cell 0.8–1.2. **Conidia** subglobose, (2.5–)3.0–3.2(–3.7) × (2.7–)3.0–3.5(–4.2) μm, L/W (0.9–)1.0–1.2(–1.3), gray-green in mass, smooth. **Chlamydospores** globose to subglobose, terminal and intercalary, observed in only one strain.

Distinguishing characters: The subglobose conidia of *T. evansii* distinguish it from the related species *T. hamatum* and *T. pubescens*.

Habitat: Endophytic in sapwood of *Cola verticillata*, *Lophira alata*, and *Theobroma gileri*.

Distribution: Cameroon, Ecuador.

Literature: Samuels and Ismaiel (2009).

Figure 6.17. *Trichoderma evansii*. **A–C,** Conidial pustules. **D–H,** Conidiophores. **I,** Conidia. **J,** Chlamydospores. **A, D, F, and G–I,** From Spezieller Nährstoffarmer agar (SNA). **B, C, E, and J,** From cornmeal dextrose agar. Scale bars: **A** = 1 mm; **B, C** = 0.5 mm; **D–F, I, J** = 20 μm; **G, H** = 10 μm. (D–F, H, and J reprinted from Samuels, G. J., and Ismaiel, A. 2009. *Trichoderma evansii* and *T. lieckfeldtiae*: Two new *T. hamatum*-like species. Mycologia 101:142-156, with permission from *Mycologia*, ©The Mycological Society of America) ➤

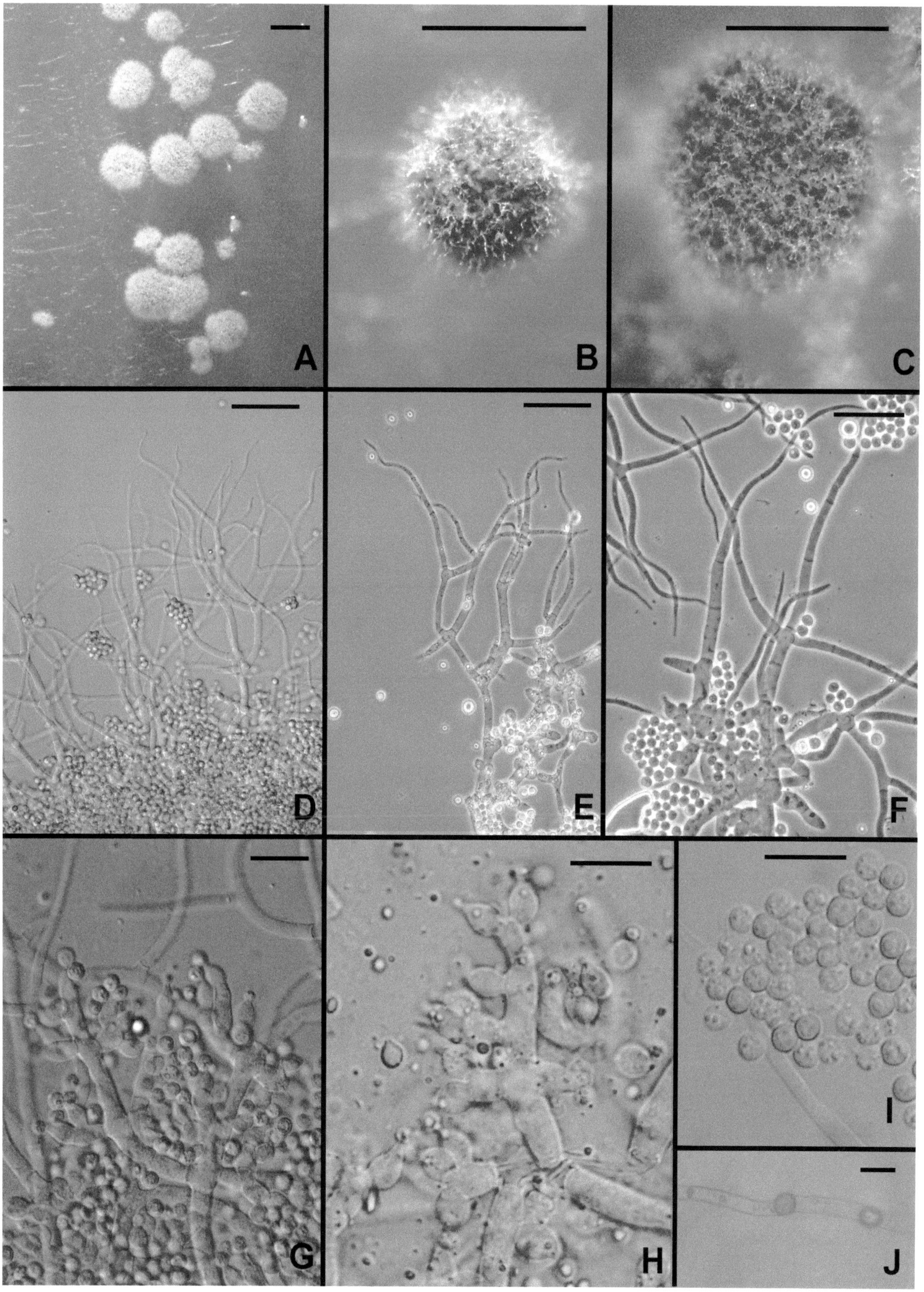
A
B
C
D
E
F
G
H
I
J

Trichoderma flagellatum

Mulaw, C. P. Kubicek & Samuels *in* Samuels et al., Fungal Diversity 55:89. 2012. Figs. 6.2B, 6.18.

Ex-type culture: CBS 130626
GenBank: *rpb2*: JN258688; *tef1*: FJ763184

Clade: Longibrachiatum

Colony radius at 25°C: PDA 65–70. **SNA** 40–45. On PDA a diffusing yellow pigment, appearing first at 30 and 35°C, later at 25°C; no distinctive odor.

Colony radius at 35°C: PDA 68–70. **SNA** 40–50.

Conidiation: Conidia forming sparingly on SNA at 25°C, more abundant at 30 and 35°C in approximately 0.25-mm-diameter pulvinate, discrete, and confluent pustules, then forming lawns of conidiophores; long, flexuous hairs arising from the pustules; conidia gray-green.

Microscopic characteristics: Branching pattern Longibrachiatum type; conidiophores arising laterally from intertwined hyphae of the pustule, typically constituting three to five levels of paired fertile branches, longest fertile branches nearest the conidiophore base, solitary phialides produced near the tip; fertile branches producing phialides directly or often producing paired secondary branches; secondary branches longest near the branching point and reduced to single phialides near the tip of the conidiophore; phialides appearing to be held in whorls; **intercalary phialides** common. **Phialides** lageniform, (3.7–)5.0–8.0(–11.5) μm long, (2.2–)2.7–3.5(–4.9) μm at the widest point, L/W (1.1–)1.6–2.9(–4.2). **Supporting cell** (1.5–)2.5–3.2(–5.0) μm wide; ratio of length of phialide to width of supporting cell (1.3–)1.7–3.2(–4.9); ratio of width of phialide to width of supporting cell (0.6–)0.9–1.3(–1.7). **Conidia** ellipsoidal to nearly oblong, (1.5–)2.0–2.7(–3.5) × (2.7–)3.0–5.0(–7.2) μm, L/W (1.2–)1.5–2.1(–2.8), green, smooth. **Chlamydospores** abundant, subglobose, terminal and intercalary, often in pairs.

Distinguishing characters: *Trichoderma flagellatum* is characterized within the Longibrachiatum clade by its whorled phialides and whiplike hairs and smooth conidia; four members of the Longibrachiatum clade are similar in this respect: *T. capillare*; *T. gillesii*; *T. konilangbra*; and *T. pseudokoningii*. Of these species, only *T. flagellatum* and *T. gillesii* fill a 9-cm-diameter petri plate within 72 h at 35°C when grown on PDA, and of these two species, only *T. flagellatum* produces diffusing yellow pigment on PDA.

Habitat: Endophytic in roots of *Coffea arabica* (coffee).

Distribution: Ethiopia, common in coffee-growing areas.

Literature: Druzhinina et al. (2012a), Samuels et al. (2012).

Figure 6.18. *Trichoderma flagellatum.* **A and B,** Conidial pustules. **C–H,** Conidiophores. **I,** Conidia. **J,** Chlamydospores. **A–J,** From Spezieller Nährstoffarmer agar (SNA). Scale bars: **A** = 1 mm; **B** = 0.5 mm; **C–G, J** = 20 μm; **H, I** = 10 μm. (G and J reproduced, by permission, from Samuels et al., 2012)

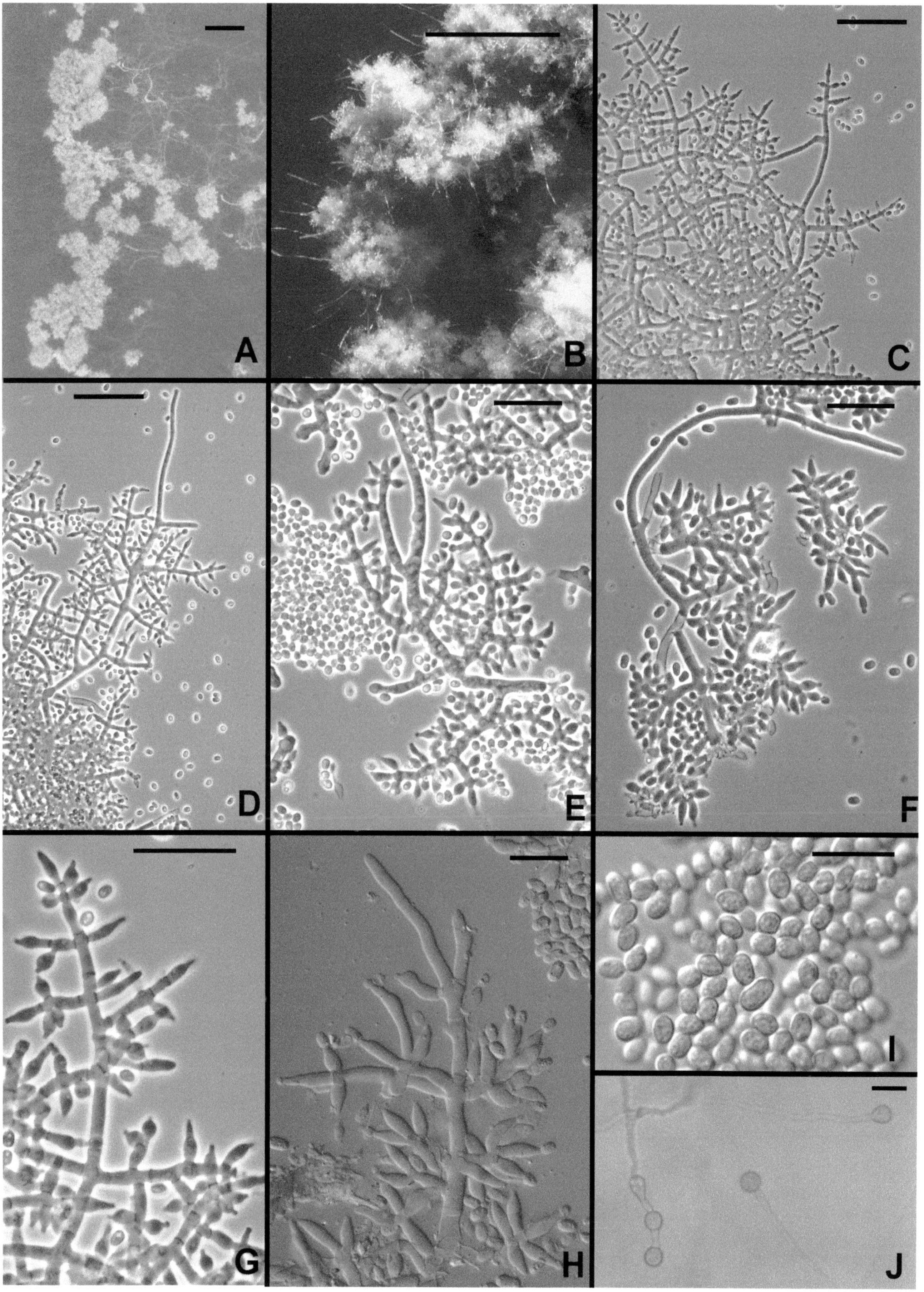
A
B
C
D
E
F
G
H
I
J

Trichoderma gamsii

Samuels & Druzhin. *in* Jaklitsch et al., Stud. Mycol. 56:168. 2006. Figs. 6.2C, 6.19.

Ex-type culture: CBS 120075
GenBank: *rpb2*: JN133561; *tef1*: DQ307541

Clade: Viride

Colony radius at 25°C: PDA 45–60. **SNA** 35–50. On PDA usually no diffusing pigment but conidia often yellow and sometimes a diffusing yellow pigment; usually no distinctive odor but sometimes a coconut odor.

Colony radius at 35°C: PDA 2–6. **SNA** 2–10.

Conidiation: Pustules 1–2 mm in diameter, forming in conspicuous concentric rings, often confluent and forming extensive flat sheets, cottony; lacking protruding sterile hairs or conidiophores; conidiophores visible within pustules when viewed with a stereomicroscope. Conidia dark green but on PDA pustules tending to be yellow.

Microscopic characteristics: Branching pattern Pyramidal type; central axis well defined, branches tending to arise in pairs and increasing in length with distance from the tip, phialides typically held in whorls; internodes between branches typically long but often short and then branches crowded, branches terminating in phialides in pairs or in compact heads of several, solitary phialides common; conidiophores occasionally terminating in a long extension that is sterile or bears one or two phialides at the tip. On rich media such as PDA and cornmeal dextrose agar (CMD) moniliform chains of phialides can be seen. **Phialides** lageniform, at most swollen in the middle (when crowded), straight, (5.2–)8.5–10.0(–18.5) μm long, (1.5–)2.2–3.0 (–4.0) μm at the widest point, L/W (1.4–)2.5–5.2(–8.4). **Supporting cell** (1.0–)1.8–3.0(–5.0) μm wide; ratio of length of phialide to width of supporting cell (1.6–)2.7–6.0(–10.0); ratio of width of phialide to width of supporting cell (0.4–)0.8–1.4(–2.3). **Intercalary phialides** infrequent. On PDA and CMD conidiophore branching irregular to a greater or lesser extent and phialides often proliferating percurrently to form long, moniliform chains. **Conidia** ellipsoidal, oblong or narrowly ovoidal, (2.0–)2.5–3.0(–3.2) × (3.2–)4.0–5.0(–6.2) μm, L/W (1.1–)1.4–1.8(–2.4), green, smooth. **Chlamydospores** typically abundant, terminal on hyphae, subglobose.

Distinguishing characters: The uniformly branched conidiophores seen on SNA and the ellipsoidal to nearly oblong conidia with a large L/W combined with aberrant conidiophores with proliferating phialides seen on richer media distinguish this species from the morphologically very similar *T. koningii*.

Habitat: Soil.

Distribution: North and south temperate, Mediterranean.

Literature: Jaklitsch et al. (2006).

Figure 6.19. *Trichoderma gamsii.* **A,** Conidial pustules. **B–H,** Conidiophores. Monilioid chains of phialides can be seen in **G** and **H**. Intercalary phialides are indicated by arrows in **E** and **F. I and J,** Conidia. **K,** Chlamydospores. **A–F, and I,** From Spezieller Nährstoffarmer agar (SNA). **G, H, J, and K,** From cornmeal dextrose agar. Scale bars: **A** = 0.5 mm; **B–H, K** = 20 μm; **I, J** = 10 μm. (D, F–H, J, and K reproduced, by permission, from Jaklitsch et al., 2006)

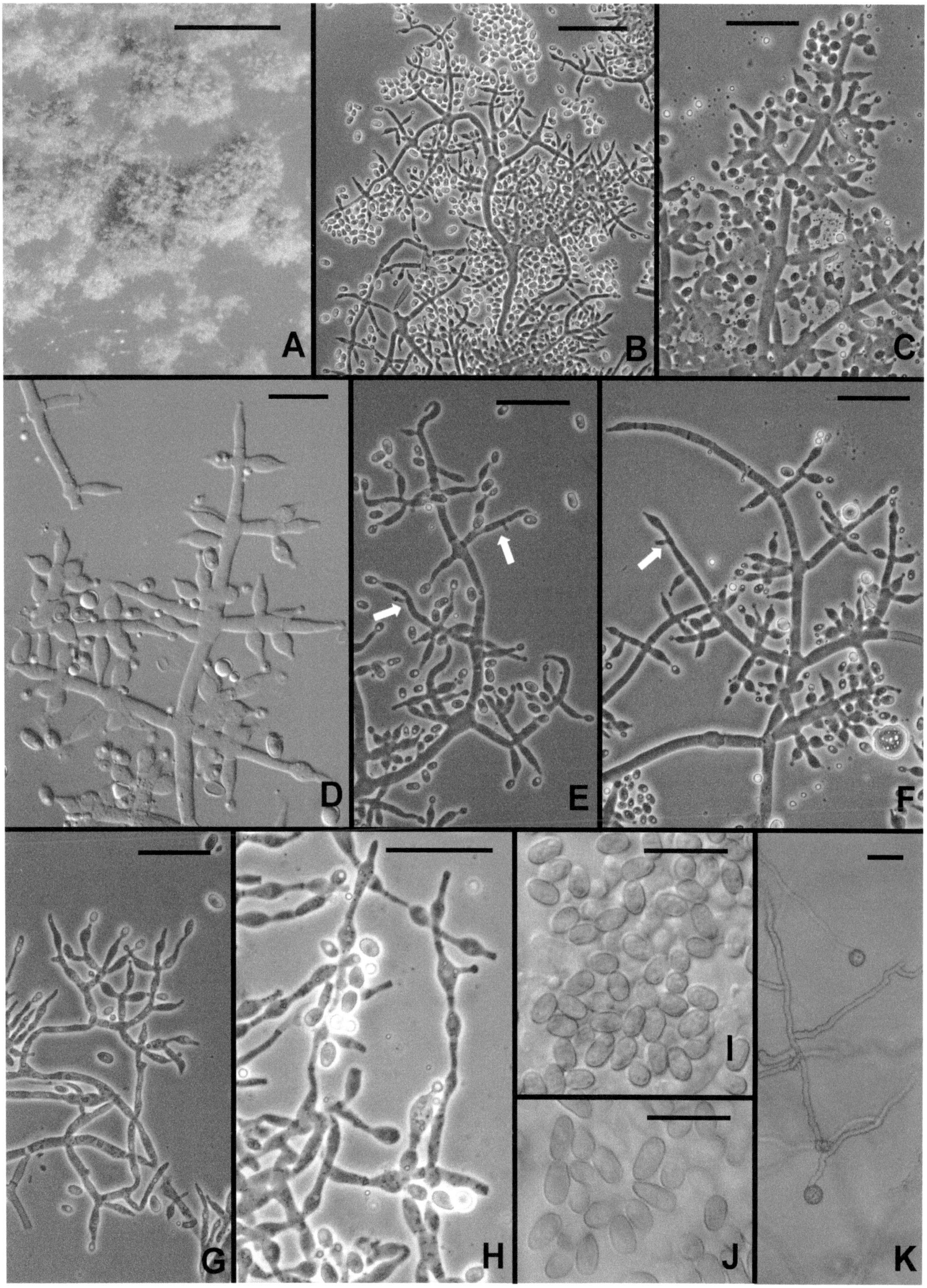
A
B
C
D
E
F
G
H
I
J
K

Trichoderma ghanense

Yoshim. Doi, Y. Abe & Sugiy. *in* Doi et al., Bull. Natl. Sci. Mus. Tokyo Ser. B (Bot.) 13:3. 1987. Figs. 6.2D, 6.20.

= *Trichoderma parceramosum* Bissett, Can. J. Bot. 69:2418. 1991 [1992].

= *Trichoderma atroviride* Bissett, Can. J. Bot. 62:930. 1984. non P. Karst. 1892.

Ex-type culture: IAM 13109 = ATCC 208858 = G.J.S. 95-137

GenBank: *rpb2*: JN175559; *tef1*: AY937423

Clade: Longibrachiatum

Colony radius at 25°C: PDA (55–)63–>70. **SNA** 70. On PDA pale yellow pigment visible in colony reverse, sometimes from above; no distinctive odor.

Colony radius at 35°C: PDA 70. **SNA** 60–70. Filling 9-cm-diameter PDA and SNA petri plates within 72 h.

Conidiation: Conidia forming in more-or-less distinct concentric rings; in some cultures obvious pustules lacking and conidia appearing to form in the aerial mycelium; pustules 0.25–0.5 mm in diameter, becoming confluent, sometimes plates flat, 4–5 mm in diameter, cottony; sterile hairs long, flexuous, arising from pustules, sometimes forming a net at the surface of pustules; conidia gray-green to dark green.

Microscopic characteristics: Branching pattern Longibrachiatum type with a strongly developed main axis and lateral branches increasing in length with distance from the tip, arising at right angles to the main axis; conidiophores fertile to the tip or varying amounts of the tip sterile and projecting as **hairs**; phialides arising from lateral branches or directly from the main axis, usually solitary, sometimes widely spaced. **Phialides** straight or hooked, subcylindrical to lageniform, (4–)6–10(–15) µm long, (1.7–)2.7–3.5(–4.5) µm at the widest point, L/W (1.3–)1.9–2.5(–6.8). **Supporting cell** (1.7–)2.5–3.2(–4.0) µm wide; ratio of length of phialide to width of supporting cell (1.3–)2.1–3.7(–6.5); ratio of width of phialide to width of supporting cell (0.8–)1.0–1.4(–1.9). **Conidia** ellipsoidal, (2.0–)2.7–4.0(–4.5) × (3.5–)5.0–6.7(–9.5) µm, L/W (1.3–)1.6–2.2(–3.2), green, tuberculate but often nearly smooth with one or a few tubercles visible. **Intercalary phialides** infrequent. **Chlamydospores** terminal and intercalary common.

Distinguishing characters: The Longibrachiatum branching pattern, relatively large, ellipsoidal, tuberculate conidia, and tropical distribution distinguish *T. ghanense*. The most morphologically similar species is *T. saturnisporum*.

Habitat: Soil.

Distribution: Tropical.

Literature: Bissett (1984), Samuels et al. (1998, 2012).

Figure 6.20. *Trichoderma ghanense*. **A and B,** Conidial pustules. **C–H,** Conidiophores. **I and J,** Conidia. **A, B, G, and I,** From cornmeal dextrose agar. **C–E and J,** From Spezieller Nährstoffarmer agar (SNA). **F and H,** From cornmeal agar. Scale bars: **A** = 1 mm; **B** = 0.5 mm; **C, D, G** = 20 µm; **E, F, H–J** = 10 µm.

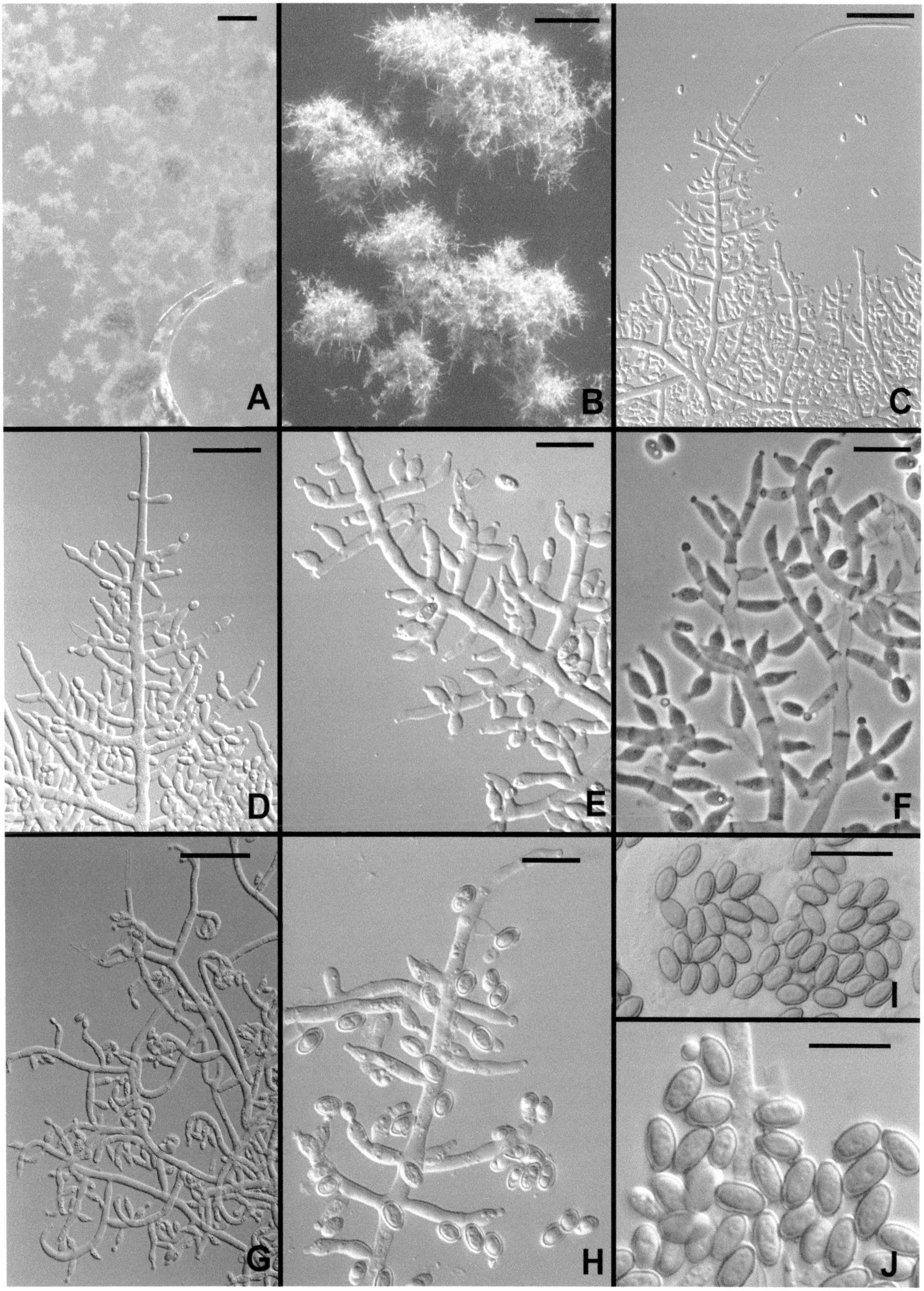
A
B
C
D
E
F
G
H
I
J

Trichoderma hamatum

(Bonord.) Bainier *in* Bainier, Bull. Soc. Mycol. France 22:131. 1906. Figs. 6.2E, 6.21.

≡ *Verticillium hamatum* Bonord. *in* Bonorden, Handb. Allgem. Mykol. p. 97. 1851.

≡ *Pachybasium hamatum* (Bonord.) Sacc. *in* Saccardo and Roumeguère, Rev. Mycol. Toulouse 7:161. 1885.

≡ *Phymatotrichum hamatum* (Bonord.) Oudem. *in* Oudemans, Ned. Kruidk. Archf. 3(Ser. 2):908. 1903.

Ex-neotype culture: DAOM 167057 = CBS 102160 (Bissett, 1991b)
GenBank: *rpb2*: AF545548; *tef1*: AF456911, AY750893

Clade: Viride

Colony radius at 25°C: PDA 50–55. **SNA** 40–47. No diffusing pigment or distinctive odor on PDA.

Colony radius at 35°C: PDA and **SNA** 0.

Conidiation: Pustules forming in concentric rings, pulvinate, compact, 0.5–1.5(–3.0) mm in diameter, often confluent; sterile hairs, sinuous to nearly helical, conspicuous, white. Conidia gray-green. A *Verticillium*-like synanamorph forming in the aerial mycelium apart from pustules.

Microscopic characteristics: Branching pattern Pachybasium type with fertile branches arising at the base of a sinuous, branched, septate, thin-walled, acute sterile extension. Fertile branches composed of one or a few broad cells, often rebranching; secondary branches typically unicellular. **Phialides** crowded at the tips of secondary branches, ampulliform, (3.7–)5.0–7.2(–10.7) µm long, (2.2–)3.2–4.0(–5.5) µm at the widest point, L/W (1.0–)1.4–2.1(–3.6). **Supporting cell** (1.5–)3.0–4.2(–6.0) µm wide; ratio of length of phialide to width of supporting cell (0.8–)0.9–2.3(–5.7); ratio of width of phialide to width of supporting cell (0.5–)0.8–1.2(–2.4). **Conidia** ellipsoidal, (2.0–)2.5–3.0(–3.5) × (3.0–)3.7–4.2(–5.7) µm, L/W (1.0–)1.2–1.6(–2.0), gray-green in mass, smooth. **Chlamydospores** terminal and intercalary, common.

Distinguishing characters: *Trichoderma hamatum* is characterized by its compact pustules from which arise branched, sterile hairs. It is most similar to *T. evansii*, *T. lieckfeldtiae*, and *T. pubescens*. If pustules are white, see *T. polysporum*. Stromata are yellow-brown or dull orange and ascospores are colorless (Jaklitsch and Voglmayr, 2015).

Habitat: Soil; commonly isolated as an endophyte from sapwood of *Theobroma cacao*.

Distribution: Cosmopolitan. *Trichoderma hamatum* is phylogenetically diverse and may eventually be divided.

Literature: Bissett (1991b), Chaverri et al. (2003a), Jaklitsch and Voglmayr (2015), Samuels and Ismaiel (2009).

Figure 6.21. *Trichoderma hamatum*. **A–C,** Conidial pustules. **D–H,** Conidiophores. **I,** Conidia. **J,** Chlamydospores. **A–J,** From cornmeal dextrose agar. Scale bars: **A** = 1 mm; **B, C** = 0.5 mm; **D–G, J** = 20 µm; **H, I** = 10 µm. ➤

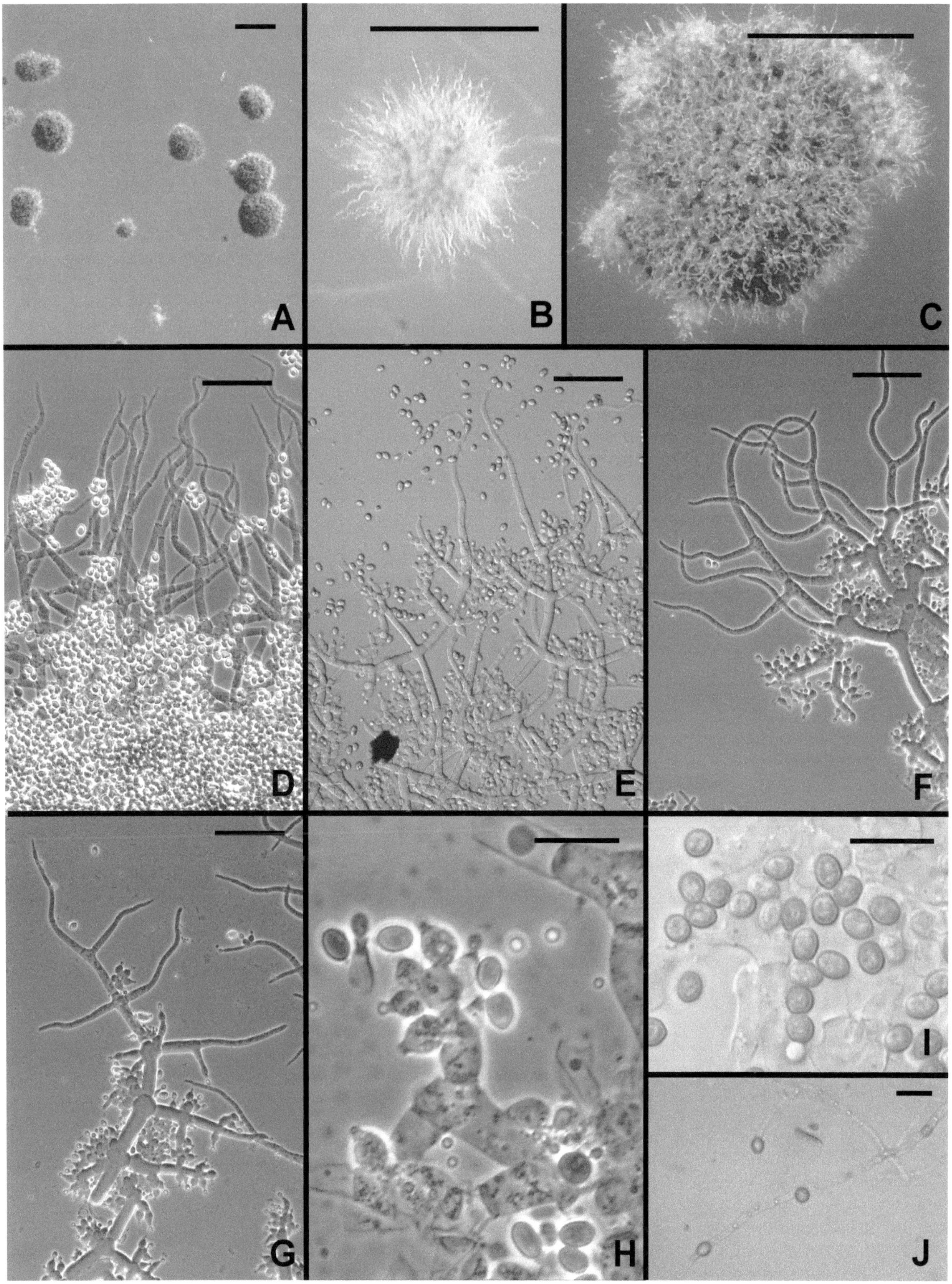
A
B
C
D
E
F
G
H
I
J

Trichoderma harzianum

Rifai, Mycol. Pap. 116:38. 1969. Figs. 6.2F, 6.22.

Ex-neotype culture: CBS 226.95 (Gams and Meyer, 1998)
GenBank: *rpb2*: AF348101; *tef1*: AF220720

Clade: Harzianum

Colony radius at 25°C: PDA 50–60. **SNA** (40–)45–52(–55). A diffusing yellow pigment typically developing within 48 h on PDA. No distinctive odor.

Colony radius at 35°C: PDA (40–)50–60. **SNA** 35–45.

Conidiation: Conidia dispersed in two or three concentric rings, conidiophores mainly arising in the aerial mycelium but also in confluent, cottony pustules beginning in the center of the colony and progressing outwardly. Conidia gray-green, appearing dry. No protruding conidiophores or sterile hairs formed.

Microscopic characteristics: Branching pattern Pyramidal type with the conidiophore terminating in hyphae of the pustule and consisting of a short main axis with a whorl of three to five diverging phialides at the tip with few (two to four) levels of usually paired, often closely spaced lateral fertile branches, each producing phialides directly or producing one to three levels of paired secondary fertile branches, each terminating in a whorl of two to four diverging phialides. Often branches not paired or conidiophores otherwise laterally asymmetrical. **Phialides** ampulliform, tending to be constricted into a narrow neck at the tip, 6.2–7.5 μm long, 3.2–3.5 μm at the widest point, L/W 1.8–2.2. **Supporting cell** 1.7–2.2 μm wide; ratio of length of phialide to width of supporting cell 2.0–2.9; ratio of widest part of phialide to width of supporting cell 1.1–1.2. **Conidia** subglobose, (2.0–)2.5–3.0(–3.7) × (2.2–)2.7–3.5(–4.2) μm, L/W (0.9–)1.0–1.2(–1.6), green, smooth. **Chlamydospores** terminal and intercalary.

Distinguishing characters: *Trichoderma harzianum* is typical of a phylogenetically complex of species (Chaverri et al., 2003b; Druzhinina et al., 2010a), all characterized by relatively small, smooth subglobose conidia, uniformly branched conidiophores dispersed in aerial mycelium or disposed in large confluent aggregates, and ampulliform phialides that often have a distinct, narrow neck. Chaverri et al. (2014) have recognized several species in the complex. The most common species are the cosmopolitan *T. afroharzianum* and *T. guizhouense*, the north temperate *T. simmonsii*, and the primarily neotropical endophytic *T. lentiforme* and *T. endophyticum*. The present description is based on *T. harzianum* in a restricted sense. The only reliable way to distinguish species in the complex is through DNA sequences, although biogeography can be helpful. Similar also is the unrelated *T. atroviride*.

Note: The report of *Hypocrea lixii* as the teleomorph of *T. harzianum* s. str. (Chaverri and Samuels, 2003) is incorrect. *Hypocrea lixii* is a distinct species within the Harzianum clade (Chaverri et al., 2015).

Habitat: Soil and wood.

Distribution: *Trichoderma harzianum* s. str. is an uncommon species mainly found in north temperate regions, Europe, and North America. The *T. harzianum* complex is very widely distributed and is the most commonly encountered "species" of *Trichoderma*.

Literature: Chaverri et al. (2003b, 2015), Druzhinina et al. (2010a), Gams and Meyer (1998), Samuels et al. (2002).

Figure 6.22. *Trichoderma harzianum*. **A and B,** Conidial pustules. **C–G,** Conidiophores. **H,** Conidia. **I,** Chlamydospores. **A–I,** From cornmeal dextrose agar. Scale bars: **A** = 1 mm; **B** = 0.5 mm; **C–G, I** = 20 μm; **H** = 10 μm.

➤

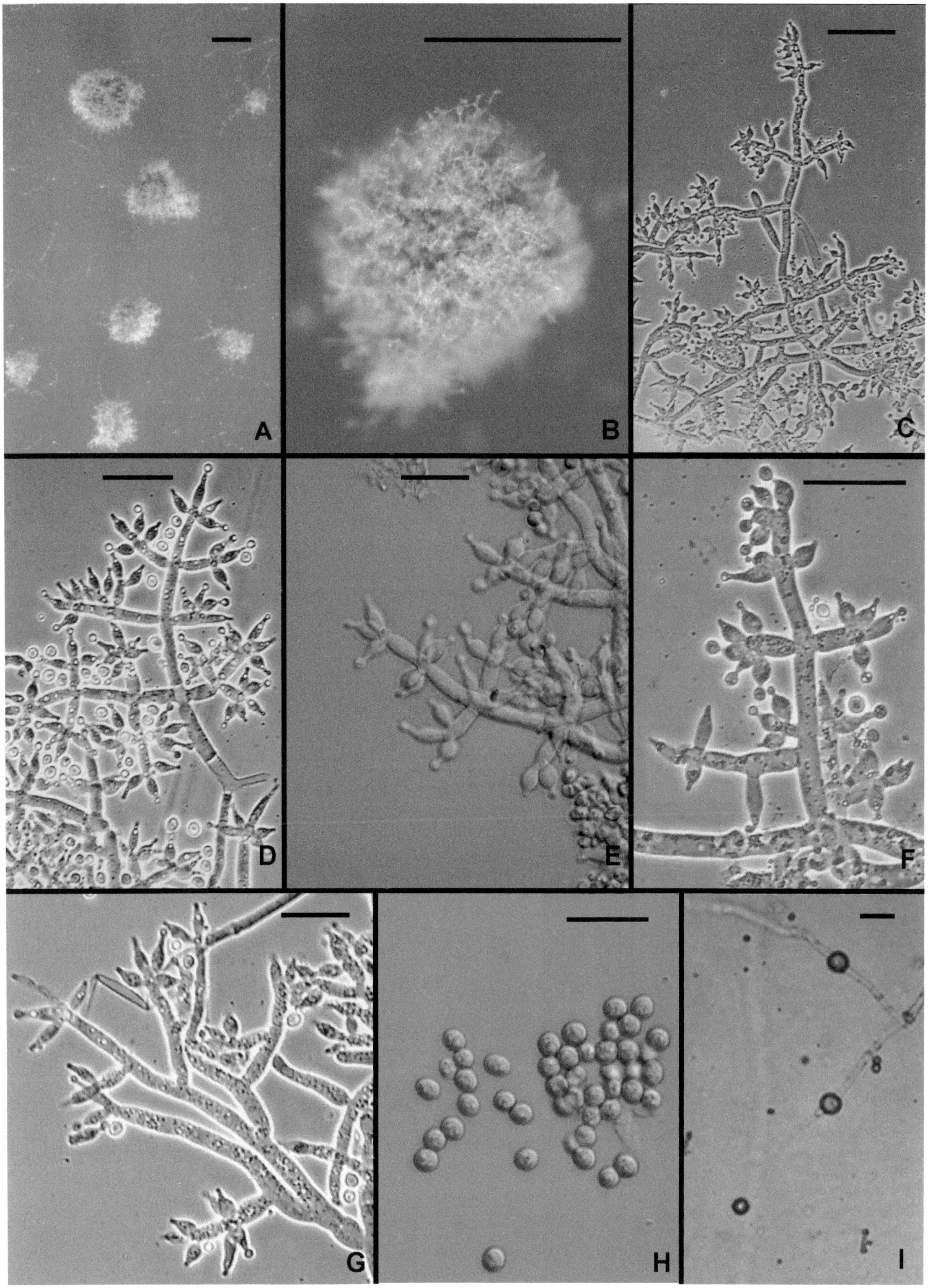
A
B
C
D
E
F
G
H
I

Trichoderma koningii

Oudem. *in* Oudemans and Koning, Arch. Néerl. Sci. Exactes Nat., Sér. 2, 7:291. 1902. Figs. 6.2G, 6.23.
= *Hypocrea koningii* Lieckf., Samuels & W. Gams *in* Lieckfeldt et al., Can. J. Bot. 76:1519. 1998.

Ex-neotype culture: CBS 457.96 = IMI 374798 (Lieckfeldt et al., 1998)
GenBank: *rpb2*: FJ442761; *tef1*: AF456909

Clade: Viride

Colony radius at 25°C: PDA (40–)50–60. **SNA** (25–)30–35. On PDA no diffusing pigment or distinctive odor noted.

Colony radius at 35°C: PDA and **SNA** <3.

Conidiation: Pustules forming abundantly in concentric rings progressing from the center of the colony, typically not well formed, loosely cottony, becoming confluent, sometimes forming flat sheets, at the advancing edge of the colony; conidiophores formed abundantly in the aerial mycelium and only slightly aggregated into pustules. Conidia dark green.

Microscopic characteristics: Branching pattern Pyramidal type with branches paired or in whorls of three and phialides held in divergent whorls at the tips of branches, often in great numbers in pseudowhorls (phialides separated from each other by a very short internode). **Sterile hairs** or projecting conidiophores not formed. **Phialides** lageniform, at most slightly swollen in the middle, straight, (4.2–)6.2–10.0(–15.5) µm long, (2.0–)2.7–3.5(–4.5) µm at the widest point, L/W (1.2–)1.8–3.6(–6.6). **Supporting cell** (1.5–)2.2–3.0 (–4.0) µm wide; ratio of length of phialide to width of supporting cell (1.5–)2.0–4.0(–6.2); ratio of widest point of phialide to width of supporting cell (0.5–) 1.0–1.4(–2.0). **Conidia** oblong, (2.0–)2.5–3.0(–3.5) × (3.0–)3.7–4.5(–4.7) µm, L/W (1.3–)1.5–1.7(–2.0), dark green, smooth. **Chlamydospores** sparingly produced, terminal on immersed hyphae, subglobose.

Distinguishing characters: *Trichoderma koningii* is morphologically similar to *T. gamsii and T. koningiopsis*. It is most easily distinguished from *T. gamsii* by the lack of yellow pigmentation and proliferating phialides on PDA and SNA. It has slightly larger conidia than *T. koningiopsis*, but the latter species is common in tropical regions and uncommon in temperate regions, whereas *T. koningii* is a species of temperate regions.

Habitat: Isolated from soil; teleomorph found on bark and wood.

Distribution: North America, Europe; not common.

Literature: Jaklitsch (2011), Lieckfeldt et al. (1998), Samuels et al. (2006a).

Figure 6.23. *Trichoderma koningii.* **A–C,** Conidial pustules. **D–H,** Conidiophores. Pseudowhorls can be seen in **G. I,** Conidia. **A and D–I,** From cornmeal dextrose agar. **B and C,** From Spezieller Nährstoffarmer agar (SNA). Scale bars: **A** = 1 mm; **B, C** = 0.5 mm; **D, F–H** = 20 µm; **E, I** = 10 µm. (D–G reproduced, by permission, from Samuels et al., 2006a) ➤

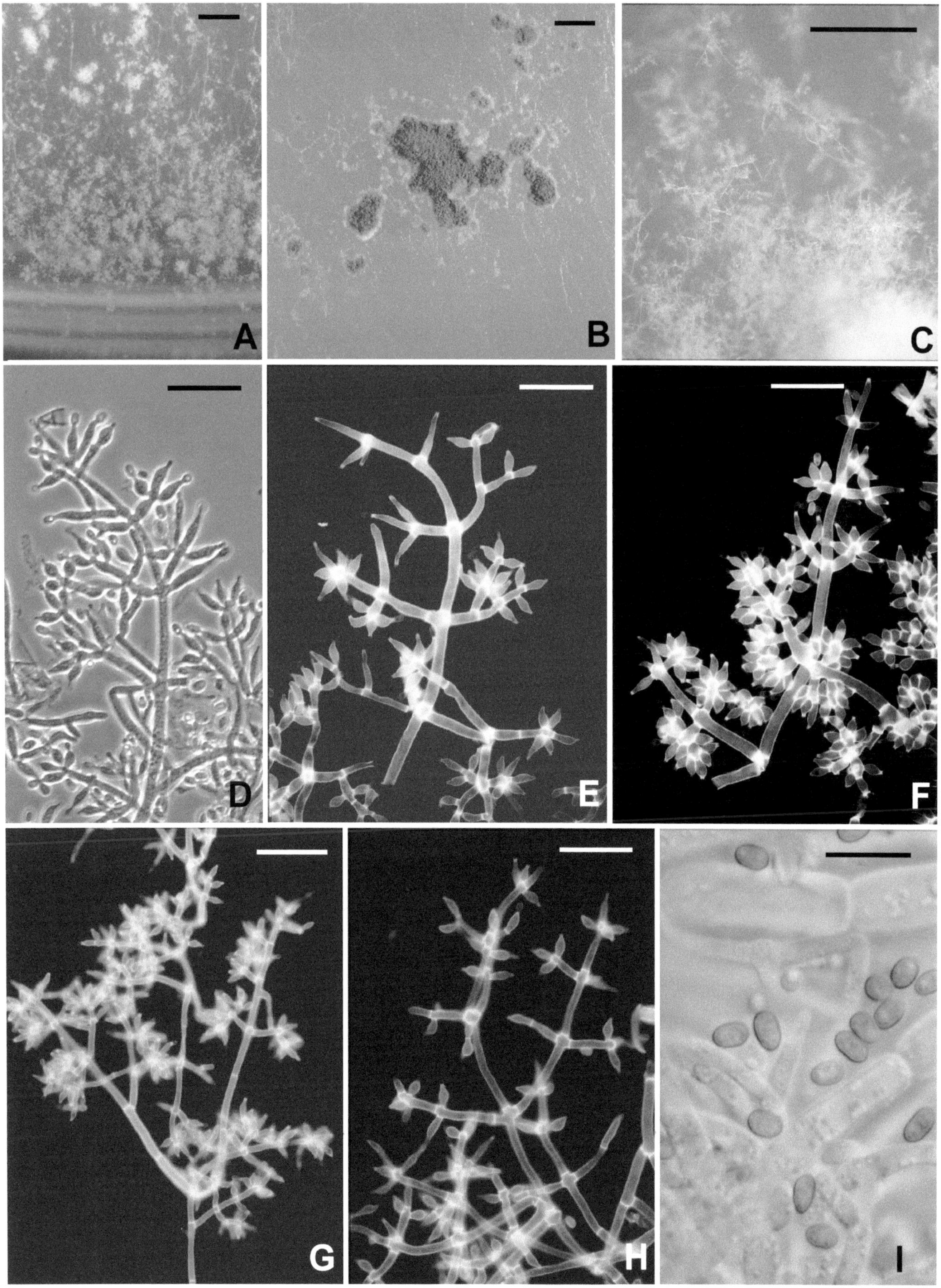
A
B
C
D
E
F
G
H
I

Trichoderma koningiopsis

Samuels, C. Suárez & H. C. Evans *in* Samuels et al., Stud. Mycol. 56:117. 2006. Figs. 6.2H, 6.24.

= *Hypocrea koningiopsis* Samuels *in* Samuels et al., Stud. Mycol. 56:117. 2006.

Ex-type culture: CBS 119075

GenBank: *rpb2*: EU241506; *tef1*: DQ284966

Clade: Viride

Colony radius at 25°C: PDA 48–67. **SNA** 40–50. On PDA no diffusing pigment; rarely a faint coconut odor.

Colony radius at 35°C: PDA 2–12. **SNA** 0–14.

Conidiation: Pustules cottony poorly formed, confluent; conidia tending to be dispersed in the aerial mycelium in concentric rings.

Microscopic characteristics: Branching pattern Pyramidal type with a conspicuous main axis having short, frequently paired lateral branches; lateral branches longer with distance from the tip, rebranching, producing phialides directly toward the tip. **Sterile hairs** or long fertile extensions of conidiophores lacking. **Phialides** on SNA lageniform, but shorter and swollen in the middle when crowded, often very closely spaced forming a pseudowhorl, (3.5–)5.5–9.2(–16.0) µm long, (2.0–)2.7–3.5(–4.5) µm at the widest point, L/W (1.0–)1.7–3.3(–6.0). **Supporting cell** (1.5–)2.2–3.0(–10.2) µm wide; ratio of length of phialide to width of supporting cell (0.8–)1.8–3.6(–6.0); ratio of widest point of phialide to width of supporting cell (0.4–)0.9–1.4(–2.1); occasionally proliferating percurrently. **Intercalary phialides** uncommon. **Conidia** ellipsoidal, (2.0–)2.5–3.0(–3.5) × (3.0–)3.5–4.0(–6.2) µm, L/W (1.0–)1.3–1.7(–2.5), green, smooth. **Chlamydospores** abundant to sparse or lacking, terminal or intercalary, globose to subglobose.

Distinguishing characters: *Trichoderma koningiopsis* is morphologically similar to *T. gamsii*, *T. koningii*, and *T. ovalisporum*. It differs from *T. koningii* in producing percurrently proliferating phialides and from *T. ovalisporum*, which has broader conidia. It differs from *T. gamsii* in having somewhat smaller conidia.

Habitat: A common endophyte of sapwood of cacao and rubber trees. Teleomorph found on decorticated hardwood.

Distribution: Pantropical and common but reaching into the Appalachian Mountains of the United States.

Literature: Samuels et al. (2006a).

Figure 6.24. *Trichoderma koningiopsis.* **A and B,** Conidial pustules. **C–J,** Conidiophores. A pseudowhorl of phialides can be seen in **H**; a short chain of proliferated phialides can be seen in **I** (arrow); an intercalary phialide can be seen in **J** (arrow). **K,** Conidia. **L,** Chlamydospores. **A,** From Spezieller Nährstoffarmer agar (SNA). **B–L,** From cornmeal dextrose agar. Scale bars: **A** = 1 mm; **B** = 0.5 mm; **C–H, J, L** = 20 µm; **I, K** = 10 µm. (A, D, J, and K reproduced, by permission, from Samuels et al., 2006a) ➤

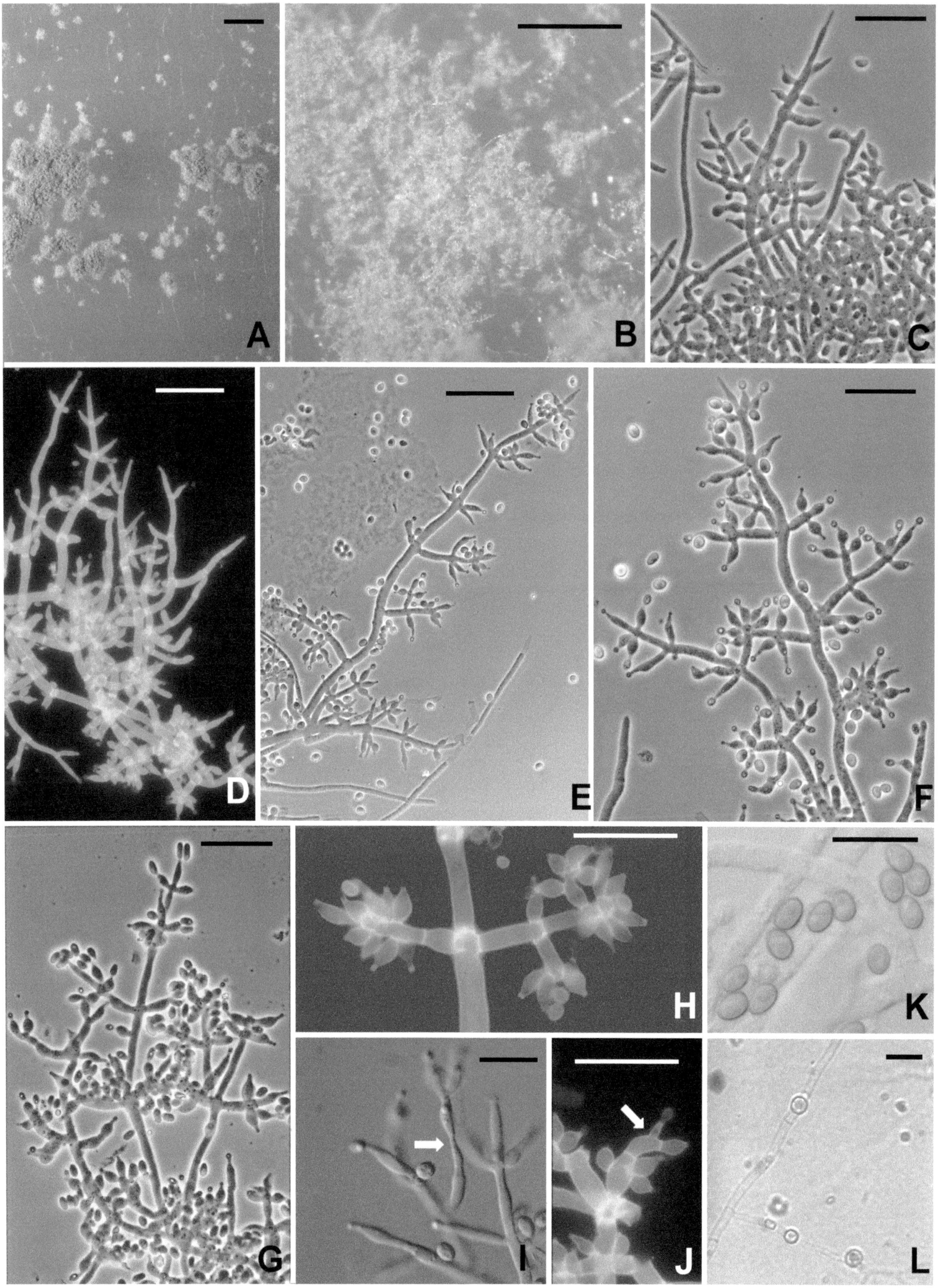
A
B
C
D
E
F
G
H
I
J
K
L

Trichoderma lieckfeldtiae

Samuels *in* Samuels and Ismaiel, Mycologia 101:149. 2009. Figs. 6.2I, 6.25.

Ex-type culture: CBS 123049
GenBank: *rpb2*: EU883562; *tef1*: EU856325

Clade: Viride

Colony radius at 25°C: PDA 60–70. **SNA** 35–45. On PDA no diffusing pigment or distinctive odor.

Colony radius at 35°C: No growth.

Conidiation: Pustules few, scattered, pulvinate to hemispherical, compact, 0.5–3.0 mm in diameter, remaining discrete, gray-green; sterile hairs, sinuous to nearly helical, conspicuous, white.

Microscopic characteristics: Branching pattern Pachybasium type with fertile branches arising at the base of sterile hairs; fertile branches composed of one or a few broad cells, often rebranching; secondary branches typically unicellular. **Sterile hairs** sinuous, branched, septate, thin walled, acute sterile. No **synanamorph** observed. **Phialides** crowded at the tips of secondary branches, flask shaped, (3.0–)5.2–5.7(–9.2) µm long, (1.5–)2.7–3.7(–4.5) µm at the widest point, L/W (0.9–)1.3–2.0(–3.5). **Supporting cell** (1.2–)2.0–3.7 (–5.5) µm wide; ratio of length of phialide to width of supporting cell (0.7–)1.3–2.6(–4.7); ratio of width of phialide to width of supporting cell (0.5–)0.8–1.6(–2.9). **Conidia** ellipsoidal, (1.7–)2.0–2.7(–3.7) × (2.5–)3.2–4.2(–6.0) µm, L/W (0.6–)1.4–1.9(–2.4), smooth, gray-green in mass. **Chlamydospores** not observed.

Distinguishing characters: *Trichoderma lieckfeldtiae* is most similar to *T. evansii*, *T. hamatum*, and *T. pubescens*. Of these, *T. pubescens* has the slowest growth rate and produces an abundant *Verticillium*-like synanamorph, as does *T. hamatum*. Neither *T. lieckfeldtiae* nor *T. evansii* produces a synanamorph. Conidia of *T. evansii* are not formed in discrete pustules on SNA, rather the pustules are poorly formed. Both *T. evansii* and *T. hamatum* produce conidia in large numbers on SNA and cornmeal dextrose agar (CMD), whereas colonies of *T. lieckfeldtiae* on CMD are sterile and far fewer conidia form on SNA. *Trichoderma evansii* is rather rare, whereas *T. hamatum*, *T. pubescens*, and *T. lieckfeldtiae* are more common.

Habitat: Soil and pseudostroma of *Moniliophthora roreri*.

Distribution: Tropical, known from Central Africa (Cameroon) and South America (Colombia, Peru).

Literature: Samuels and Ismaiel (2009).

Figure 6.25. *Trichoderma lieckfeldtiae*. **A and B,** Conidial pustules. **C–H,** Conidiophores. **I,** Conidia. **A–C and E–G,** From Spezieller Nährstoffarmer agar (SNA). **D, H, and I,** From cornmeal dextrose agar. Scale bars: **A, B** = 0.5 mm; **C–E** = 20 µm; **F–I** = 10 µm. (C and F reprinted from Samuels, G. J., and Ismaiel, A. 2009. *Trichoderma evansii* and *T. lieckfeldtiae*: Two new *T. hamatum*-like species. Mycologia 101:142-156, with permission from *Mycologia*, ©The Mycological Society of America) ➤

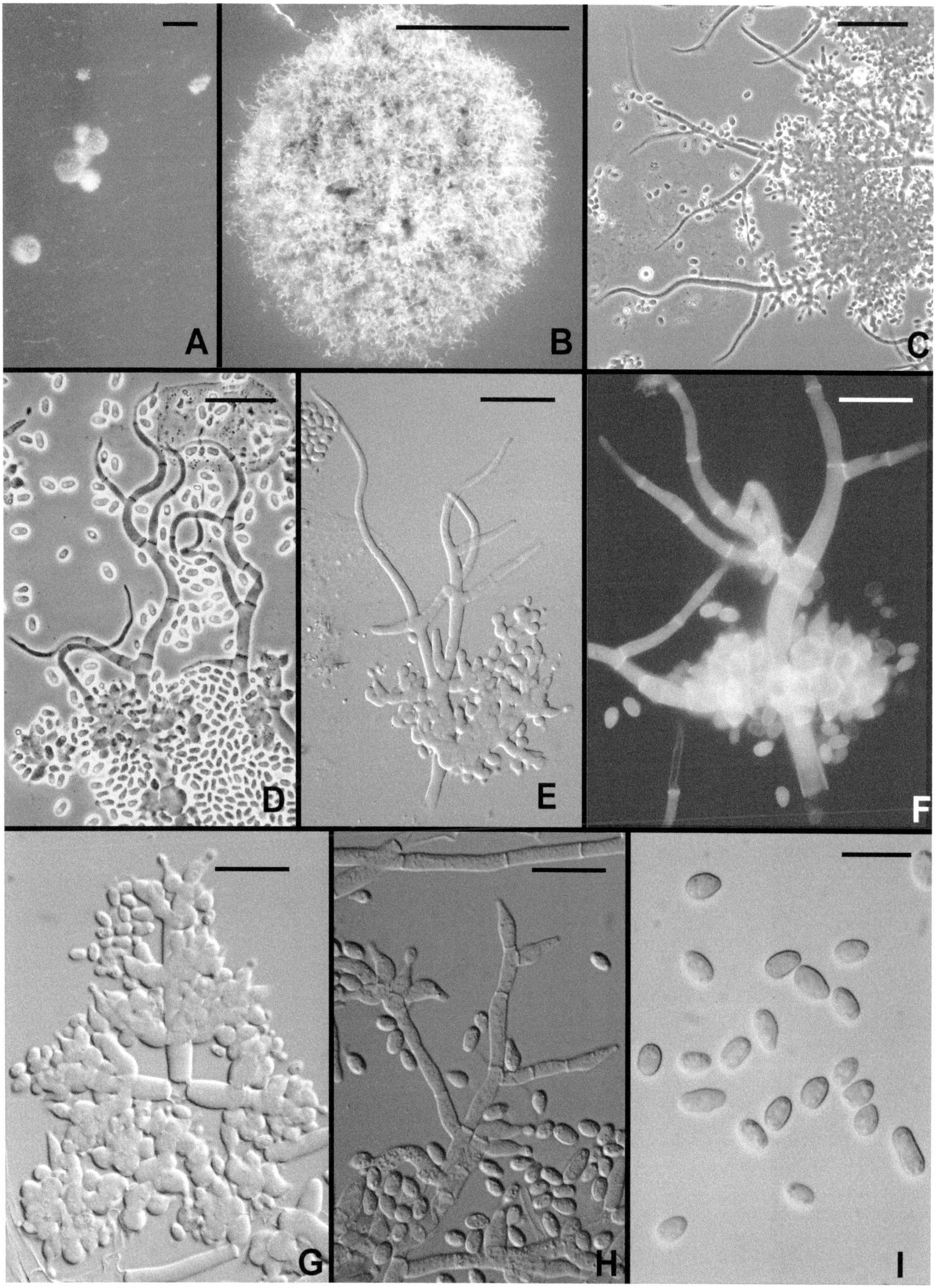
A
B
C
D
E
F
G
H
I

Trichoderma longibrachiatum

Rifai, Mycol. Pap. 116:42. 1969. Figs. 6.2J, 6.26.

Ex-type culture: ATCC 18648 = CBS 816.68 = DAOM 267674
GenBank: *rpb2*: DQ087242; *tef1*: AY937412

Clade: Longibrachiatum

Colony radius at 25°C: PDA (55–)62–70. **SNA** (50–)42–56(–65). On PDA a diffusing yellow pigment; no distinctive odor.

Colony radius at 35°C: PDA (62–)68–70. **SNA** (38–)43–57(–63). **At 40°C: PDA** (24–)34–54(–62).

Conidiation: Pustules abundant, uniformly dispersed, sometimes in concentric rings, often forming a continuous lawn, minute, <0.25 mm in diameter, becoming confluent, cottony or wooly, dark green; completely fertile conidiophores visible.

Microscopic characteristics: Branching pattern Longibrachiatum type, fertile branches typically consisting of several levels of branching, branches near the tip bearing a single phialide and not rebranched, branches distal to the tip longer and rebranching; **sterile hairs** not seen; single or ropes of conidiophores giving the pustules a spiky or aspirate aspect. **Phialides** cylindrical to lageniform, often swollen in the middle, solitary, often terminating in a single cell and subtended by an intercalary phialide, straight, hooked or sinuous, (3.5–)5.5–10.0(–17.2) μm long, (1.0–)2.2–3.2(–4.5) μm at the widest point, L/W (1.1–)1.8–3.7(–7.3). **Supporting cell** (1.5–)2.2–3.2(–5.0) μm wide; ratio of length of phialide to width of supporting cell (1.0–)2.0–3.8(–8.1); ratio of width of phialide to width of supporting cell (0.5–)0.8–1.2(–2.0). **Intercalary phialides** (arrows) abundant. **Conidia** ellipsoidal, (2.0–)2.5–3.0(–4.0) × (2.0–)3.5–4.5(–6.2) μm, L/W (0.9–)1.3–1.9(–2.7), green, smooth. **Chlamydospores** generally abundant on cornmeal dextrose agar (CMD), terminal and then subglobose to globose or intercalary and then conforming to the shape of the hyphal cell in which they form.

Distinguishing characters: *Trichoderma longibrachiatum* is essentially morphologically indistinguishable from *T. reesei* and *T. parareesei.* Unlike *T. reesei*, on SNA conidia of *T. longibrachiatum* form in abundance within 96 h under light. *Trichoderma longibrachiatum* tends to have more sinuous or hooked phialides than *T. reesei.*

Habitat: Soil.

Distribution: Cosmopolitan.

Literature: Bissett (1984), Gams and Bissett (1998), Samuels et al. (1998, 2012).

Figure 6.26. *Trichoderma longibrachiatum.* **A and B,** Conidial pustules. Note the spiky conidiophores. **C–I,** Conidiophores. Intercalary phialides indicated by arrows in **E** and **H. J,** Conidia. **A, C, G, and I,** From cornmeal agar. **B, D, E, and H,** From cornmeal dextrose agar. **F and J,** From Spezieller Nährstoffarmer agar (SNA). Scale bars: **A, B** = 0.5 mm; **C, D, G, I** = 20 μm; **E, F, H, J** = 10 μm.

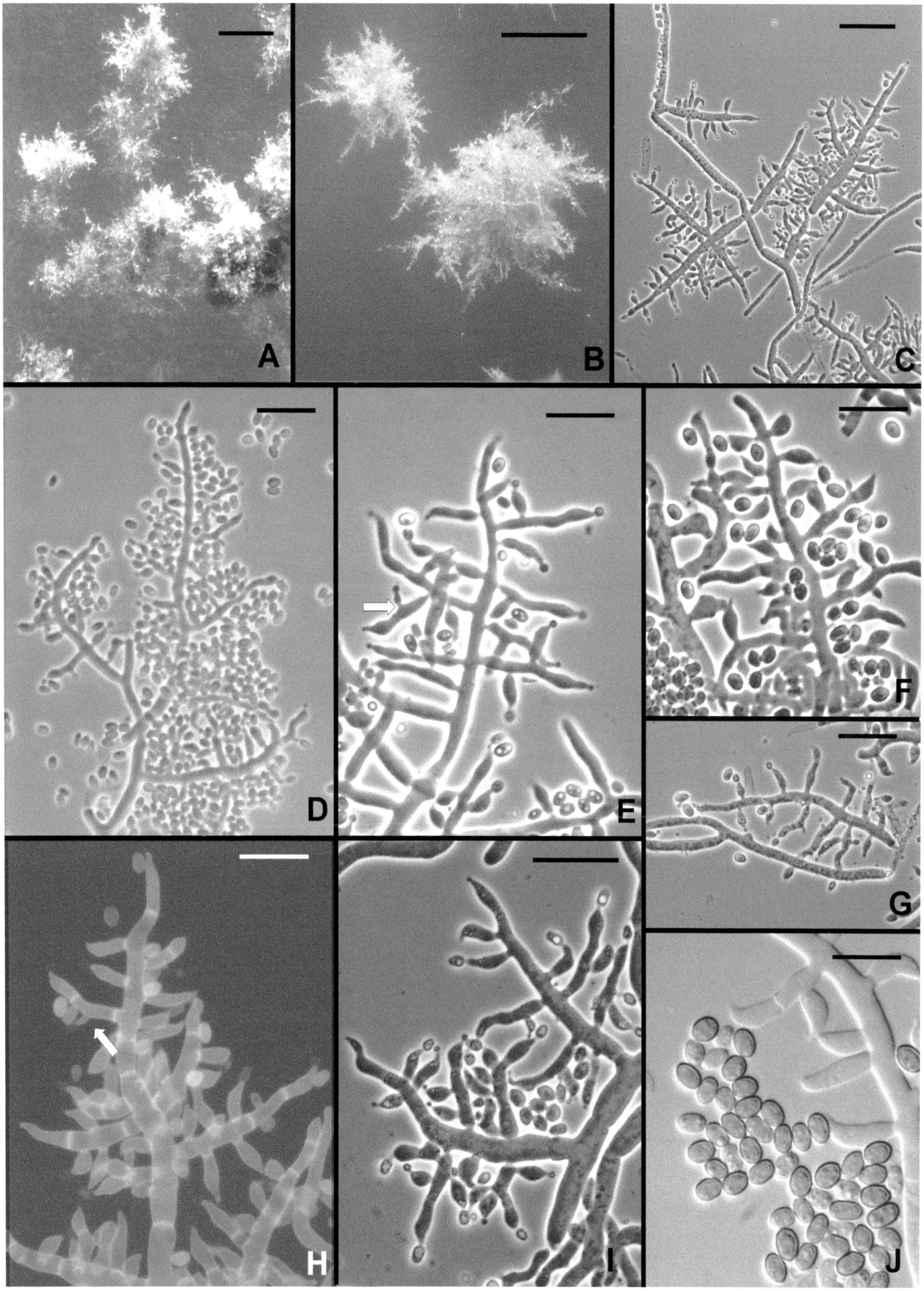
A
B
C
D
E
F
G
H
I
J

Trichoderma longipile

Bissett, Can. J. Bot. 69:2395. 1991 [1992] [as *longipilis*]. Figs. 6.2K, 6.27.
= *Hypocrea longipilosa* Jaklitsch, Stud. Mycol. 63:62. 2009.

Ex-type culture: DAOM 177227 (CBS 120953, ex-type culture of *Hypocrea longipilosa*)
GenBank: *rpb2*: AF545550; *tef1*: AY937430 (FJ860643, CBS 120953 *H. longipilosa*)

Clade: Strictipile

Colony radius at 25°C: PDA 30–45. **SNA** 25–30(–45). On PDA no diffusing pigment or distinctive odor, diffused conidia forming within 7 days at 25°C under light; pustules forming only after 14 days.

Colony radius at 35°C: PDA <30. **SNA** <15.

Conidiation: Conidia forming within 2 weeks, effused throughout the colony in the aerial mycelium, conidiophores *Verticillium*-like and in pustules. Pustules abundant, loose, to 2 mm in diameter, often confluent to 10 mm, grayish green; producing conspicuous long hairs.

Microscopic characteristics: Branching pattern Pachybasium type; fertile branches arising at right angles toward the base of hairs, usually only one or a few cells in length, producing unicellular secondary branches; primary and secondary branches terminating in one or several phialides, increasing in length and more profusely branched with distance from the tip. **Hairs** straight or undulate, unbranched or branching, often anastomosing, sterile or bearing a few terminal phialides, otherwise acute. **Phialides** ampulliform, (4.0–)4.7–7.0(–9.5) µm long, (3.0–)3.2–4.5(–5.0) µm at the widest point, L/W (1.0–)1.2–1.9(–2.7). **Supporting cell** (2.7–)3.0–4.7(–5.7) µm wide; ratio of length of phialide to width of supporting cell (1.0–)1.1–2.0(–2.5); ratio of width of phialide to width of supporting cell (0.6–)0.8–1.2(–1.5). **Conidia** oblong, (1.2–)2.5–3.2(–4.0) × (4.2–)4.5–5.5(–7.0) µm, L/W (1.3–)1.5–2.1(–2.4), green, smooth. **Chlamydospores** terminal and intercalary.

Distinguishing characters: *Trichoderma longipile* is a northern species characterized by long, branched, acute fertile or sterile hairs. It is most similar to *T. oblongisporum*, which has somewhat smaller conidia and unbranched hairs with rounded tips, and the more common *T. strictipile*, which has smaller conidia (2.2–3.0 × 2.7–4.0 µm). *Trichoderma longipile* and *T. strictipile* are more commonly encountered as their teleomorphs than they are isolated directly from soil or other substrates. The reddish color of the stroma is unusual (Jaklitsch, 2009).

Habitat: Wood and soil. The teleomorph is known from a single collection found in Sweden on a decorticated branch of *Corylus avellana*.

Distribution: North temperate/boreal (Canada, Sweden).

Literature: Bissett (1991b), Gams and Bissett (1998), Jaklitsch (2009).

Figure 6.27. *Trichoderma longipile*. **A and B,** Conidial pustules. **C,** Sterile hairs arising from a pustule. **D–H,** Conidiophores. **I and J,** Phialides. **K,** Conidia. **L,** Chlamydospores. **A–L,** From Spezieller Nährstoffarmer agar (SNA). Scale bars: **A** = 1 mm; **B** = 0.5 mm; **C–H, J, L** = 20 µm; **I, K** = 10 µm.

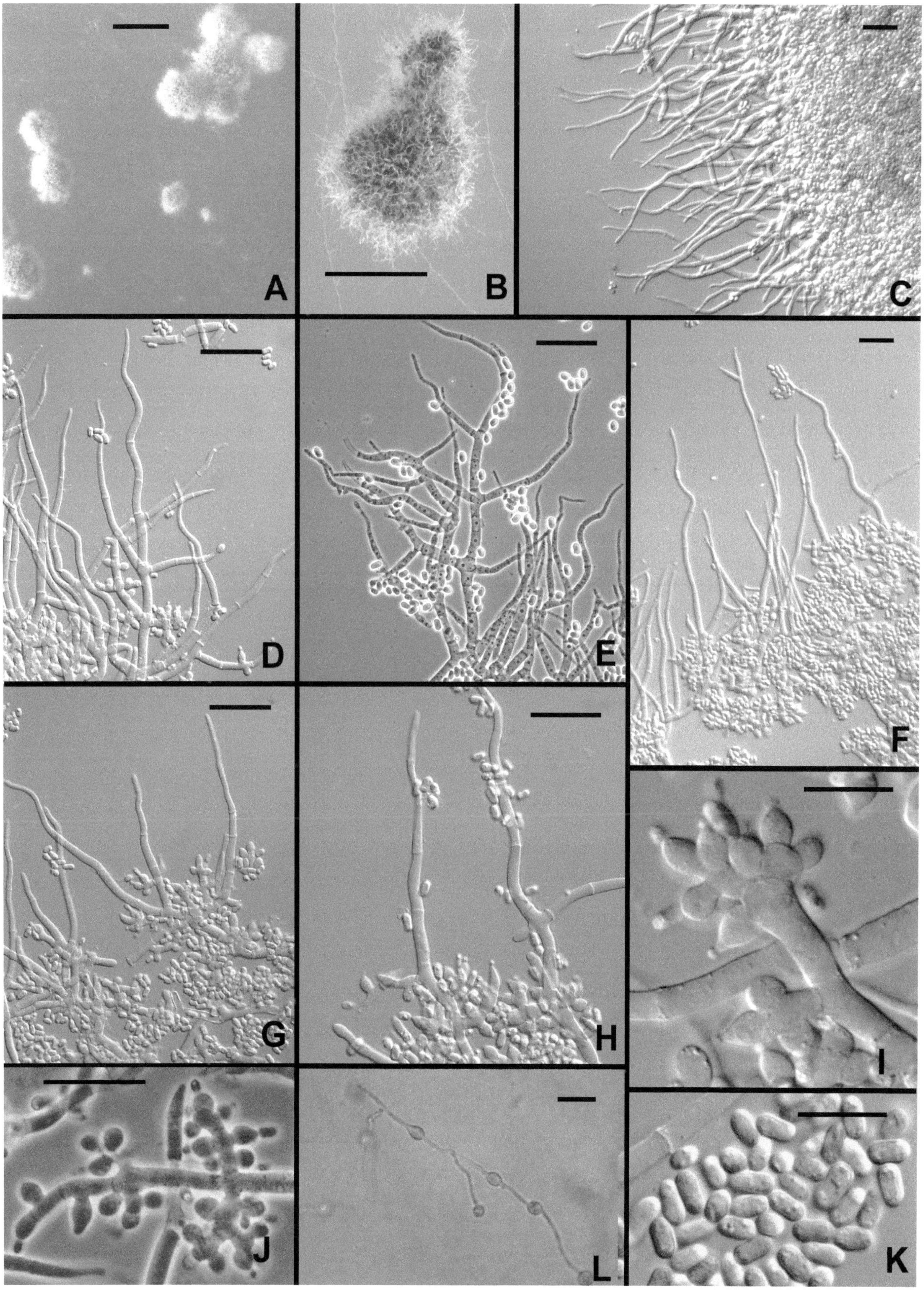
A
B
C
D
E
F
G
H
I
J
L
K

Trichoderma minutisporum

Bissett, Can. J. Bot. 69:2396. 1991 [1992]. Figs. 6.2L, 6.28.

= *Hypocrea minutispora* B.-S. Lu, Fallah & Samuels *in* Lu et al., Mycologia 96:335. 2004.

Ex-type culture: DAOM 167069 = CBS 341.93
GenBank: *rpb2*: KJ665314; *tef1*: AY750883

Clade: Polysporum

Colony radius at 25°C: PDA and **SNA** 25–40. On PDA no diffusing pigment or distinctive odor.

Colony radius at 35°C: PDA and **SNA** <5.

Conidiation: Pustules abundant or few, scattered, discrete or sometimes adjacent pustules becoming confluent, pulvinate, 0.5–2 mm in diameter, appearing papillate or granular because of the conidia held in cabbagelike heads, sterile hairs lacking, grayish green, green. On PDA conidia forming within (7–)10 days under light at 25°C, covering the entire surface with densely disposed green pustules.

Microscopic characteristics: Branching pattern Pachybasium type; typical conidiophores with a well-developed main axis, fertile branches arising at right angles to the main axis, typically paired, arising at close intervals, increasing in length from the tip to the base of the conidiophore, rebranching; branches arising near the tip of the conidiophore and secondary branches usually one or a few cells long, arising at right angles, paired, terminating in a crowded head of phialides, the main axis of the conidiophore and the fertile branches appearing wide with respect to the phialides; often conidiophore extending to a greater or lesser distance and terminating in one or a few phialides; occasionally conidiophores having a Pyramidal-type branching pattern with relatively long phialides in whorls; conidiophores often produced in aerial mycelium apart from pustules. **Phialides** typically ampulliform less frequently lageniform, (3.0–)4.5–7.0(–11.2) μm long, (2.2–)3.0–3.7(–4.5) μm at the widest point, L/W (1.0–)1.3–2.1 (–4.4). **Supporting cell** conspicuously wide, (2.0–)3.0–4.2(–5.5) μm wide; ratio of length of phialide to width of supporting cell (0.7–)1.1–2.0(–3.8); ratio of width of phialide to width of supporting cell (0.6–)0.8–1.1(–1.5). **Conidia** ellipsoidal, (1.7–)2.0–2.7(–3.2) × (2.5–)2.7–4.0 (–5.5) μm, L/W (1.0–)1.2–1.5(–2.0), green, smooth. **Chlamydospores** not observed.

Distinguishing characters: *Trichoderma minutisporum* is characterized by discrete pustules (on SNA) that lack sterile hairs or conspicuously long and terminally fertile conidiophores and that appear mealy or papillate, the result of uniformly branched conidiophores that are fertile over their entire length, the conidia forming in dry heads at the end of each branch. The tendency to form discrete pustules is reduced on cornmeal dextrose agar (CMD), where conidiation tends to be effused. Bissett (1991b) described the conidia as "bright green"; Jaklitsch (2011) described them as "bluish to medium green" in nature. The phylogenetic diversity shown for this species by Lu et al. (2004a) suggests that *T. minutisporum* as it is now recognized is a species complex. Jaklitsch and Voglmayr (2015) distinguished *T. minutisporum* (United States) and *T. europaeum* (Europe) primarily on phylogenetic differences, and cultures illustrated for European collections of this species by Jaklitsch (2011) are different from those illustrated here for a North American collection, supporting a taxonomic distinction between Europe and North America.

Habitat: Bark, soil. The teleomorph is the most common European species of the genus having hyaline ascospores, common on wood of *Fagus sylvatica*, less common on conifers (Jaklitsch, 2011).

Distribution: North temperate (Europe, North America), rare or absent in the tropics.

Literature: Bissett (1991b), Jaklitsch (2011), Jaklitsch and Voglmayr (2015), Lu et al. (2004a).

Figure 6.28. *Trichoderma minutisporum.* **A and B,** Conidial pustules. **C–I,** Conidiophores. **J,** Phialides. **K,** Conidia. **A,** From cornmeal dextrose agar. **B–K,** From Spezieller Nährstoffarmer agar (SNA). Scale bars: **A** = 1 mm; **B** = 0.5 mm; **C–E, G–I** = 20 μm; **F, J, K** = 10 μm.

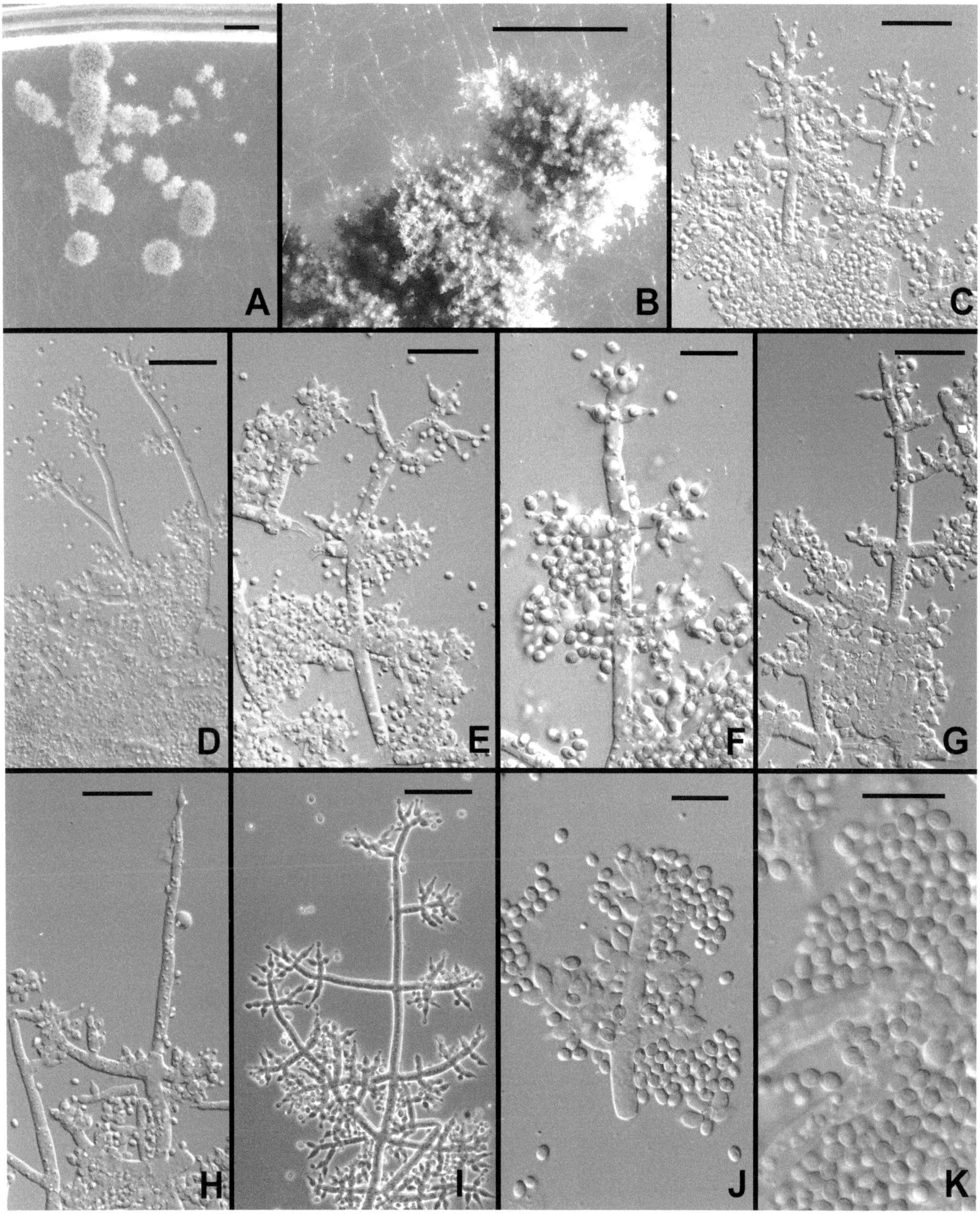
A
B
C
D
E
F
G
H
I
J
K

Trichoderma orientale

(Samuels & Petrini) Jaklitsch & Samuels *in* Jaklitsch and Voglmayr, Mycotaxon 126:151. 2014. Figs. 6.3A, 6.29.
≡ *Hypocrea orientalis* Samuels & Petrini *in* Samuels et al., Stud. Mycol. 41:30. 1998.

Ex-type culture: CBS 130428
GenBank: *rpb2*: JN175522; *tef1*: EU401581

Clade: Longibrachiatum

Colony radius at 25°C: PDA 65–70. **SNA** (20–)35–50(–70). On PDA often intense yellow diffusing pigment, no distinctive odor.

Colony radius at 35°C: PDA 70 (filling 9-cm-diameter petri plate). **SNA** 60–70. At 40°C typically filling a 9-cm-diameter petri plate within 72 h.

Conidiation: Pustules forming in concentric rings, minute, <0.5 mm in diameter, poorly formed and often reduced to one or a few conidiophores, discrete or confluent and then forming cottony lawns, green; individual pustules with a spiky or aspirate aspect resulting from long, erect fascicles or ropes of conidiophores arising from within pustules; sterile hairs lacking. Conidiophores also arising from aerial hyphae.

Microscopic characteristics: Branching pattern Longibrachiatum type; conidiophores arising as fertile branches of hyphae in pustules with a distinct main axis; main axis short and relatively little branched or longer and extensively rebranched with five or more levels of secondary branches; secondary branches increasing in length with distance from the tip of the main axis and rebranching; phialides typically solitary, especially for several levels below the tips of all branches. **Phialides** lageniform, less commonly ampulliform, straight, hooked or sinuous, (3.2–)6.5–10.5(–15.7) μm long, (2.0–)2.5–3.5(–4.5) μm at the widest point, L/W (1.3–)1.9–3.9(–5.8). **Supporting cell** (1.5–)2.2–3.7(–5.5) μm wide; ratio of length of phialide to width of supporting cell (1.2–)1.9–3.9(–6.4); ratio of width of phialide to width of supporting cell (0.5–)0.8–1.2(–2.1). **Intercalary phialides** (arrows) common. **Conidia** oblong to ellipsoidal, (2.0–)2.5–3.5(–5.2) × (3.2–)3.7–6.2(–10.5) μm, L/W (1.1–)1.3–2.5(–4.9), green, smooth. **Chlamydospores** terminal and intercalary, subglobose to clavate.

Distinguishing characters: *Trichoderma orientale* is a typical member of the Longibrachiatum clade. Its closest morphological comparison is with *T. gracile*, *T. longibrachiatum*, and *T. reesei*. It has a somewhat faster growth rate than the rare *T. gracile* on PDA and SNA. It is morphologically indistinguishable from the more common *T. longibrachiatum*.

Habitat: Soil, found once as an endophyte in sapwood of *Theobroma cacao*; known to cause invasive mycoses of humans. The teleomorph forms on wood.

Distribution: Pantropical; teleomorph known from China and Sardinia, possibly New Zealand.

Literature: Druzhinina et al. (2008, 2012a), Kuhls et al. (1997), Samuels et al. (1998, 2012).

Figure 6.29. *Trichoderma orientale.* **A and B,** Conidial pustules. Note the spiky aspect from projecting conidiophores in **B. C–H,** Conidiophores. Intercalary phialides indicated by arrows in **G. I,** Conidia. **J,** Chlamydospores. **A–C and F–J,** From Spezieller Nährstoffarmer agar (SNA). **D and E,** From cornmeal dextrose agar. Scale bars: **A** = 1 mm; **B** = 0.5 mm; **C–F, J** = 20 μm; **G–I** = 10 μm. (A, C, E, G, and I reproduced, by permission, from Samuels et al., 2012) ➤

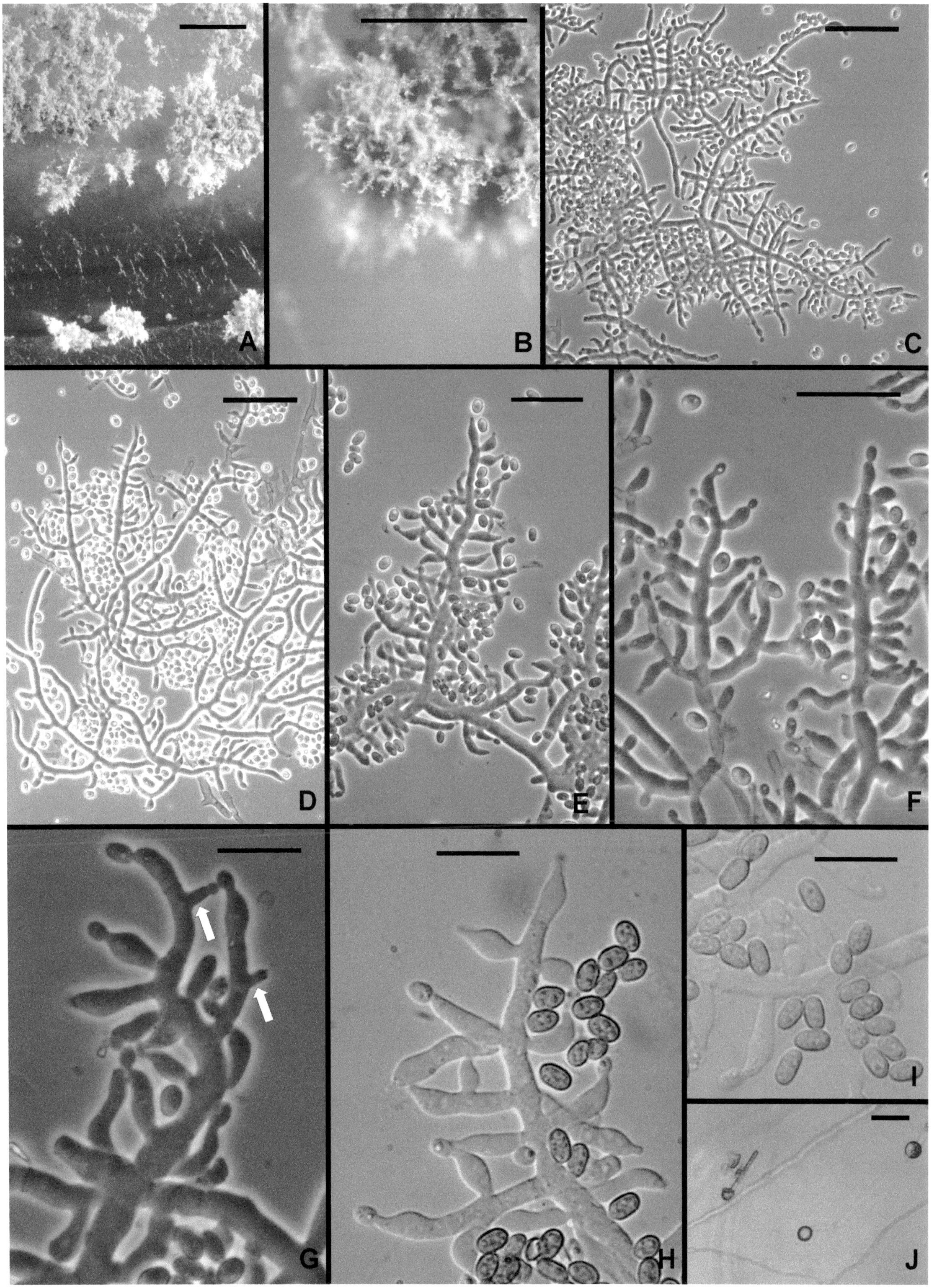
A
B
C
D
E
F
G
H
I
J

Trichoderma ovalisporum

Samuels & Schroers *in* Holmes et al., Mycol. Prog. 3:204. 2004. Figs. 6.3B, 6.30.

Ex-type culture: CBS 113300 = DAOM 232078 = IMI 385808
GenBank: *rpb2*: FJ442742; *tef1*: AY376037

Clade: Viride

Colony radius at 25°C: PDA 60–70. **SNA** 30–40(–45). On PDA no diffusing pigment or distinctive odor.

Colony radius at 35°C: PDA and **SNA** 0–10(–25).

Conidiation: Pustules forming in more-or-less distinct concentric rings, discrete, to 2 mm in diameter or often extensive, undefined and confluent, cottony, lacking protruding sterile hairs or conidiophores, dark green.

Microscopic characteristics: Branching pattern Pyramidal type; conidiophores terminating in intertwined hyphae of the pustules, composed of a well-defined central axis with up to six levels of fertile lateral branches; lateral branches relatively short, often arising from swollen nodes in the main axis, increasing somewhat in length with distance from the tip of the main axis but usually not rebranching with more than two secondary branches, often paired or arising in whorls of a few phialides; often solitary phialides arising in whorls from the main axis especially near the tip. **Phialides** typically held in divergent whorls of three or more, often in pairs or solitary and then often closely spaced and appearing to be arranged in whorls (pseudowhorls); lageniform, (4.0–)5.7–10.0(–14.7) μm long, (2.0–)2.7–3.7(–4.5) μm at the widest point, L/W (1.2–)1.6–3.5(–6.8). **Supporting cell** (1.7–)2.0–4.5(–10.7) μm wide; ratio of length of phialide to width of supporting cell (0.9–)1.7–3.7(–6.4); ratio of width of phialide to width of supporting cell (0.4–)0.8–1.4(–1.8). **Intercalary phialides** infrequent. **Conidia** ovoidal to subglobose, (1.7–)2.7–3.2(–4.5) × (2.0–)3.0–4.0(–4.5) μm, L/W (0.8–)1.1–1.3(–1.6), green, smooth. **Chlamydospores** scattered, subglobose.

Distinguishing characters: *Trichoderma ovalisporum* is distinguished by its uniformly branched conidiophores that have a strongly developed main axis and by its dark green, smooth, ovoidal to subglobose conidia. It is distinguished from *T. asperelloides* and *T. asperellum* by its smaller conidia and less uniformly branched conidiophores; from *T. atroviride*, which has subglobose to globose conidia; and from the unrelated *T. harzianum* by the more extensively branched conidiophores and longer phialides in *T. ovalisporum*.

Habitat: Soil, endophytic in sapwood of *Theobroma* species, isolated from wood of the liana *Banisteriopsis caapi*.

Distribution: Known from Brazil, Ecuador, Panama, Peru, Vietnam; probably pantropical.

Literature: Holmes et al. (2004).

Figure 6.30. *Trichoderma ovalisporum*. **A–C,** Conidial pustules. **D–H,** Conidiophores. Pseudowhorls of phialides can be seen in **G** and **H. I,** Conidia. **A–C, G, and H,** From cornmeal dextrose agar. **D–F and I,** From Spezieller Nährstoffarmer agar (SNA). Scale bars: **A** = 1 mm; **B, C** = 0.5 mm; **D–H** = 20 μm; **I** = 10 μm. (F reproduced, by permission, from Samuels et al., 2006a) ➤

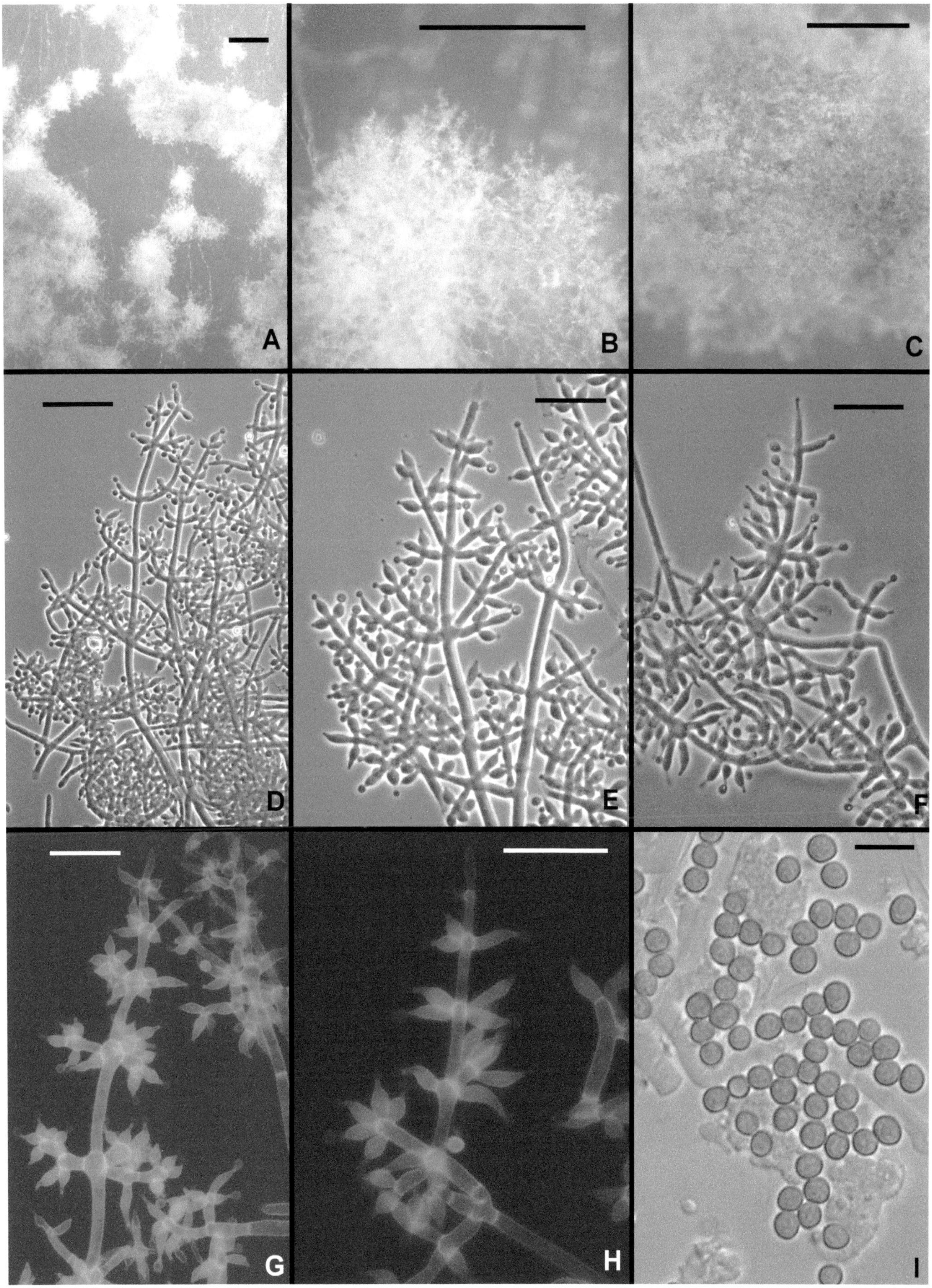
A
B
C
D
E
F
G
H
I

Trichoderma paucisporum

Samuels, C. Suárez & K. Solis *in* Samuels et al., Mycol. Res. 110:390. 2006. Figs. 6.3C, 6.31.

Ex-type culture: CBS 118645
GenBank: *rpb2*: FJ150787; *tef1*: DQ109540

Clade: Viride

Colony radius at 25°C: PDA and **SNA** 18–25.

Colony radius at 35°C: PDA and **SNA** 0–5.

Conidiation: Colonies on SNA nearly invisible with scant aerial mycelium in the center of the colony, no diffusing pigment, sometimes a strong odor of coconut detected, producing conidia in minute pustules around the margin; conidia remaining hyaline or slowly becoming green. On half-strength PDA sometimes producing conidia in broad, felty bands; conidia remaining yellow or becoming yellow-green. Colonies on PDA after 96 h off white to ivory with scant, felty aerial mycelium, pale yellow diffusing pigment, a coconut odor or no distinctive odor, sterile.

Microscopic characteristics: Branching pattern *Verticillium*- or *Gliocladium*-like or variously branched, with a more-or-less well-developed main axis having lateral branches tending to be paired or disposed in divergent whorls of up to three, solitary phialides often arising directly from the main axis; primary branches producing phialides directly singly and in whorls of three, terminating in one to three phialides in a whorl. **Phialides** lageniform and slightly swollen in the middle or tapering uniformly from base to tip, (6.0–)6.5–12.5 (–21.5) µm long, (2.0–)2.5–3.5 µm at the widest point, L/W (1.0–)1.2–1.4. **Supporting cell** 2–3 µm wide; ratio of length of phialide to width of supporting cell 3.2–5.2; ratio of width of phialide to width of supporting cell 0.8–1.2. **Conidia** ellipsoidal to broadly ellipsoidal, 2.5–3.5 × 3–5 µm, L/W 1.0–1.6, green, smooth. **Chlamydospores** sometimes produced abundantly on SNA, cornmeal dextrose agar (CMD), and half-strength PDA; terminal or less frequently intercalary in hyphae, becoming thick walled, subglobose to ellipsoidal.

Distinguishing characters: *Trichoderma paucisporum* is characterized by its *Verticillium*- or *Gliocladium*-like conidiophores, the sparse production of conidia, which are yellowish or weakly green, and the coconut odor which is especially noted on PDA colonies. It is atypical in the genus but is morphologically similar to *T. effusum* and *T. scalesiae*. *Trichoderma paucisporum* is most closely related to *T. theobromicola* and is known only from the type locality.

Habitat: Isolated from pods of *Theobroma cacao* infected with *Moniliophthora roreri* lying on leaf litter.

Distribution: Ecuador; known only from the type locality.

Literature: Samuels et al. (2006b).

Figure 6.31. *Trichoderma paucisporum*. **A–I,** Conidiophores. **J,** Conidia. **K,** Chlamydospores. **A–K,** From Spezieller Nährstoffarmer agar (SNA). Scale bars: **A–I, K** = 20 µm; **J** = 10 µm.

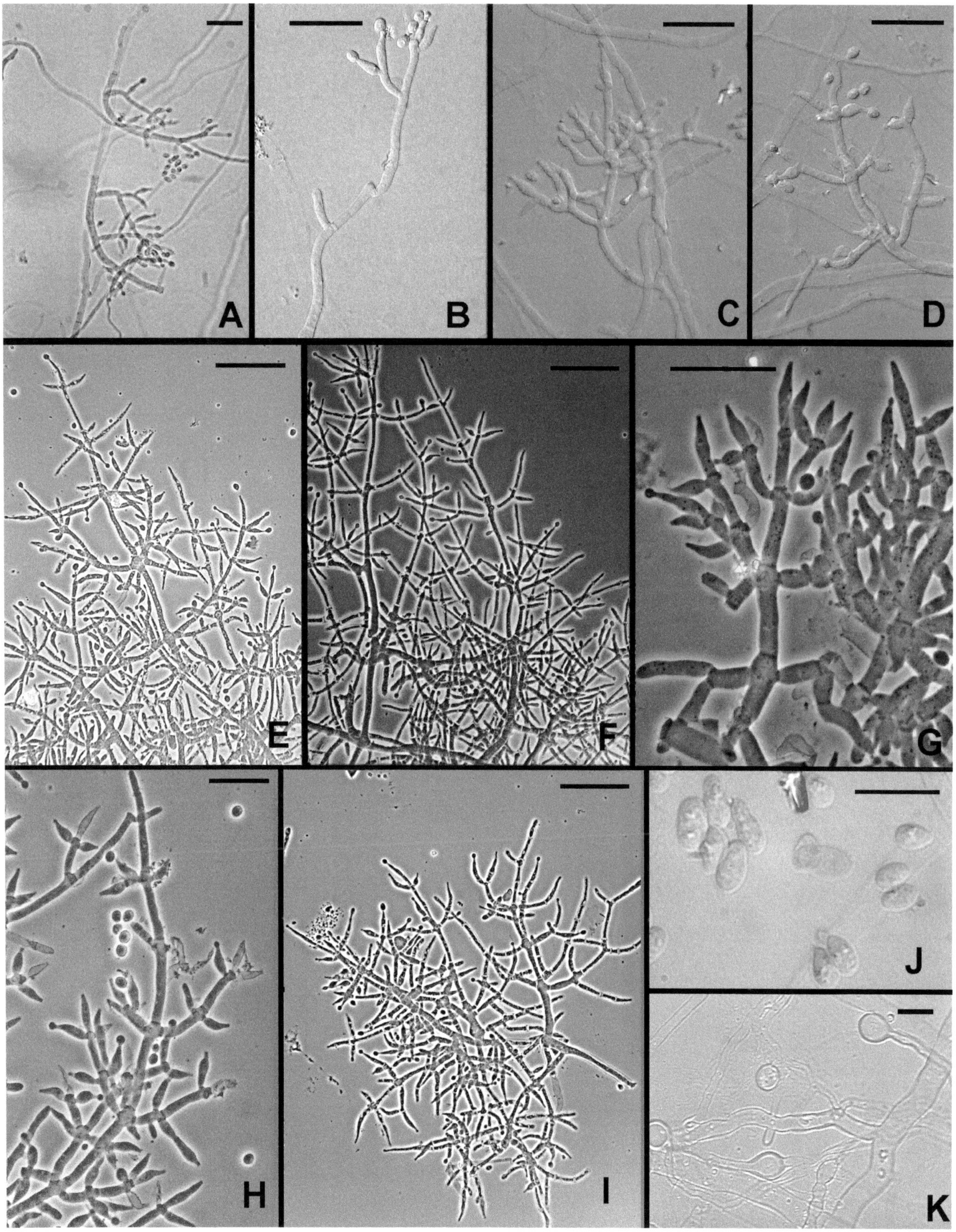
A
B
C
D
E
F
G
H
I
J
K

Trichoderma pleuroti

S. H. Yu & M. S. Park *in* Park et al., Mycobiology 34:111. 2006 [as *pleurotum*]. Figs. 6.3D, 6.32.

Ex-type culture: KCTC 26314 = CNUMH 501 = CBS 124387
GenBank: *rpb2*: HM142372; *tef1*: HM142382

Clade: Harzianum

Colony radius at 25°C: PDA 35–50. **SNA** 30–45. No diffusing pigment or distinctive odor on PDA.

Colony radius at 35°C: PDA (0–)20–30. **SNA** (0–)15–25.

Conidiation: Conidiophores forming abundantly throughout the colony in the aerial mycelium; fascicles of conidiophores not formed; individual conidiophores visible, fertile at the tips of branches. Conidia held in moist heads but not in glistening drops, slowly turning green.

Microscopic characteristics: Branching pattern Gliocladium type; conidiophores more-or-less discrete, (35–)60–120(–330) µm long, (2–)5–7(–10) µm at the base, with a distinct branched stipe, each branch terminating in a penicillate head of phialides. Conidiophores producing three to four short fertile branches or solitary phialides from the tip and one to three levels of fertile branches below the tip; primary branches (metulae) arising directly from the conidiophores at an acute angle with respect to the conidiophores, 20–25 µm long, 5.5–6.5 µm wide, producing a penicillus of three to five phialides from the tip or producing a second level of branches; second-level branches 15–20 µm long, 4.5–6.0 µm wide, producing a penicillus of phialides or less frequently producing a third level of branches; third-level branches 8–10 µm long, 4–5 µm wide, producing a penicillus of phialides or rarely producing a fourth level of branches and corresponding penicillus of phialides; lateral branches arising at or near right angles to the stipe, often increasing in length with distance from the tip of the stipe. **Phialides** lageniform, constricted at the tip, (5–)6–10(–12) µm long, (1.5–)2.5–3.5(–4.5) µm at the widest point, L/W (1.6–)1.9–3.3(–5.3). **Supporting cell** 2.5–3.5 µm wide; ratio of length of phialide to width of supporting cell 1.2–3.2; ratio of width of phialide to width of supporting cell 0.6–1.2. **Conidia** ellipsoidal to nearly oblong, (1.5–)2.0–2.7(–3.7) × (2.7–)3.2–4.2(–5.5) µm, L/W (1.0–)1.4–2.0(–2.3), green, smooth. **Chlamydospores** not observed (subglobose to ellipsoidal in the original description).

Distinguishing characters: *Trichoderma pleuroti* is readily recognizable by penicillately branched, *Gliocladium*-like conidiophores but can be confused with *T. virens*. It differs from *T. virens* in the absence of chlamydospores and by its smaller conidia.

Habitat: Soil; often found in cultivation substrate of oyster mushroom (*Pleurotus ostreatus*).

Distribution: Europe, North America, Korea; probably cosmopolitan north temperate.

Literature: Komoń-Zelazowska et al. (2007), Park et al. (2006).

Figure 6.32. *Trichoderma pleuroti*. **A and B,** Conidiation. Note the discrete conidiophores that bear wet heads of conidia. **C–H,** Conidiophores. **I,** Conidia. **A, C, and E–I,** From Spezieller Nährstoffarmer agar (SNA). **B and D,** From cornmeal dextrose agar. Scale bars: **A, B** = 0.5 mm; **C–H** = 20 µm; **I** = 10 µm. ➤

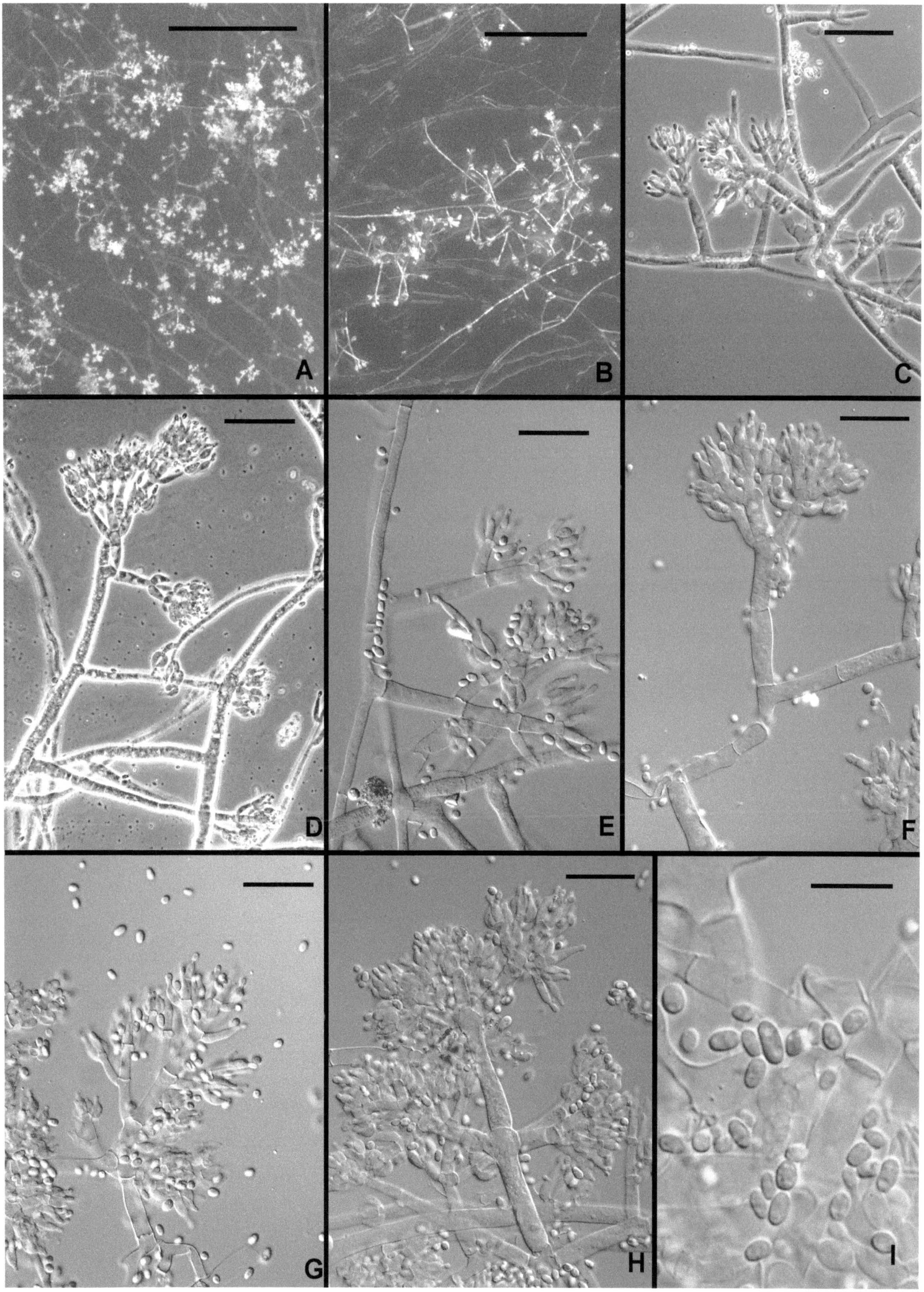
A
B
C
D
E
F
G
H
I

Trichoderma pleuroticola

S. H. Yu & M. S. Park *in* Park et al., Mycobiology 34:112. 2006. Figs. 6.3E, 6.33.

Ex-type culture: CNUMH 601 = CBS 124383
GenBank: *rpb2*: HM142371; *tef1*: HM142381

Clade: Harzianum

Colony radius at 25°C: PDA (33–)45–61(–62). **SNA** (42–)55–65(–68). No pigment or distinctive odor on PDA.

Colony radius at 35°C: PDA and **SNA** ≤10 mm.

Conidiation: Pustules abundant, mainly around the periphery of the colony, loosely cottony, becoming confluent and forming sheets, gray-green; conidia appearing dry.

Microscopic characteristics: Branching pattern Pyramidal type; phialides held in divergent whorls, infrequently solitary. **Sterile hairs** not formed but conidiophores often elongated and bearing one or a few whorls of phialides at the tip of the elongation. **Phialides** ampulliform when crowded, and less swollen, lageniform when less crowded, (4.7–)5.2–7.2(–10.0) μm long, (2.5–)2.7–3.7(–4.5) μm at the widest point, L/W (1.4–)1.6–3.2. **Supporting cell** (2.0–)2.5–2.7(–4.7) μm wide; ratio of length of phialide to width of supporting cell (1.3–)1.5–2.5(–3.1); ratio of width of phialide to width of supporting cell (0.7–)0.8–1.3(–2.0). **Conidia** subglobose, (2.0–)2.2–3.0(–3.2) × 2.5–3.2(–3.5) μm, L/W 0.9–1.5(–1.6), green, smooth. **Chlamydospores** few, subglobose to ellipsoidal.

Distinguishing characters: *Trichoderma pleuroticola* is morphologically consistent with other members of the Harzianum clade. The long extensions of conidiophores are somewhat distinctive, as is the relatively slow growth rate. Morphologically it is very similar to *T. aggressivum*, which is found only on commercial *Agaricus bisporus*, and to the tropical species *T. amazonicum*.

Habitat: Soil and wood, found also on commercial *Pleurotus* substrate as the cause of green mold disease, and on basidiomata of *Armillaria* species as biological control.

Distribution: North America, Europe, New Zealand, Cameroon; common.

Literature: Komoń-Zelazowska et al. (2007), Park et al. (2006).

Figure 6.33. *Trichoderma pleuroticola.* **A and B,** Conidial pustules. **C–H,** Conidiophores. **I,** Conidia. **J,** Chlamydospores. **A, B, D, and G–I,** From cornmeal dextrose agar. **C, E, F, and J,** From Spezieller Nährstoffarmer agar (SNA). Scale bars: **A, B** = 0.5 mm; **C–H, J** = 20 μm; **I** = 10 μm.

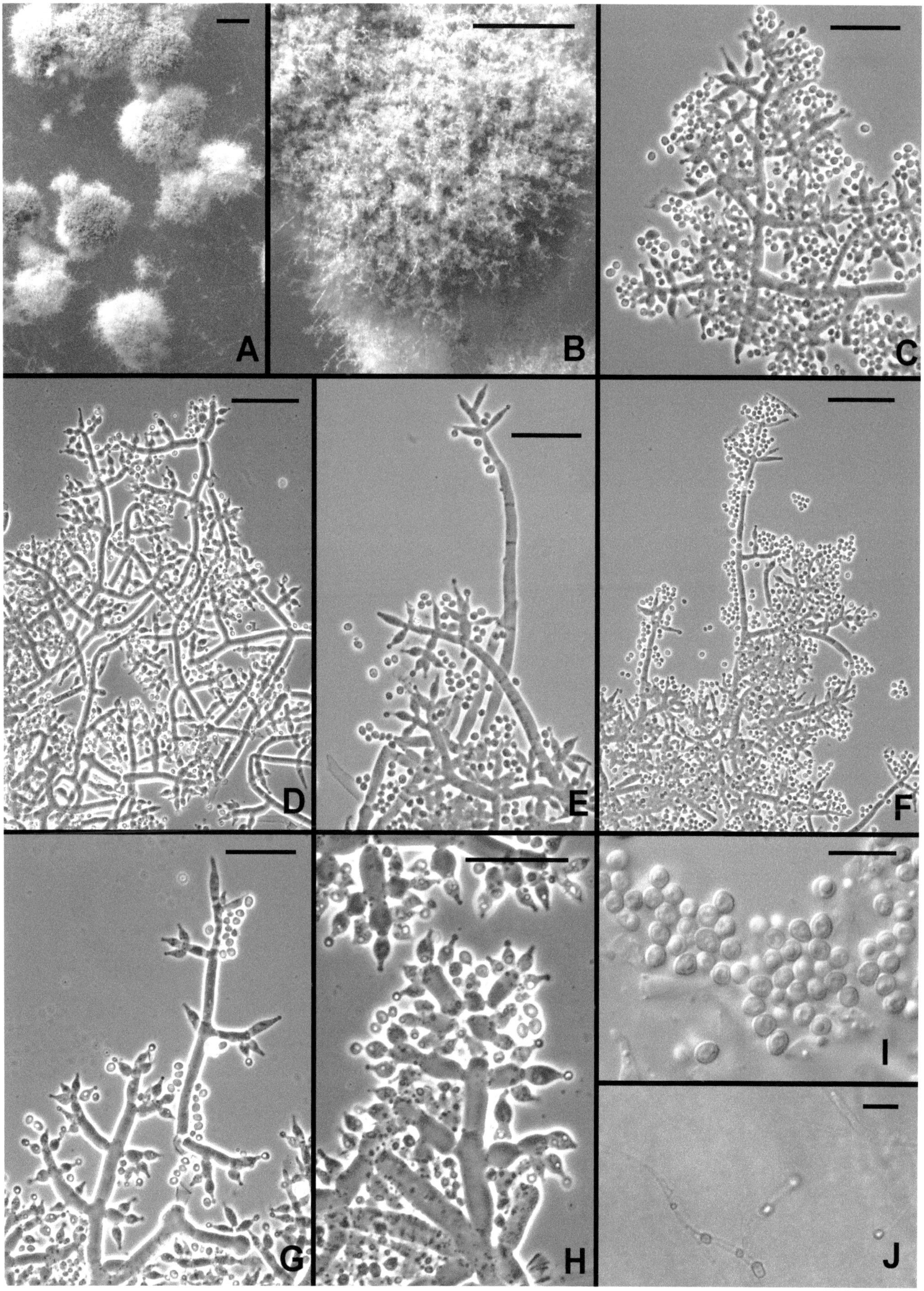
A
B
C
D
E
F
G
H
I
J

Trichoderma polysporum

(Link) Rifai *in* Rifai, Mycol. Pap. 116:18. 1969. Figs. 6.3F, 6.34.

≡ *Sporotrichum polysporum* Link, Mag. Ges. Naturforsch. Freunde Berlin 8:34. 1816. Fr., Syst. Mycol. 3:424. 1834.
= *Trichoderma croceum* Bissett, Can. J. Bot. 69:2379. 1991 [1992].
= *Hypocrea pachybasioides* Yoshim. Doi, Bull. Natl. Sci. Mus. Tokyo 15:685. 1972.
= *Trichoderma stellatum* (B. S. Lu, Druzhin. & Samuels) Jaklitsch & Voglmayr *in* Jaklitsch and Voglmayr, Mycotaxon 126:153. 2014.
≡ *Hypocrea stellata* B. S. Lu, Druzhin. & Samuels *in* Lu et al., Mycologia 96:333. 2004.

Typical culture: ATCC 18650 = CBS 820.68
GenBank: *rpb2*: FJ179613; *tef1*: AY750886

Clade: Polysporum

Colony radius at 25°C: PDA and **SNA** 18–30. On PDA no diffusing pigment or distinctive odor.

Colony radius at 35°C: PDA and **SNA** <6.

Conidiation: Pustules slow to form, discrete, scattered, pulvinate, 1–2.5 mm in diameter, hairs with a helical, corkscrew aspect arising abundantly from pustules; conidia white or less frequently ivory, yellow, or brownish.

Microscopic characteristics: Branching pattern Pachybasium type; fertile branches arising at the base of sterile hairs. **Sterile hairs** conspicuously spiraled, white, roughened or smooth, unbranched or little branched, acute, infrequently bearing one or a few phialides at the apex. Fertile branches arising at right angles from the base of hairs, often paired, usually one or a few broad cells in length, producing phialides directly or in grapelike heads of up to several at the tips of branches. No **synanamorph** observed. **Phialides** ampulliform, (3.5–)4.2–6.5(–10.0) µm long, (2.0–)2.7–3.7(–5.0) µm at the widest point, L/W (1.0–)1.3–2.1(–4.2). **Supporting cell** (2.2–)3.0–3.5(–6.5) µm wide; ratio of length of phialide to width of supporting cell (0.7–)1.1–2.0(–4.0), ratio of width of phialide to width of supporting cell (0.5–)0.7–1.1(–1.5). **Conidia** ellipsoidal, (1.5–)2.0–2.7(–3.0) × (2.0–)2.7–4.0(–9.0) µm, L/W (0.9–)1.2–1.8(–4.3), white in mass, smooth. **Chlamydospores** common, terminal and intercalary, subglobose.

Distinguishing characters: *Trichoderma polysporum* is characterized by the formation of compact, white pustules with spiraled hairs. Other species that have colorless (white in mass) conidia include *T. piluliferum* (on wood of *Betula* species) and *T. placentula* (on culms of *Juncus* species), both in northern Europe (Jaklitsch, 2011), and *Hypocrea parapilulifera* and *H. stellata* in New Zealand (Lu et al., 2004a). None of these species produces pustules with spiraled hairs.

Habitat: Soil. Teleomorph found on wood; in Japan associated with shiitake production.

Distribution: Common in north and south temperate regions in hardwood forests.

Literature: Bissett (1991b), Doi (1972), Gams and Bissett (1998), Jaklitsch (2011), Komatsu (1976), Lu et al. (2004a).

Figure 6.34. *Trichoderma polysporum.* **A and B,** White conidial pustules. Coiled hair can be seen in **B** (arrow). **C,** Sinuous to coiled sterile hairs arising from pustules. **D–F,** Conidiophores. **G and H,** Phialides. **I,** Conidia. **J,** Chlamydospores. **A–J,** From Spezieller Nährstoffarmer agar (SNA). Scale bars: **A** = 0.5 mm; **B** = 0.25 mm; **C–F, J** = 20 µm; **G–I** = 10 µm. ➤

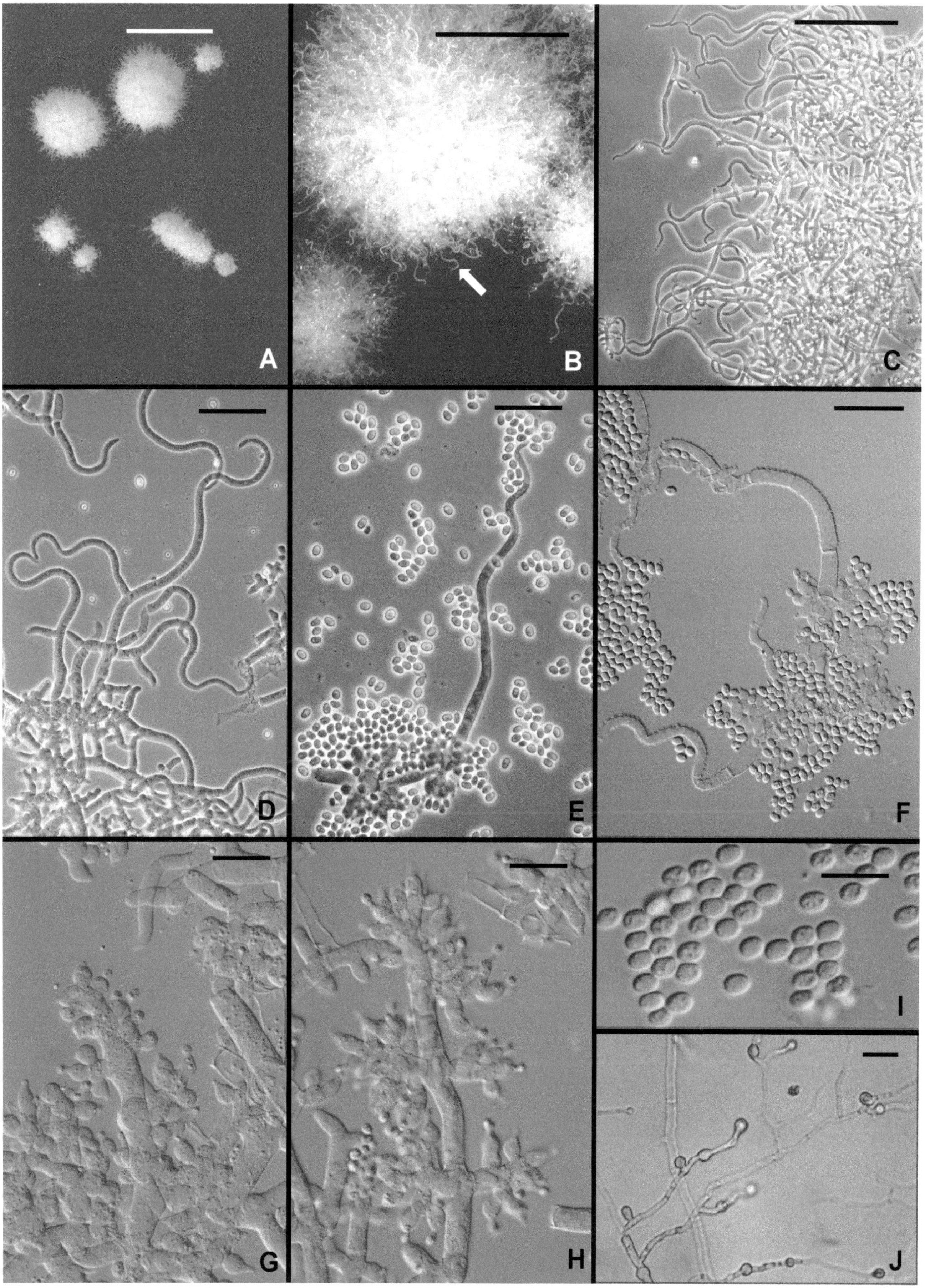
A
B
C
D
E
F
G
H
I
J

Trichoderma pseudokoningii

Rifai, Mycol. Pap. 116:45. 1969. Figs. 6.3G, 6.35.

= *Hypocrea pseudokoningii* Samuels & Petrini *in* Samuels et al., Stud. Mycol. 41:36. 1998.

Ex-type culture: CBS 408.91 = DAOM 167678
GenBank: *rpb2*: AY750886; *tef1*: AY937429

Clade: Longibrachiatum

Colony radius at 25°C: PDA 55–70. **SNA** 40–55. On PDA at most a pale yellow pigment, no distinctive odor.

Colony radius at 35°C: PDA 60–70. **SNA** 40–60(–70).

Conidiation: Pustules few in number, minute, <0.25 mm in diameter, poorly developed; many conidiophores formed throughout the colony in aerial hyphae; individual conidiophores not visible in pustules, long sinuous sterile white hairs arising in pustules; conidia green.

Microscopic characteristics: Branching pattern mainly Longibrachiatum type but with many whorls of divergent phialides. **Conidiophores** consisting of a distinct central axis, often richly branched, solitary phialides arising singly for a short distance below the tip of the main axis and secondary branches; secondary branches increasing in length with distance from the tip of the main axis, often terminating in a whorl of three or more diverging phialides. **Phialides** narrowly ampulliform, often somewhat swollen below the neck, (4–)6–10(–15) µm long, (2.0–)2.5–3.5(–4.5) µm at the widest point, L/W (1.2–)1.9–3.5(–5.9). **Supporting cell** (1.7–)2.7–3.2(–4.2) µm wide; ratio of length of phialide to width of supporting cell (1.5–)2.0–4.0(–6.7); ratio of width of phialide to width of supporting cell (0.6–)0.8–1.4(–2.1). **Intercalary phialides** not seen. **Conidia** oblong to narrowly ellipsoidal, (1.7–)2.0–3.0(–5.0) × (2.7–)3.5–5.5(–9.7) µm, L/W (1.1–)1.4–2.0(–3.3), green, smooth. **Chlamydospores** abundant or not, terminal or intercalary.

Distinguishing characters: *Trichoderma pseudokoningii* is separated from typical members of the Longibrachiatum clade by the conspicuous divergent whorls of phialides, absence of intercalary phialides, and rather large conidia. Other members of the Longibrachiatum clade having similar conidiophores are described by Samuels et al. (2012). *Trichoderma pseudokoningii* is similar to *T. citrinoviride*/*H. schweinitzii* and *H. andinensis* in having a black stroma.

Habitat: Wood (teleomorph), forest soil (anamorph).

Distribution: A south temperate species; common as its teleomorph in parts of Australia and New Zealand; rare outside of Australasia. Although this species is often reported in the literature, it is rarely isolated directly from the substrate and these reports should be viewed with suspicion.

Literature: Bissett (1984), Gams and Bissett (1998), Samuels et al. (1998, 2012).

Figure 6.35. *Trichoderma pseudokoningii*. **A and B,** Conidial pustules. **C–J,** Conidiophores. **K,** Conidia. Note the whiplike extensions of conidiophores in **B** and **C**. **A–K,** From Spezieller Nährstoffarmer agar (SNA). Scale bars: **A** = 1 mm; **B** = 250 µm; **C–F, H, I** = 20 µm; **G, J, K** = 10 µm. ➤

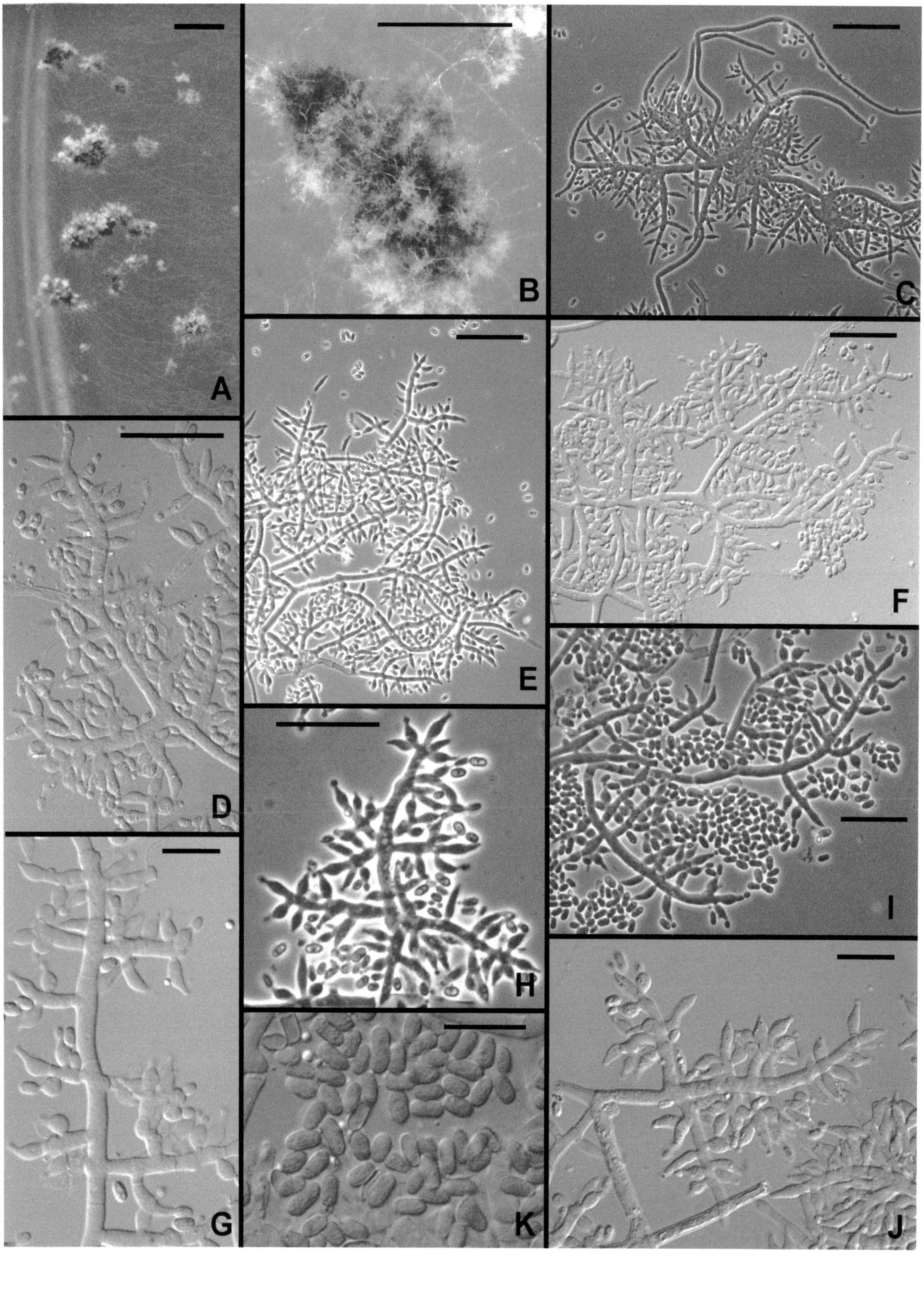
A
B
C
D
E
F
G
H
I
J
K

Trichoderma pubescens

Bissett, Can. J. Bot. 69:2405. 1991 [1992]. Figs. 6.3H, 6.36.

Ex-type culture: DAOM 166162
GenBank: *rpb2*: EU248613; *tef1*: AY750887

Clade: Viride

Colony radius at 25°C: PDA 25–40. **SNA** 22–25. On PDA no pigment or distinctive odor.

Colony radius at 35°C: No growth.

Conidiation: Conidial pustules forming around the original inoculum and sometimes in concentric rings, discrete and pulvinate, to 2 mm in diameter, or confluent and forming extensive flat areas of varying extent, grayish green; sterile hairs extending beyond the surface of the pustule.

Microscopic characteristics: Branching pattern Pachybasium type with phialides arising from fertile branches at the base of sterile hairs. **Sterile hairs** conspicuously extending beyond the surface of the pustule, acute, sinuous, septate, branched. Fertile branches arising at right angles from the base of the hair, at most a few cells in length, rebranching; secondary branches typically a single cell in length, producing two to several phialides in a terminal head. **Phialides** ampulliform, (3.7–)4.7–6.7(–7.5) μm long, (2.7–)3.2–4.0(–5.0) μm at the widest point, L/W (1.1–)1.3–1.9(–2.3). **Supporting cell** (1.7–)3.0–4.5(–5.2) μm wide; ratio of length of phialide to width of supporting cell (0.7–)1.1–2.1(–4.0); ratio of width of phialide to width of supporting cell (0.7–)0.8–1.2(–2.2). **Conidia** ellipsoidal to nearly oblong, (1.5–)2.2–2.7(–3.2) × (3.2–)4.0–4.7(–5.2) μm, L/W (1.3–)1.4–2.0(–3.0), gray-green, smooth. **Chlamydospores** sporadically produced, intercalary, subglobose to clavate. **Synanamorph** *Verticillium*-like, forming in aerial mycelium on cornmeal dextrose agar (CMD).

Distinguishing characters: *Trichoderma pubescens* is characterized by its Pachybasium-type branching pattern, acute branched sterile hairs, and inability to grow at 35°C. Similar species are *T. hamatum*, *T. longipile*, *T. oblongisporum*, and *Hypocrea semiorbis*.

Habitat: Soil; also known as an endophyte of sapwood of *Cola* species.

Distribution: United States (North Carolina), Peru, Cameroon; apparently rare.

Literature: Bissett (1991b, description), Chaverri et al. (2003a, description), Degenkolb et al. (2006a, peptaibiotics), Hoyos-Carvajal et al. (2009a, distribution).

Figure 6.36. *Trichoderma pubescens.* **A–C,** Conidial pustules. Note the hairs extending beyond the edge of the pustule in **C. D–H,** Conidiophores. **I,** Phialides. **J,** Synanamorph conidiophores. **K,** Conidia. **L,** Chlamydospores. **A–L,** From Spezieller Nährstoffarmer agar (SNA). Scale bars: **A, B** = 1 mm; **C** = 150 μm; **D–H, J, L** = 20 μm; **I, K** = 10 μm.

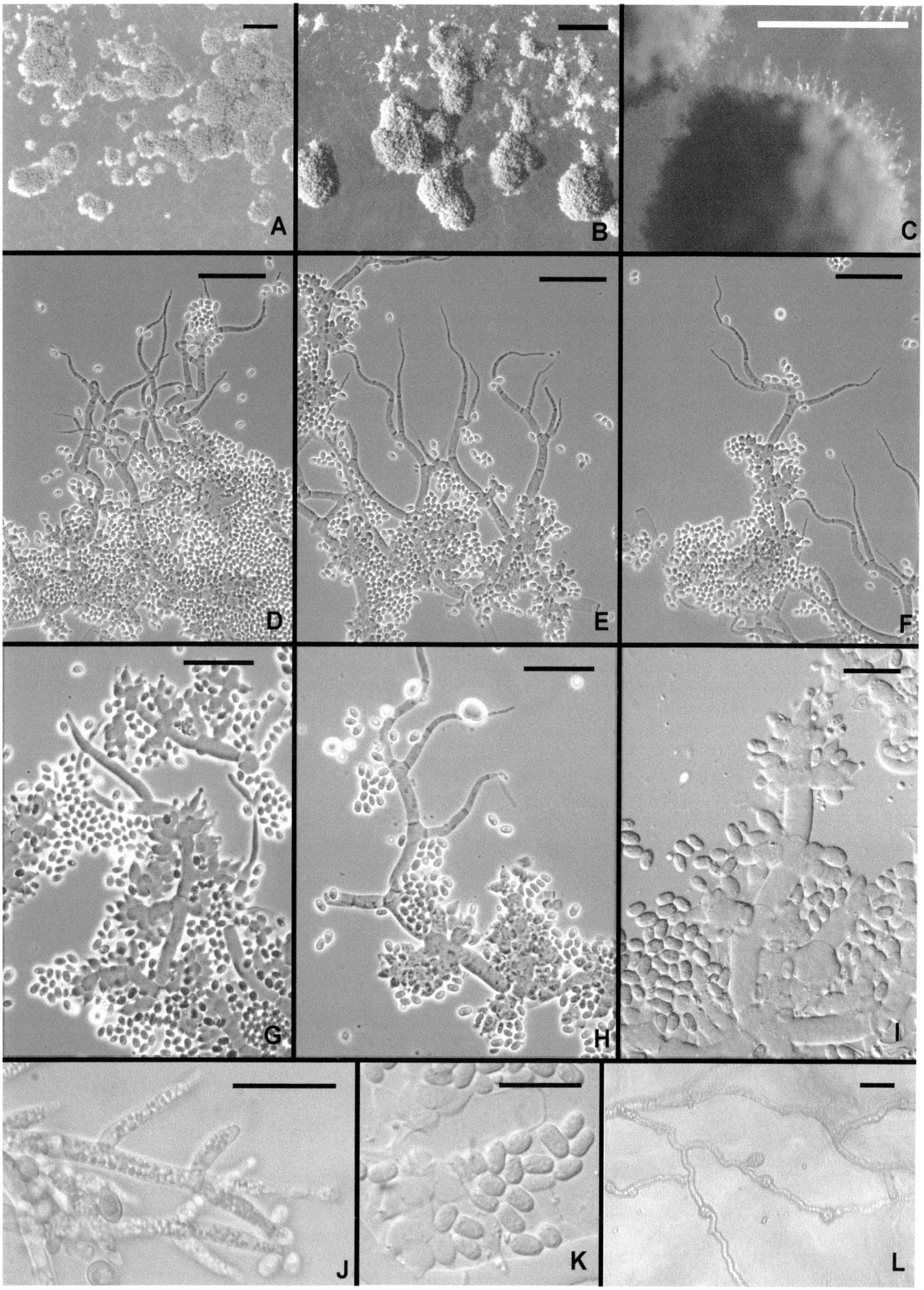
A
B
C
D
E
F
G
H
I
J
K
L

Trichoderma reesei

E. G. Simmons *in* Bigelow and Simmons, Abstracts, 2nd Int. Mycol. Congr. M-Z:618. 1977. nom cons. prop.
Figs. 6.3I, 6.37.

= *Hypocrea jecorina* Berk. & Broome *in* Berkeley and Broome, J. Linn. Soc. Bot. 14:112. 1873.

Ex-type culture: QM 6a = ATCC 13631 = IMI 192654 = CBS 383.78 = DAOM 167654
GenBank: *rpb2*: HM182969; *tef1*: AF401004

Clade: Longibrachiatum

Colony radius at 25°C: PDA (40–)54–70. **SNA** (34–)43–55. On PDA a diffusing yellow pigment, no distinctive odor; colony surface with a radiating aspect.

Colony radius at 35°C: PDA (45–)62–70. **SNA** (32–)45–70. **At 40°C after 72 h: PDA** 25–55. **SNA** 15–35.

Conidiation: Conidia just beginning to form; after 10 days pustules numerous, uniformly dispersed throughout the colony, minute, <0.25 mm in diameter, remaining discrete or at most slightly confluent; cottony or wooly, completely fertile conidiophores visible, long sinuous sterile white hairs visible especially in young pustules; ropes of conidiophores give the pustules a spiky or aspirate aspect; conidia dark green.

Microscopic characteristics: Branching pattern Longibrachiatum type, fertile branches typically long, composed of several levels of branching, the branches near the tip monophialidic, the branches distal to the tip longer and rebranching; **sterile hairs** seen in young pustules. **Phialides** cylindrical to lageniform, often swollen in the middle, solitary, often terminating in a single cell and subtended by an intercalary phialide, straight, (3.5–)5.2–9.2(–16.0) μm long, (1.7–)2.5–3.2 (–4.5) μm at the widest point, L/W (1.2–)1.5–3.6(–6.1). **Supporting cell** (1.5–)2.0–3.2(–5.5) μm wide; ratio of length of phialide to width of supporting cell (1.1–) 2.1–3.7(–7.1); ratio of width of phialide to width of supporting cell (0.3–)0.9–1.4(–1.9). **Intercalary phialides** abundant, conspicuous. **Conidia** ellipsoidal, (1.7–)2.5–3.0(–4.0) × (2.0–)3.5–4.5(–6.2) μm, L/W (0.7–)1.2–1.8(–2.6), green, smooth. **Chlamydospores** typically abundant but sometimes lacking, globose to subglobose, terminal or intercalary.

Distinguishing characters: *Trichoderma reesei* is morphologically indistinguishable from *T. longibrachiatum* and *T. parareesei*. Unlike *T. longibrachiatum*, conidia of *T. reesei* only form sparingly on SNA at 25°C under light within 1 week.

Habitat: Originally isolated from canvas fabric, found also in soil. The teleomorph occurs on wood.

Distribution: Until recently this species was known only from the original isolation, and this culture was utilized in many biotechnological articles because of its ability to produce cellulase. *Trichoderma reesei* is a tropical species that is rarely isolated directly from substrate; it is common as its teleomorph in tropical countries. Druzhinina et al. (2010b) described *T. parareesei*, a closely related and essentially morphologically indistinguishable but apparently strictly asexual species.

Literature: Druzhinina et al. (2010b), Samuels et al. (1998, 2012). The genome of *T. reesei* was sequenced by Diener et al. (2004) and Kubicek et al. (2011). The vast enzymological literature is summarized in Kubicek and Harman (1998).

Figure 6.37. *Trichoderma reesei.* **A–C,** Pustules. Note the spiky aspect of pustules in **B** and **C**. **D–I,** Conidiophores. **J,** Phialides. Intercalary phialides are indicated by arrows in **E, F, I,** and **J**. **K,** Conidia. **L,** Chlamydospores. **A–D, G–J, and L,** From cornmeal dextrose agar. **E, F, and K,** From cornmeal agar. Scale bars: **A** = 1 mm; **B** = 0.5 mm; **C** = 150 μm; **D–G, I, L** = 20 μm; **H, J, K** = 10 μm.

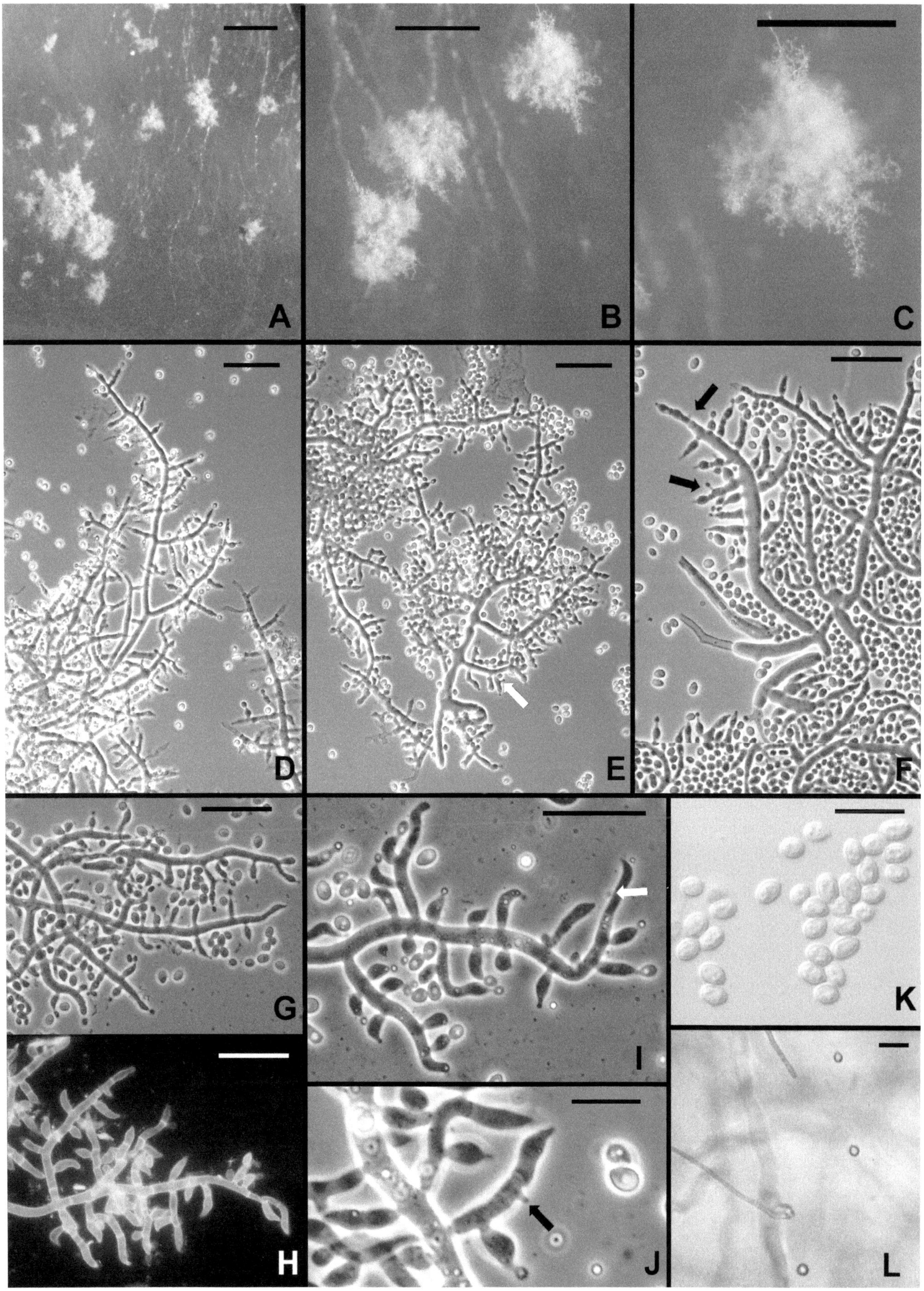
A
B
C
D
E
F
G
H
I
J
K
L

Trichoderma rossicum

Bissett, C. P. Kubicek & Szakács *in* Bissett et al., Can. J. Bot. 81:578. 2003. Figs. 6.3J, 6.38.

Ex-type culture: DAOM 230011
GenBank: *rpb2*: HC342288; *tef1*: AY937441

Clade: Stromaticum

Colony radius at 25°C: PDA 45–55. **SNA** 25–32. On PDA no diffusing pigment or distinctive odor.

Colony radius at 35°C: No growth.

Conidiation: Pustules forming in concentric rings, 0.25–0.5 mm in diameter, hemispherical, discrete or confluent, green, easily removed from the agar, sterile hairs sinuous, abundant, pustule appearing wooly.

Microscopic characteristics: Branching pattern typically Pachybasium type with fertile branches arising at the base of hairs. Fertile branches composed of one or a few broad cells, sometimes rebranching; secondary branches typically unicellular; phialides arising directly from and terminating in primary and secondary fertile branches; solitary or several densely clustered. **Sterile hairs** sinuous branched, septate, thin walled, acute. **Phialides** flask shaped to subglobose with a short constricted neck, (4.5–)5.0–7.0(–7.7) µm long, (2.7–)3.5–4.5(–4.7) µm at the widest point, L/W (1.2–)1.3–1.9(–2.2). **Supporting cell** (3.5–)4.0–5.2(–6.0) µm wide; ratio of length of phialide to width of supporting cell (0.8–)1.0–1.6(–2.0); ratio of width of phialide to width of supporting cell (0.6–)0.7–1.1. **Conidia** oblong, (2.2–)2.7–3.0(–3.2) × (4.0–)4.2–5.0(–5.7) µm, L/W (1.4–)1.5–1.7(–2.0), green, smooth. **Chlamydospores** not observed (few chlamydospores observed in the protolog). No **synanamorph** observed.

Distinguishing characters: *Trichoderma rossicum* is a species of temperate regions characterized by its wooly pustules. Unlike most *Trichoderma* species, its growth rate at 30°C is markedly slower than at 25 or 20°C. It is morphologically similar to the ubiquitous *T. hamatum*, which has shorter conidia and a *Verticillium*-like synanamorph on SNA. Compare also with *T. evansii*, *T. lieckfeldtiae*, and *T. pubescens*.

Habitat: Soil.

Distribution: North temperate (Austria, Siberia).

Literature: Bissett et al. (2003), Grosch et al. (2006, as *T. viride*, antagonistic to *Rhizoctonia solani*), Samuels et al. (2012).

Figure 6.38. *Trichoderma rossicum*. **A–C,** Conidial pustules. Note the hairs extending beyond the edge of the pustule in **B** and **C**. **D–G,** Conidiophores. **H–J,** Phialides. **K,** Conidia. **A,** From cornmeal dextrose agar. **B–K,** From Spezieller Nährstoffarmer agar (SNA). Scale bars: **A–C** = 1 mm; **D–J** = 20 µm; **K** = 10 µm. (E–G, J, and K reproduced from Samuels et al., [2011 online] 2012, with kind permission of Springer Science+Business Media)

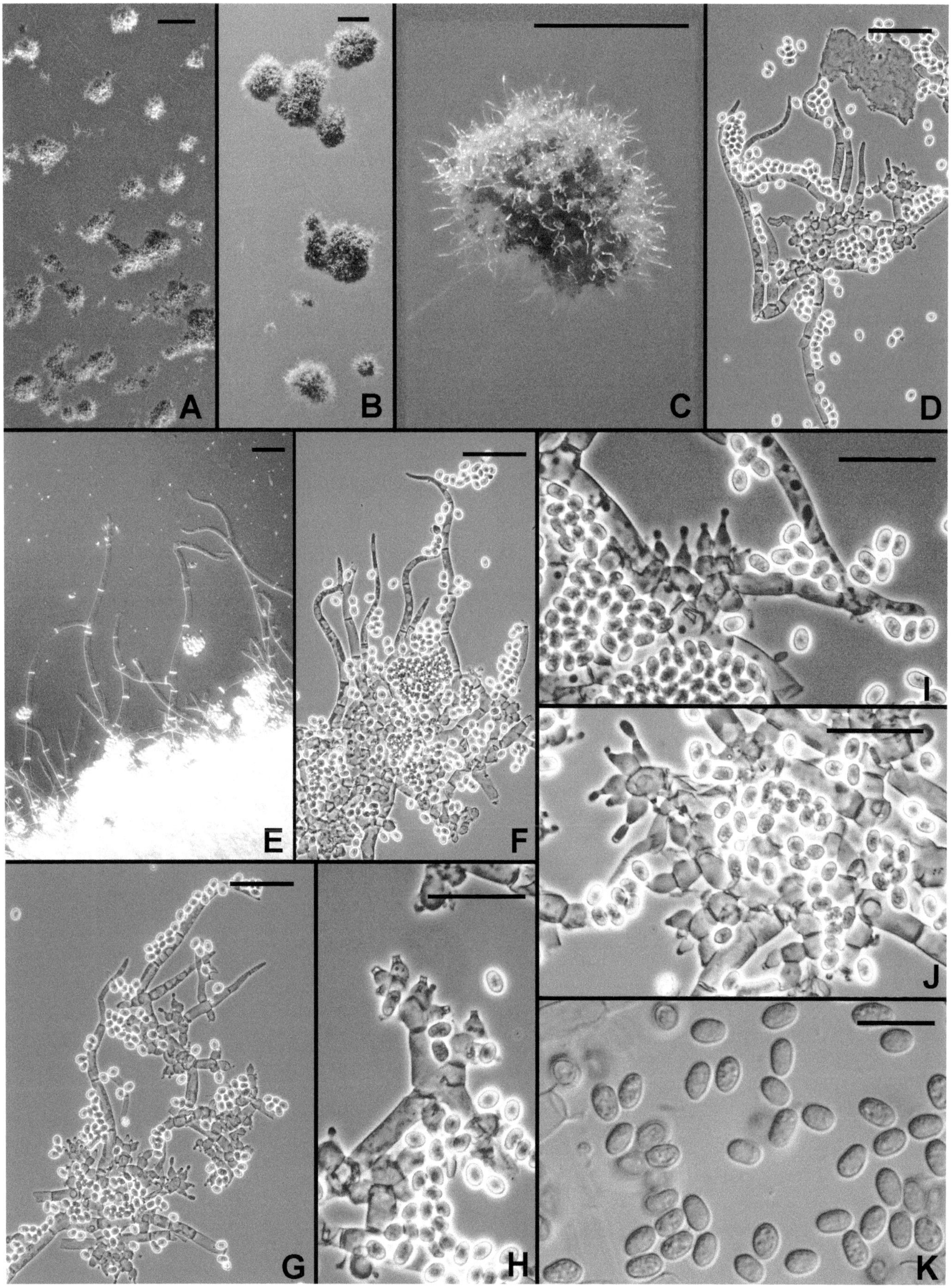
A
B
C
D
E
F
I
J
G
H
K

Trichoderma saturnisporopsis

Samuels & Jaklitsch *in* Samuels et al., Fungal Diversity 55:103. 2012. Figs. 6.3K, 6.39.

Ex-type culture: CBS 130751
GenBank: *rpb2*: DQ857348; *tef1*: JQ685869

Clade: Longibrachiatum

Colony radius at 25°C: PDA 40–70. **SNA** 15–50. On PDA sometimes a pale yellow diffusing pigment, no distinctive odor.

Colony radius at 35°C: PDA 20–65. **SNA** 10–30.

Conidiation: Pustules forming in one or two concentric rings beginning in the center of the colony; flat to hemispherical, <1 mm in diameter, dark green, becoming confluent; hairs often evident in pustules.

Microscopic characteristics: Branching pattern not well defined, sometimes Pyramidal type; conidiophore sometimes composed of a wide discernable central axis with paired lateral branches, the branches increasing in length from the tip, each branch rebranching to produce solitary phialides or convergent or divergent whorls of phialides; sometimes fertile branches arising singly and at irregular intervals along hyphae of the pustule. The tip of the conidiophore often elongating to form a hair; **hairs** stiff, erect, straight, septate, with blunt ends. **Phialides** mostly solitary; sometimes phialides densely clustered in convergent heads at the tips of short branches of hyphae; lageniform to ampulliform, straight, widest below the middle, (4.0–)5.7–10.5(–14.0) μm long, (2.2–)3.0–3.7(–5.5) μm at the widest point, L/W (1.3–)1.6–3.2(–5.5). **Supporting cell** (1.7–)2.2–3.2(–4.5) μm wide; ratio of length of phialide to width of supporting cell (1.5–)2.0–4.0(–5.0); ratio of width of phialide to width of supporting cell (0.5–)0.8–1.8(–3.0). **Intercalary phialides** rare. **Conidia** broadly ellipsoidal, (2.5–)3.2–4.0(–4.5) × (3.7–)4.2–5.0(–6.0) μm, L/W (1.0–)1.1–1.5(–1.9), green, typically conspicuously tuberculate, less frequently tubercles few. **Chlamydospores** uncommon, terminal and intercalary, globose, ellipsoidal or pyriform.

Distinguishing characters: The tuberculate, subglobose conidia and irregular branching of conidiophores distinguish *T. saturnisporopsis*. The most morphologically similar species are the mainly tropical *T. ghanense* and *T. saturnisporum*, both of which also have tuberculate conidia. *Trichoderma saturnisporopsis* grows faster than *T. saturnisporum*. The only two known strains of *T. saturnisporopsis* have very different rates of growth, and colonies of the strain having the slowest growth rate appear to be abnormal, having highly dissected margins (Samuels et al., 2012).

Habitat: Soil.

Distribution: North temperate regions, Europe, North America; apparently rare.

Literature: Samuels et al. (2012).

Figure 6.39. *Trichoderma saturnisporopsis*. **A,** Conidial pustules. **B–G,** Conidiophores. Pyramidal-type branching can be seen in **G. H,** Phialides. **I,** Conidia. **J,** Chlamydospores. **A–J,** From Spezieller Nährstoffarmer agar (SNA). Scale bars: **A** = 0.5 mm; **B–E, G, J** = 20 μm; **F, H, I** = 10 μm. (Reproduced, by permission, from Samuels et al., 2012)

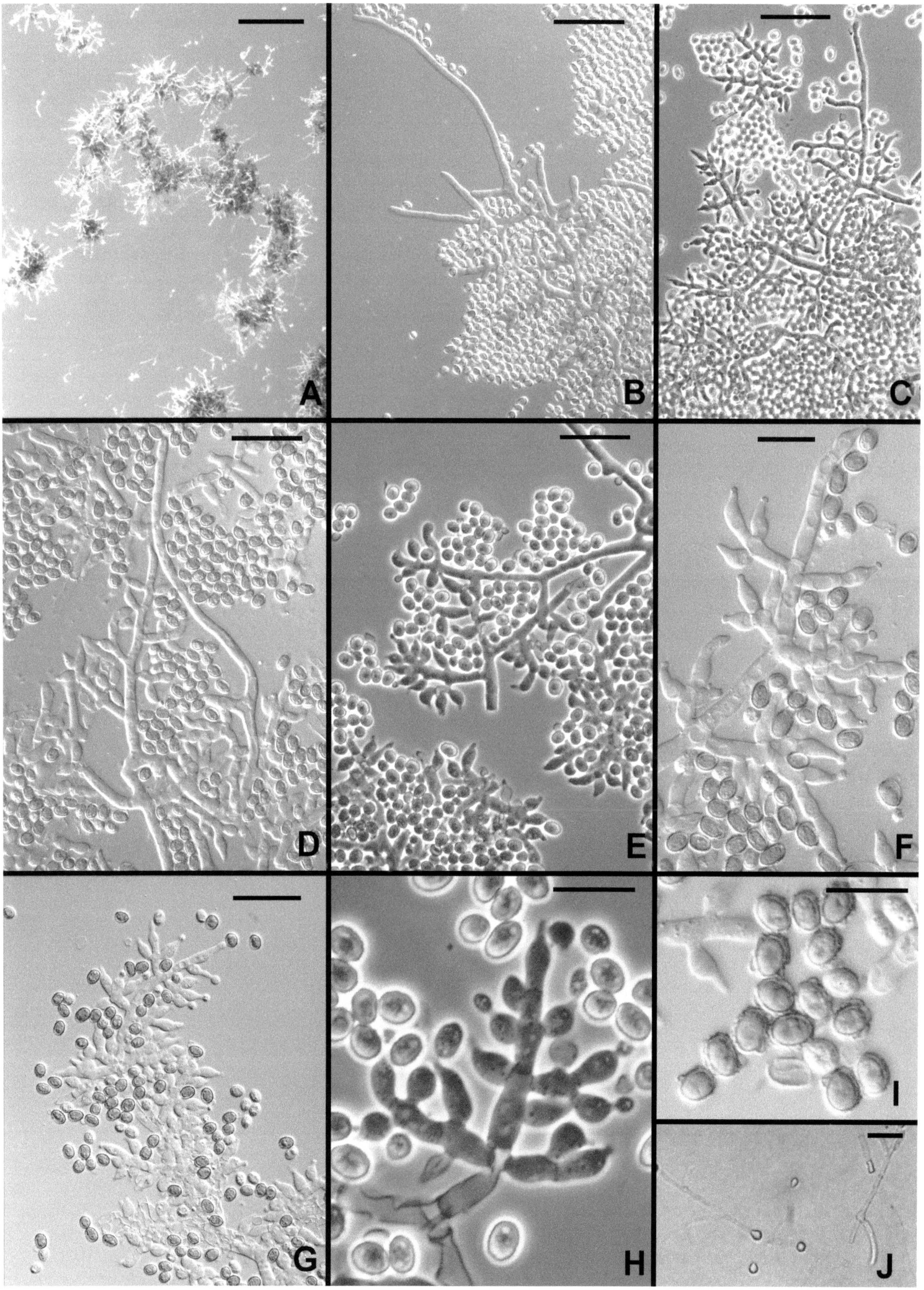
A
B
C
D
E
F
G
H
I
J

Trichoderma saturnisporum

Hammill, Mycologia 62:112. 1970. Figs. 6.3L, 6.40.

Ex-type culture: ATCC 18903 = CBS 330.70 = IMI 146852
GenBank: *rpb2*: DQ087243; *tef1*: AY937414 (CBS 886.72)

Clade: Longibrachiatum

Colony radius at 25°C: PDA 20–35. **SNA** 20–25(–30). On PDA yellow diffusing pigment visible in colony reverse. No distinctive odor.

Colony radius at 35°C: PDA 30–45. **SNA** 15–20.

Conidiation: Pustules uniformly dispersed throughout the colony, sometimes in concentric rings, minute, 0.25–1 mm in diameter, gray-green, typically cottony, typically with short sterile hairs protruding.

Microscopic characteristics: Branching pattern not well defined, in part Longibrachiatum type, main axis visible only in protruding conidiophores; pustules formed of frequently branched, intertwined, sinuous hyphae. **Phialides** terminating and arising along the length of hyphae of the pustule, typically solitary, less frequently in whorls of two or three, lageniform, (5.0–)7.0–11.5(–15.0) µm long, (2.2–)3.0–4.0(–5.0) µm at the widest point, L/W (1.1–)1.8–3.6(–6.1). **Supporting cell** (2.0–)2.5–4.0(–5.5) µm wide; ratio of length of phialide to width of supporting cell (1.2–)1.8–3.8(–7.3); ratio of width of phialide to width of supporting cell (0.7–)0.8–1.5(–2.3). **Conidia** broadly ellipsoidal, (2.2–)3.0–3.5(–4.0) × (3.0–)3.7–4.5(–5.0) µm, L/W (0.9–)1.2–1.6(–2.2), gray-green, tuberculate. **Intercalary phialides** infrequent. **Chlamydospores** terminal and intercalary, subglobose, typically abundant.

Distinguishing characters: The tuberculate, subglobose conidia, irregular branching of conidiophores, and temperate (north and south) distribution distinguish *T. saturnisporum*. The most morphologically similar species are *T. ghanense*, mainly a tropical species, and the rare temperate species *T. saturnisporopsis*.

Habitat: Soil.

Distribution: North and south temperate regions.

Literature: Bissett (1984), Gams and Bissett (1998), Samuels et al. (1998, 2012).

Figure 6.40. *Trichoderma saturnisporum.* **A and B,** Conidial pustules. **C–H,** Conidiophores. **I,** Conidia (optical view in top portion; surface view in bottom portion). **A–E and H,** From cornmeal agar. **F, G, and I,** From Spezieller Nährstoffarmer agar (SNA). Scale bars: **A, B** = 0.5 mm; **C, G–I** = 10 µm; **D–F** = 20 µm.

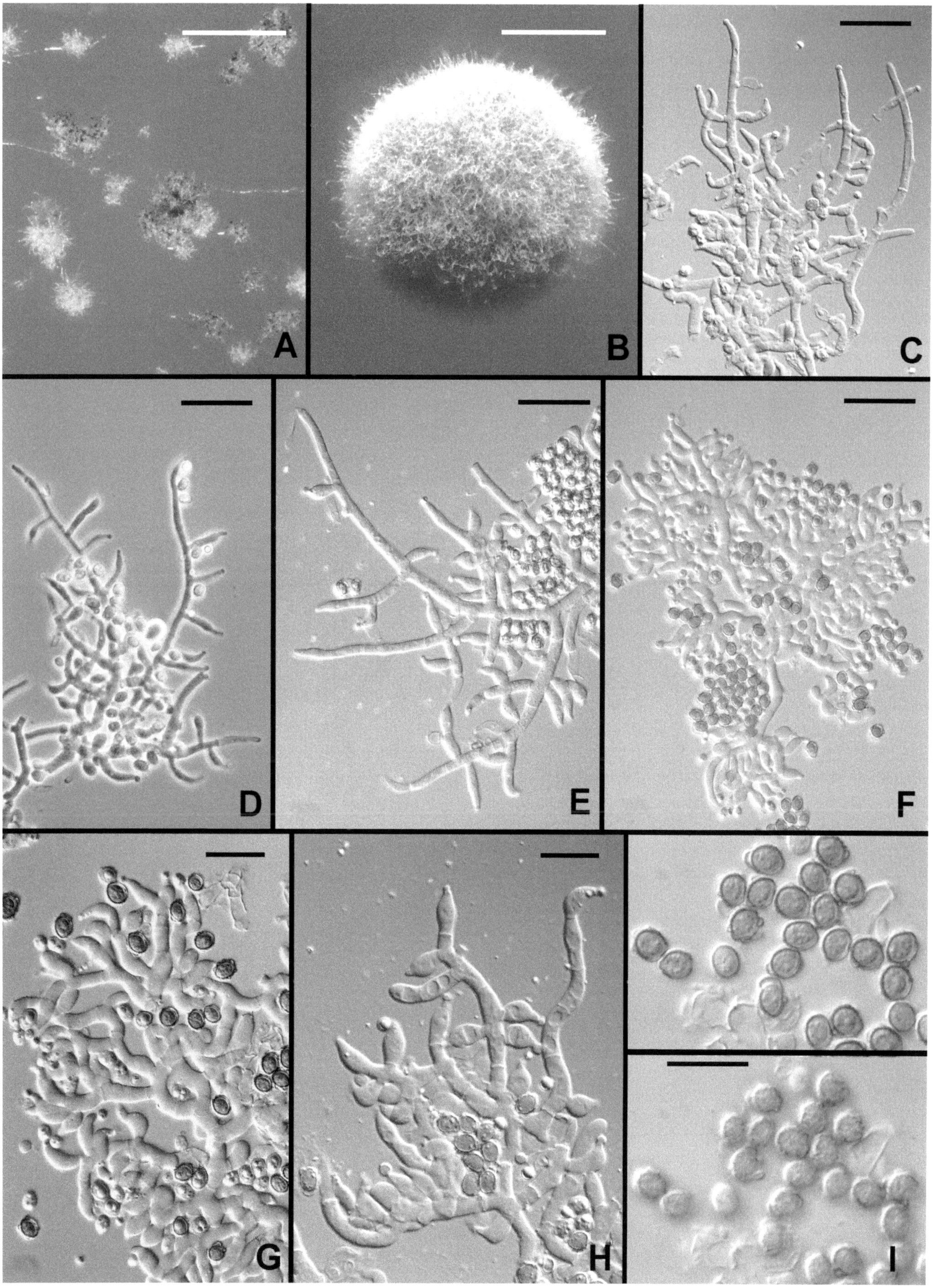
A
B
C
D
E
F
G
H
I

Trichoderma spirale

Bissett, Can. J. Bot. 69:2408. 1991 [1992]. Figs. 6.4A, 6.41.

Ex-type culture: DAOM 183974
GenBank: *rpb2*: FJ442694; *tef1*: AY750890

Clade: Virens

Colony radius at 25°C: PDA and **SNA** (29–)30–55(–60), slightly slower on SNA. A yellow diffusing pigment developing in some colonies on PDA; no distinctive odor on PDA.

Colony radius at 35°C: PDA and **SNA** 0–10(–23).

Conidiation: Pustules few to abundant, forming in more-or-less visible concentric rings, to 3 mm in diameter, gray-green; sterile hairs spiraled or whiplike, abundant, white. Conidiophores with glistening green conidial masses arising abundantly in the aerial mycelium. A *Verticillium*-like synanamorph with glistening heads of green conidia conspicuous, even if no pustules are formed.

Microscopic characteristics: Branching pattern Pachybasium type; fertile branches arising at right angles from the base of hairs, increasing in length and more profusely branched with distance from the tip. **Sterile hairs** typically spiraled (especially in fresh cultures) or sinuous, branched, septate, thin walled, acute, sometimes terminating in a single phialide. **Phialides** arising in clusters of few to many at the tip of fertile branches and singly along the length of the branch, ampulliform, (3.5–)4.2–6.7(–14.0) μm long, (2.7–)3.2–4.2(–5.0) μm at the widest point, L/W (1.0–)1.1–1.9(–4.2). **Supporting cell** (2.2–)3.0–4.2(–5.0) μm wide; ratio of length of phialide to width of supporting cell (0.8–)1.0–2.0(–4.8); ratio of width of phialide to width of supporting cell (0.7–)0.8–1.2(–1.5). **Conidia** oblong to narrowly ellipsoidal, (1.5–)2.0–2.7(–3.5) × (3.0–)3.7–4.2(–5.0) μm, L/W (1.2–)1.5–1.9(–2.3), gray-green, smooth. **Chlamydospores** terminal and intercalary, common.

Distinguishing characters: *Trichoderma spirale* is characterized by the abundant spiraled white hairs arising from pustules, the *Verticillium*-like synanamorph, and the abundant chlamydospores. The synanamorph is sometimes the only conidial form in a petri dish. *Trichoderma fertile* can be distinguished from *T. spirale* by its straight, generally unbranched conidiophore elongation. *Trichoderma fertile* has slightly smaller conidia and slightly longer phialides.

Habitat: Soil; it is commonly isolated from sapwood of *Theobroma cacao* and other *Theobroma* species. Isolated from surface-sterilized seed of *Cecropia insignis* incubated in forest soil (as *Trichoderma* species GenBank accession nos. FJ612747, FJ612715, FJ612701, FJ612790; see also U'ren et al., 2009).

Distribution: Cosmopolitan.

Literature: Bissett (1991b), Chaverri et al. (2003a), Gams and Bissett (1998).

Figure 6.41. *Trichoderma spirale*. **A and B,** Conidial pustules. **C–J,** Conidiophores. **K,** Conidia. **L,** Chlamydospores. **M,** Synanamorph. **A–M,** From Spezieller Nährstoffarmer agar (SNA). Scale bars: **A** = 1 mm; **B** = 250 μm; **C–H, L** = 20 μm; **I–K, M** = 10 μm.

➤

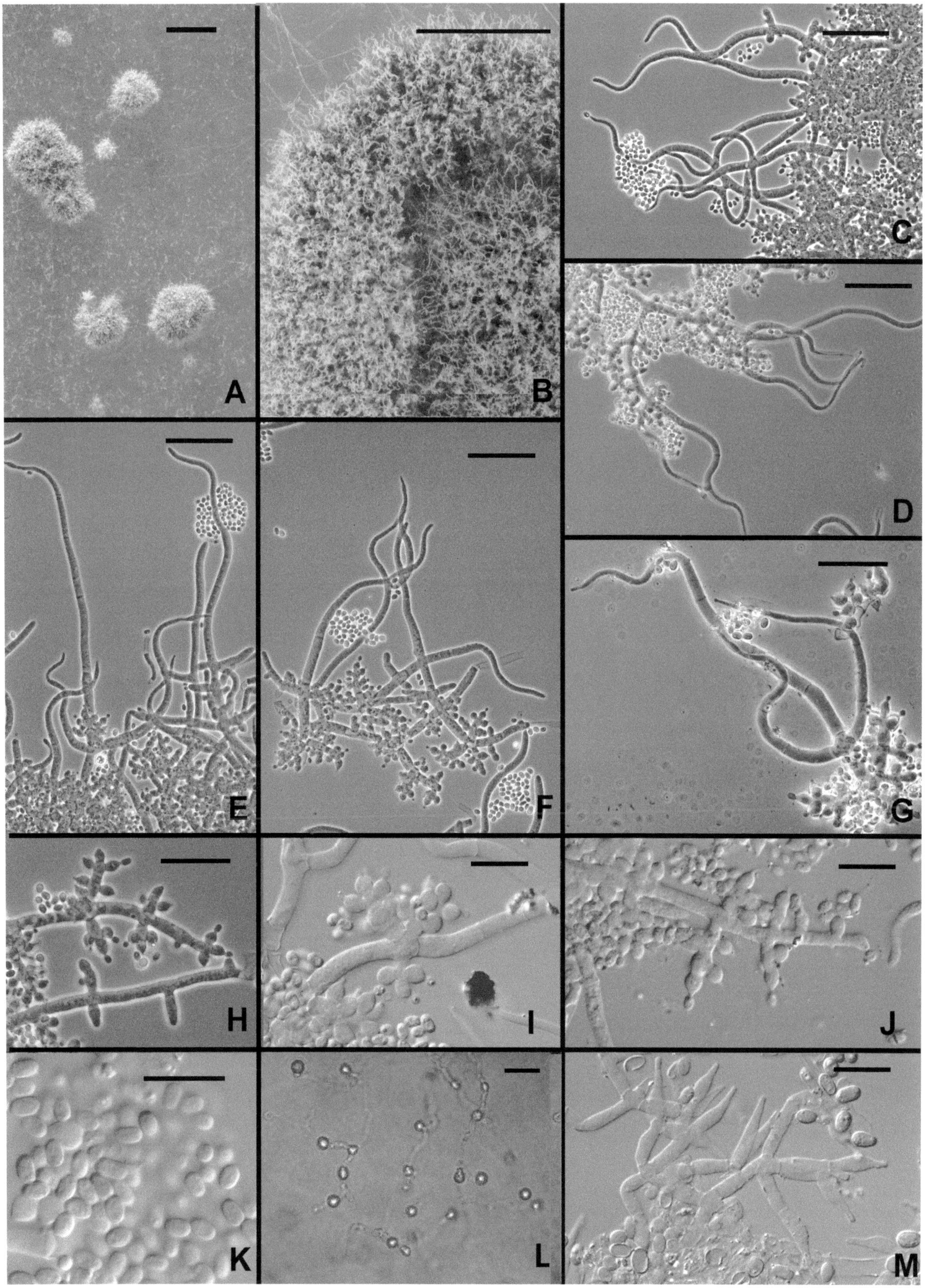
A
B
C
D
E
F
G
H
I
J
K
L
M

Trichoderma stilbohypoxyli

Samuels & Schroers *in* Samuels et al., Stud. Mycol. 56:128. 2006. nom cons. prop. Figs. 6.4B, 6.42.
= *Hypocrea stilbohypoxyli* B. S. Lu & Samuels *in* Lu and Samuels, Sydowia 55:265. 2003.

Ex-type culture: CBS 992.97 = ATCC MYA 2970 = DAOM 231834
GenBank: *rpb2*: EU341805; *tef1*: DQ109546

Clade: Viride

Colony radius at 25°C: PDA (16–)33–55(–65). **SNA** (18–)24–40(–42). On PDA typically a diffusing yellow pigment; no distinctive odor.

Colony radius at 35°C: PDA and **SNA** <5.

Conidiation: Conidia forming in more-or-less continuous, loosely cottony lawns in concentric rings, breaking into loosely cottony pustules at the edges; pustules 0.5–1 mm in diameter, lacking projecting conidiophores or sterile hairs, grayish green.

Microscopic characteristics: Branching pattern Pyramidal type; fertile branches uniformly spaced, frequently paired or verticillately whorled; phialides held in divergent whorls. **Phialides** lageniform to ampulliform, often sharply constricted at the neck, not greatly constricted at the base, (4.2–)5.5–9.0(–12.5) μm long, (2.0–)2.7–3.5(–4.5) μm wide at the widest point, L/W (1.2–)1.4–3.1(–5.6). **Supporting cell** (1.6–)2.2–3.2 (–5.0) μm wide; ratio of length of phialide to width of supporting cell (1.2–)1.7–3.7(–7.2); ratio of widest point of phialide to width of supporting cell (1.2–)1.2–1.4(–1.9). **Intercalary phialides** occasionally formed. **Conidia** ellipsoidal to broadly ellipsoidal, smooth, (1.7–)2.5–3.0 (–3.5) × (2.5–)3.0–4.0(–5.0) μm, L/W (0.9–)1.1–1.5(–1.9), green, smooth. **Chlamydospores** present or absent, subglobose, terminal and intercalary in hyphae.

Distinguishing characters: *Trichoderma stilbohypoxyli* is morphologically similar to several species in the Viride clade. Its conidia are somewhat shorter than those of *T. koningiopsis* and broader than those of *T. ovalisporum*; however, unlike *T. stilbohypoxyli*, neither of these species produces a diffusing yellow pigment on PDA.

Habitat: Bark, soil, on ascomycetes on wood; endophytic in sapwood of *Fagus sylvatica* and *Theobroma* species. Teleomorph on stomata of Xylariaceae, Puerto Rico, United Kingdom, Europe.

Distribution: South America (Argentina, Brazil, Ecuador), Ghana, United Kingdom, Europe.

Literature: Jaklitsch (2011), Lu and Samuels (2003), Samuels et al. (2006a).

Figure 6.42. *Trichoderma stilbohypoxyli*. **A and B,** Conidial pustules. See the more-or-less free conidiophores in **B**. **C–G,** Conidiophores. An intercalary phialide can be seen in **F** (arrow). **H,** Conidia. **I,** Chlamydospores. **A–I,** From cornmeal dextrose agar. Scale bars: **A** = 1 mm; **B** = 0.5 mm; **C, D, I** = 20 μm; **E–H** = 10 μm. (H reproduced, by permission, from Samuels et al., 2006a)

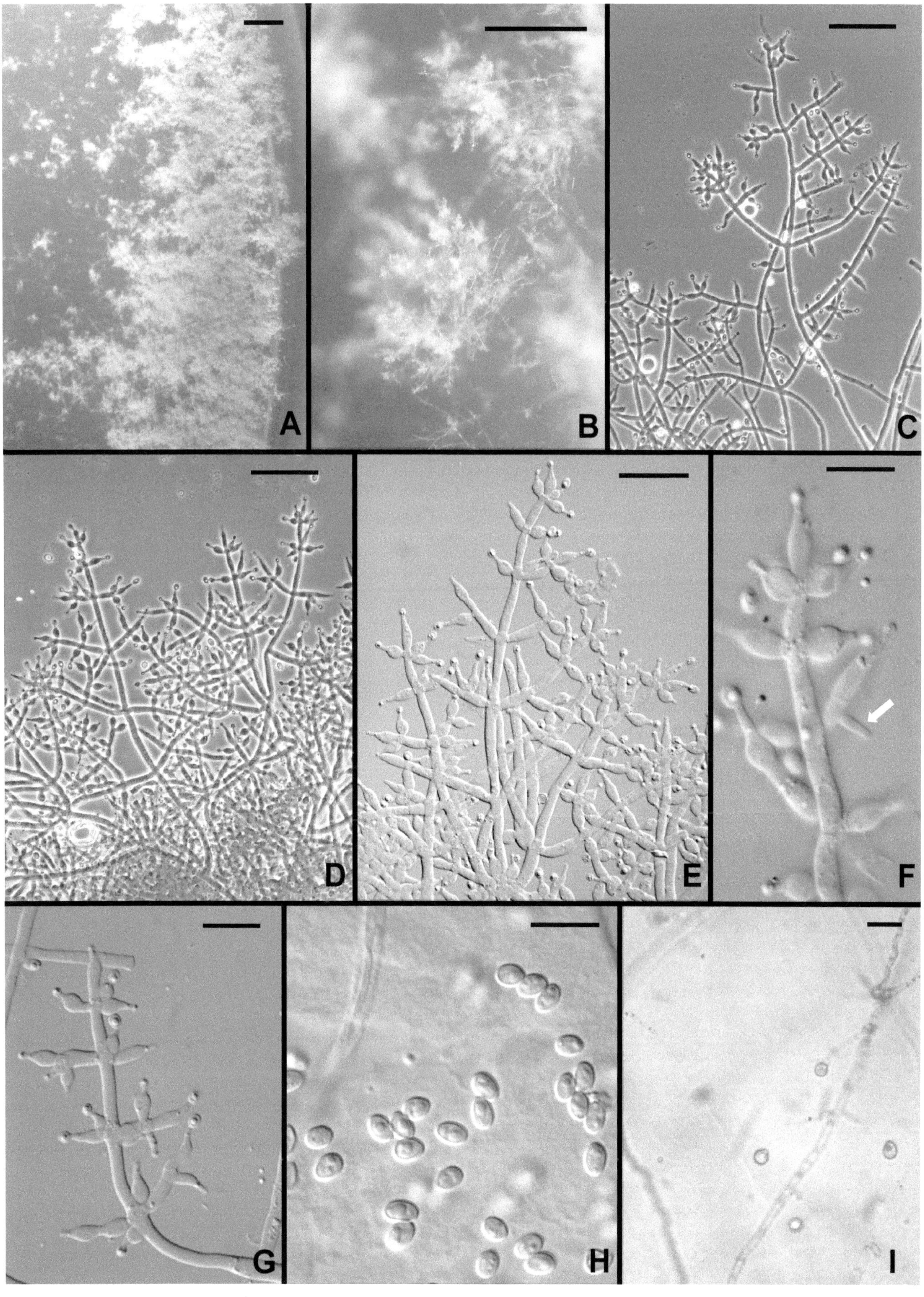
A
B
C
D
E
F
G
H
I

Trichoderma strigosum

Bissett, Can. J. Bot. 69:2411. 1991 [1992]. Figs. 6.4C, 6.43.

Ex-type culture: DAOM 1661214 = CBS 348.93
GenBank: *rpb2*: AF545556; *tef1*: AY376057

Clade: Viride

Colony radius at 25°C: PDA (20–)48–55. **SNA** 24–40. On PDA no diffusing pigment, less frequently with a pale yellow diffusing pigment; often with a strong coconut odor.

Colony radius at 35°C: PDA (0–)10–35. **SNA** 0–10.

Conidiation: Pustules mainly forming at the periphery of the colony, discrete, 0.5–1 mm in diameter, hemispherical, becoming dark green, sometimes conidia yellow; hairs typically conspicuous, straight, relatively short, terminating in one or a few phialides. No synanamorph.

Microscopic characteristics: Branching pattern Pachybasium type but Pyramidal-type conidiophores also forming within pustules. Pachybasium-type conidiophores composed of fertile branches arising at the base of the hairs. **Hairs** stiff, straight, septate, unbranched or infrequently branched, bearing one or a few phialides at the tip, sometimes more-or-less widely spaced, paired branches arising below the tip along the length of the hair, often with long distances between branches; fertile branches arising from near the base of the hair often three or more cells in length and rebranched, increasing in length with distance from the tip of the hair, sometimes composed of a single cell from which a terminal cluster of phialides arises; phialides typically held in divergent whorls at the tips of branches. **Phialides** lageniform, (4.2–)5–10(–13) µm long, (2.0–)2.5–3.5(–4.0) µm at the widest point, L/W (1.1–)1.6–3.0(–5.5). **Supporting cell** (1.7–)2.5–3.5 (–5.0) µm wide; ratio of length of phialide to width of supporting cell (1.3–)1.6–3.2(–4.5); ratio of width of phialide to width of supporting cell (0.6–)0.9–1.3(–2.0). **Conidia** oblong to narrowly ellipsoidal, (1.5–)2.0–2.7 (–3.5) × (3.2–)3.7–5.0(–8.0) µm, L/W (1.2–)1.5–2.1, green, smooth. **Chlamydospores** terminal and intercalary, absent or abundant.

Distinguishing characters: *Trichoderma strigosum* is characterized by the stiff, terminally fertile extensions of the conidiophores (hairs), combined with lageniform phialides that are typically more than twice as wide as the cell that produced them and the coconut odor frequently detected in PDA cultures. The closest relative of *T. strigosum* is *T. strigosellum*, a pantropical species (López-Quintero et al., 2013).

Habitat: Soil; reported as an endophyte from sapwood of *Theobroma* species and from roots of *Anthocleista nobilis* in Brazil (GenBank accession nos. EU718073, EU718074; C. Douanla-Meli and E. Langer, *unpublished data*).

Distribution: North, Central, and South America, Africa (Cameroon).

Literature: Bissett (1991b), Chaverri et al. (2003a), Gams and Bissett (1998).

Figure 6.43. *Trichoderma strigosum.* **A and B,** Conidial pustules. Note the protruding, terminally fertile hairs at the periphery of the pustule in **B. C–K,** Conidiophores and phialides. Note the straight, relatively short hairs in **C, H,** and **I.** Pyramidal-type branching can be seen in **K. L,** Conidia. **M,** Chlamydospores. **A and C,** From cornmeal dextrose agar. **B and D–M,** From Spezieller Nährstoffarmer agar (SNA). Scale bars: **A** = 5 mm; **B** = 100 µm; **C–F, H–K, M** = 20 µm; **G, L** = 10 µm. ➤

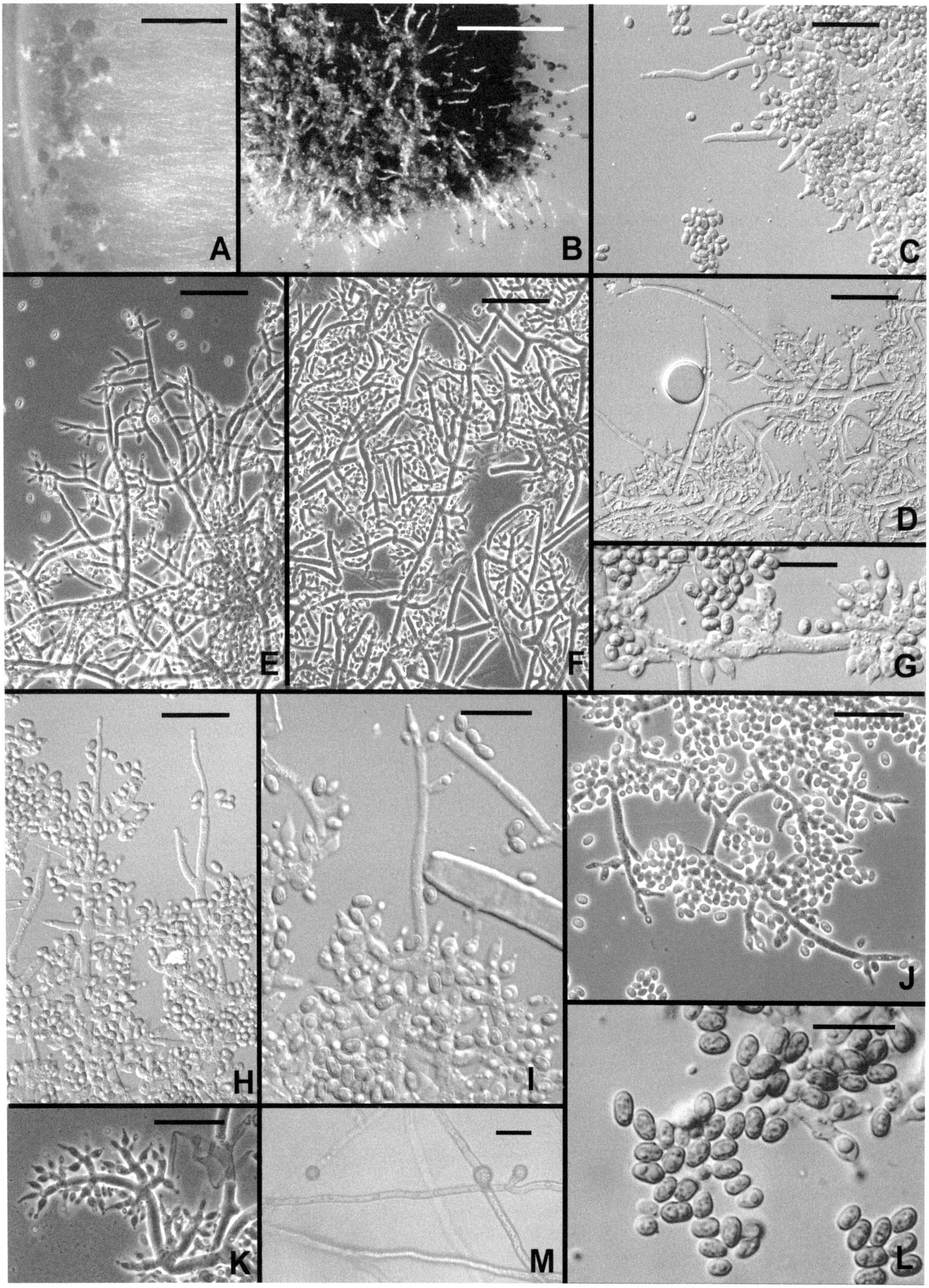
A
B
C
E
F
D
G
H
I
J
K
M
L

Trichoderma stromaticum

Samuels & Pardo-Schulth. *in* Samuels et al., Mycol. Res. 104:762. 2000. Figs. 6.4D, 6.44.
= *Hypocrea stromatica* J. L. Bezerra, J. C. B. Costa & C. N. Bastos *in* Bezerra et al., Fitopatol. Bras. 28:409. 2003.

Ex-type culture: ATCC 204426 = CBS 101875
GenBank: *rpb2*: HQ342245; *tef1*: AY937418

Clade: Stromaticum

Colony radius at 25°C: PDA (20–)30–50. **SNA** (5–)15–25(–30). On PDA no distinctive diffusing pigment or distinctive odor.

Colony radius at 35°C: PDA and **SNA** 0(<10).

Conidiation: Pustules typically abundant in a wide band around the periphery, less frequently uniformly distributed throughout the colony, approximately 0.5 mm in diameter, hemispherical, very easily dislodged from the agar, beginning as a white knot, usually producing conidia only from the top and often only partially fertile, conidia dark green, often yellow. Hairs arising abundantly from the pustule, stiff, awl-like, producing conidia from the tip. On PDA pustules often covering the entire surface of the colony, very easily dislodged. On SNA *Verticillium*-like synanamorph with glistening green conidial masses arising abundantly in the aerial mycelium.

Microscopic characteristics: Branching pattern Pachybasium type; the pustule formed of short, wide hyphae terminating in fertile branches and hairs. Fertile branches arising from the base of hairs usually in dense clusters, no more than a few cells in length, usually producing unicellular secondary branches, each branch terminating in one to three phialides. **Hairs** very broad at the base, tapering to an acute tip, rigid, septate; one or a few phialides arising from the tip. **Phialides** ampulliform, (3.5–)5.0–8.0(–17) μm long, (1–)3–5(–10) μm at the widest point, L/W (0.9–)1.3–2.1(–4.2). **Supporting cell** (3.0–)3.5–5.5(–9.0) μm wide; ratio of length of phialide to width of supporting cell (0.7–)1.1–1.9(–3.1); ratio of width of phialide to width of supporting cell (0.3–)0.7–1.1(–1.5). **Conidia** ellipsoidal, (1.7–)2.5–3.0(–3.7) × (3.0–)3.7–4.5(–7.2) μm, L/W (1.0–)1.4–1.8(–2.3), green, smooth (warted in scanning electron microscopy [SEM], de Melo and Faull, 2004). **Chlamydospores** abundant, terminal and intercalary, often aggregating to form large masses. **Synanamorph** *Verticillium*-like.

Distinguishing characters: *Trichoderma stromaticum* is characterized by the formation of discrete pustules that are formed of short, wide, hyphae that readily disarticulate in microscope crush mounts, short awl-like hairs, and the fact that the pustules are very easily dislodged from the colony surface.

Habitat: Constantly associated with *Moniliophthora perniciosa* on *Theobroma cacao*, on dead cacao tissue infected with *Moniliophthora perniciosa* (dry brooms) or fruit bodies of *Moniliophthora perniciosa*. Collections from Peru were found on the pseudostroma of *Moniliophthora roreri*. Isolated rarely as an endophyte from the trunk of *Theobroma cacao*.

Distribution: South America (Brazil, Colombia, Ecuador, Peru). DNA detected in soil in the Canary Islands (Zachow et al., 2009).

Literature: Bezerra et al. (2003, teleomorph), de Melo and Faull (2004), Medeiros et al. (2010), Samuels et al. (2000, [2011 online] 2012), Zachow et al. (2009).

Figure 6.44. *Trichoderma stromaticum*. **A–D,** Conidial pustules. Note that only part of the pustule is forming conidia in **B** and **C**. **E,** *Verticillium*-like synanamorph. **F–J,** Conidiophores. Note the short, broad stromatic cells that produce conidiophores in **F**. Note the awl-like hairs in **G–J**. **K,** Phialides. **L,** Conidia. **M,** Chlamydospores. **N,** *Verticilllium*-like synanamorph. **A and J,** From potato dextrose agar. **B, D–H, K, L, and N,** From Spezieller Nährstoffarmer agar (SNA). **C,** From cornmeal agar. **I and M,** From cornmeal dextrose agar. Scale bars: **A–D** = 0.5 mm; **E** = 150 μm; **F–J, M, N** = 20 μm; **K, L** = 10 μm. (C, E–H, and K reproduced from Samuels et al., [2011 online] 2012, with kind permission of Springer Science+Business Media)

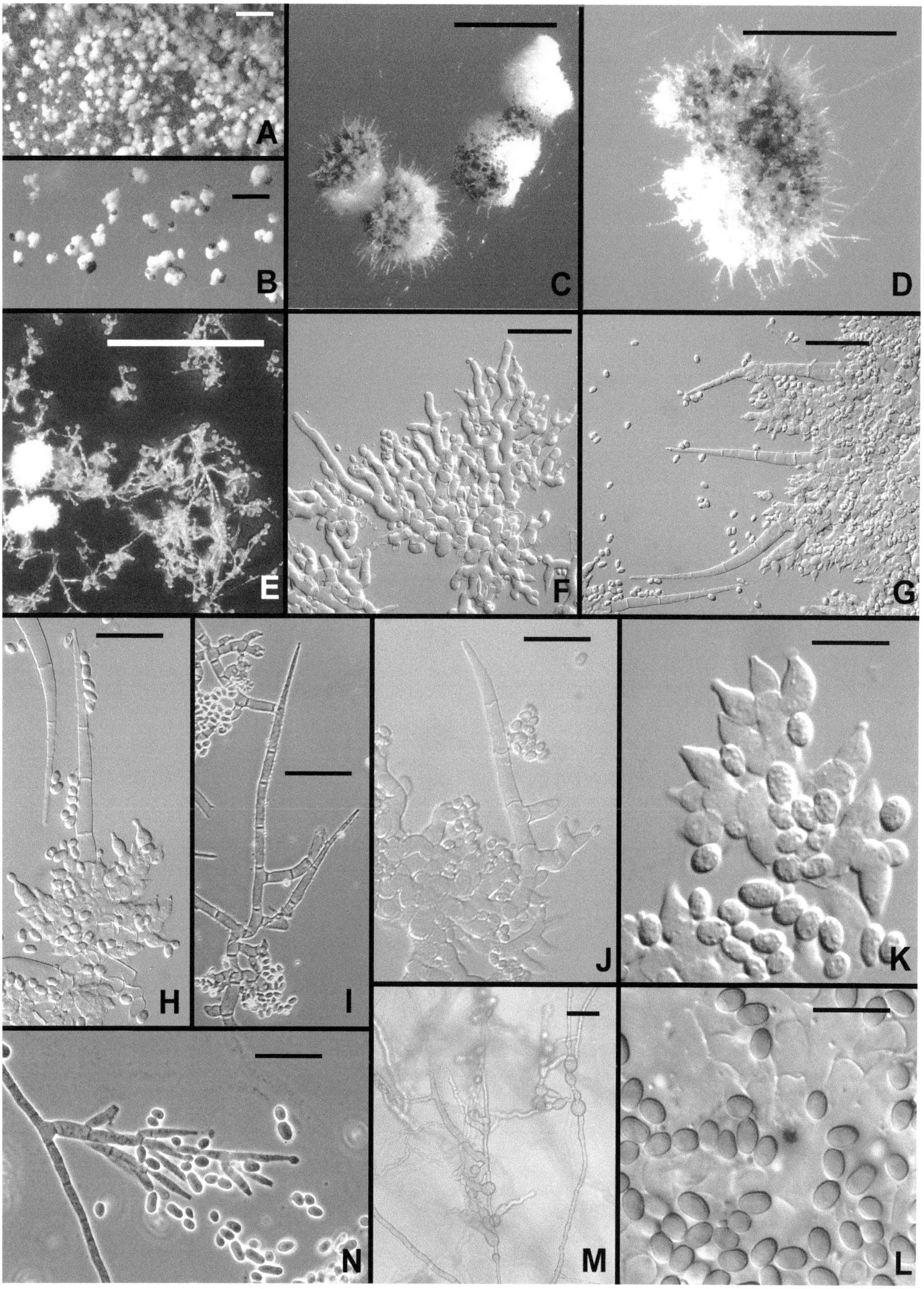
A
B
C
D
E
F
G
H
I
J
K
L
M
N

Trichoderma tomentosum

Bissett, Can. J. Bot. 69:2412. 1991 [1992]. Figs. 6.4E, 6.45.

Ex-type culture: DAOM 178713a
GenBank: *rpb2*: AF545557; *tef1*: AY750882

Clade: Tomentosum

Colony radius at 25°C: PDA 35–55. **SNA** 35–40. No diffusing pigment or distinctive odor on PDA.

Colony radius at 35°C: PDA and **SNA** 0(<10).

Conidiation: Pustules few, scattered, hemispherical, 1–2 mm in diameter, gray-green. Hairs fine, whiplike, arising abundantly from each pustule, sterile or fertile near the tip. Abundant conidial production from mononematous conidiophores in the scant aerial mycelium, conidia held in drops of green liquid. The production of pustules can be suppressed and only mononematous conidiophores are observed.

Microscopic characteristics: Branching pattern Pachybasium type; conidiophores consisting of a very long, sinuous, branched, septate, sterile or fertile hair with fertile branches arising at right angles near its base; fertile branches broad, 3–4 μm wide, composed of two or three levels of closely spaced, frequently paired secondary branches; secondary branches increasing in length with distance from the tip of the primary fertile branch, rebranching to form unicellular tertiary branches or solitary phialides; all ultimate branches terminating in three or more crowded phialides. **Phialides** ampulliform, (3.7–)4.2–6.2(–10.2) μm long, (2.7–)3.0–3.7(–4.0) μm at the widest point, L/W (1.1–)1.2–1.6(–2.1). **Supporting cell** (2.0–)2.7–3.7(–4.5) μm wide; ratio of length of phialide to width of supporting cell (1.0–)1.1–2.5(–4.7); ratio of width of phialide to width of supporting cell (0.8–)0.9–1.3(–1.6). **Conidia** ellipsoidal, (1.5–)2.0–2.5 × (2.5–)3.0–3.7(–4.0) μm, L/W (1.1–)1.4–1.8(–2.2), green, smooth. **Synanamorph** *Verticillium*-like. **Chlamydospores** typically abundant, terminal and intercalary.

Distinguishing characters: *Trichoderma tomentosum* is most closely related to *T. velutinum*, which has smaller conidia. Conidial measurements given here are somewhat greater than was given in the original description (Bissett, 1991b).

Habitat: Soil, roots, wood of conifers and dicotyledonous trees. Uncultured genomic DNA of a *Trichoderma* species isolated from fungus combs of termite nests in China is most likely *T. tomentosum* (GenBank accession nos. FN812786, FN812788, FN812792, FN812797; Q. Qian and H. A. Wen, *unpublished data*).

Distribution: North temperate (North America, Europe, China), reported from Guatemala (Hoyos-Carvajal et al., 2009a) and Brazil (Lopes et al., 2012), but possibly rare in tropics.

Literature: Bissett (1991b), Chaverri and Samuels (2003).

Figure 6.45. *Trichoderma tomentosum*. **A and B,** Conidial pustules. *Verticillium*-like synanamorph can be seen in **B** (upper left). **C,** Hairs. **D–H,** Conidiophores. **I,** Phialides. **J,** Conidia. **K,** Chlamydospores. **L and M,** *Verticillium*-like synanamorph. **A, B, D–F, and J–M,** From Spezieller Nährstoffarmer agar (SNA). **C and G–I,** From cornmeal dextrose agar. Scale bars: **A** = 1 mm; **B** = 0.5 mm; **C–E, G, H, K, L** = 20 μm; **F, I, J, M** = 10 μm.

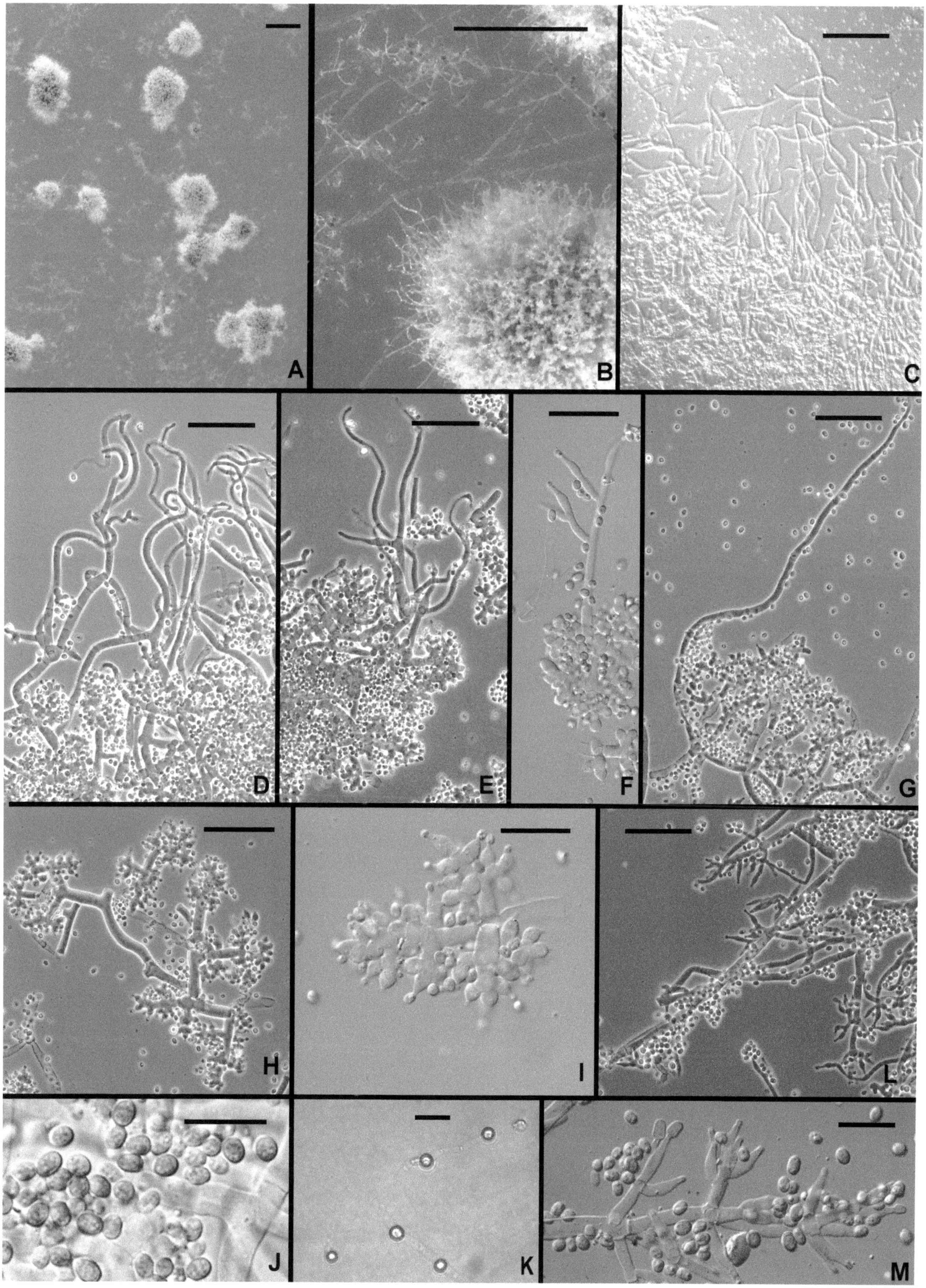
A
B
C
D
E
F
G
H
I
L
J
K
M

Trichoderma velutinum

Bissett, C. P. Kubicek & Szakács *in* Bissett et al., Can. J. Bot. 81:579. 2003. Figs. 6.4F, 6.46.

Ex-type culture: DAOM 230013 = TUB F784
GenBank: *rpb2*: FN812788; *tef1*: AF510444

Clade: Lone lineages

Colony radius at 25°C: PDA 55–65. **SNA** 40–50. No distinctive pigment or distinctive odor on PDA.

Colony radius at 35°C: PDA and **SNA** 2–35.

Conidiation: Pustules beginning to form around the periphery of the colony, to 0.5 mm in diameter, becoming confluent and forming a nearly continuous thick lawn, difficult to remove from the agar, yellow-green. Hairs abundant, whiplike, arising from the pustules, producing conidia from the tip. Synanamorph *Verticillium*-like, observed on malt agar (Bissett et al., 2003).

Microscopic characteristics: Branching pattern Pachybasium type; fertile branches arising at right angles from the base of hairs, increasing in length and more profusely branched with distance from the tip. **Hairs** flexuous, branched or unbranched, septate, acute, sterile or bearing a few terminal phialides. Fertile branches arising near the base of the hairs, one to a few cells in length, often rebranching, bearing phialides singly and in terminal clusters of several. **Phialides** ampulliform, 4.0–7.2 μm long, 2.7–4.5 μm at the widest point, L/W 1.5. **Supporting cell** approximately 3 μm wide; ratio of length of phialide to width of supporting cell 1.6; ratio of width of phialide to width of supporting cell approximately 0.7. **Conidia** ellipsoidal, 1.5–2.5 × 2.7–4.0 μm, L/W 1.6, gray-green or yellow-green, smooth. **Chlamydospores** abundant, terminal and intercalary.

Distinguishing characters: *Trichoderma velutinum* is morphologically similar to *T. tomentosum* and *T. cerinum*, both of which have small conidia. *Trichoderma cerinum* (25°C after 72 h: PDA 45–50 mm, SNA 35–45 mm) and *T. tomentosum* have a somewhat slower rate of growth than *T. velutinum*. In its original description, the hairs of *T. velutinum* were described as being sterile but in at least one culture of this species we have observed that solitary phialides are often found at the tips of hairs. Bissett et al. (2003) included this species in a broadly conceived Pachybasium clade where its closest relatives appear to be *T. cerinum* and *T. harzianum* s. lat. but its systematic position is unresolved.

Habitat: Soil.

Distribution: Europe (Austria), Nepal; uncommon.

Literature: Bissett et al. (2003).

Figure 6.46. *Trichoderma velutinum*. **A and B,** Conidial pustules. Hairs can be seen at the periphery of pustules in **B**. **C–G,** Conidiophores. **H,** Phialides. **I,** Conidia. **A–I,** From cornmeal dextrose agar. Scale bars: **A** = 1 mm; **B** = 250 μm; **C–F** = 20 μm; **G–I** = 10 μm.

➤

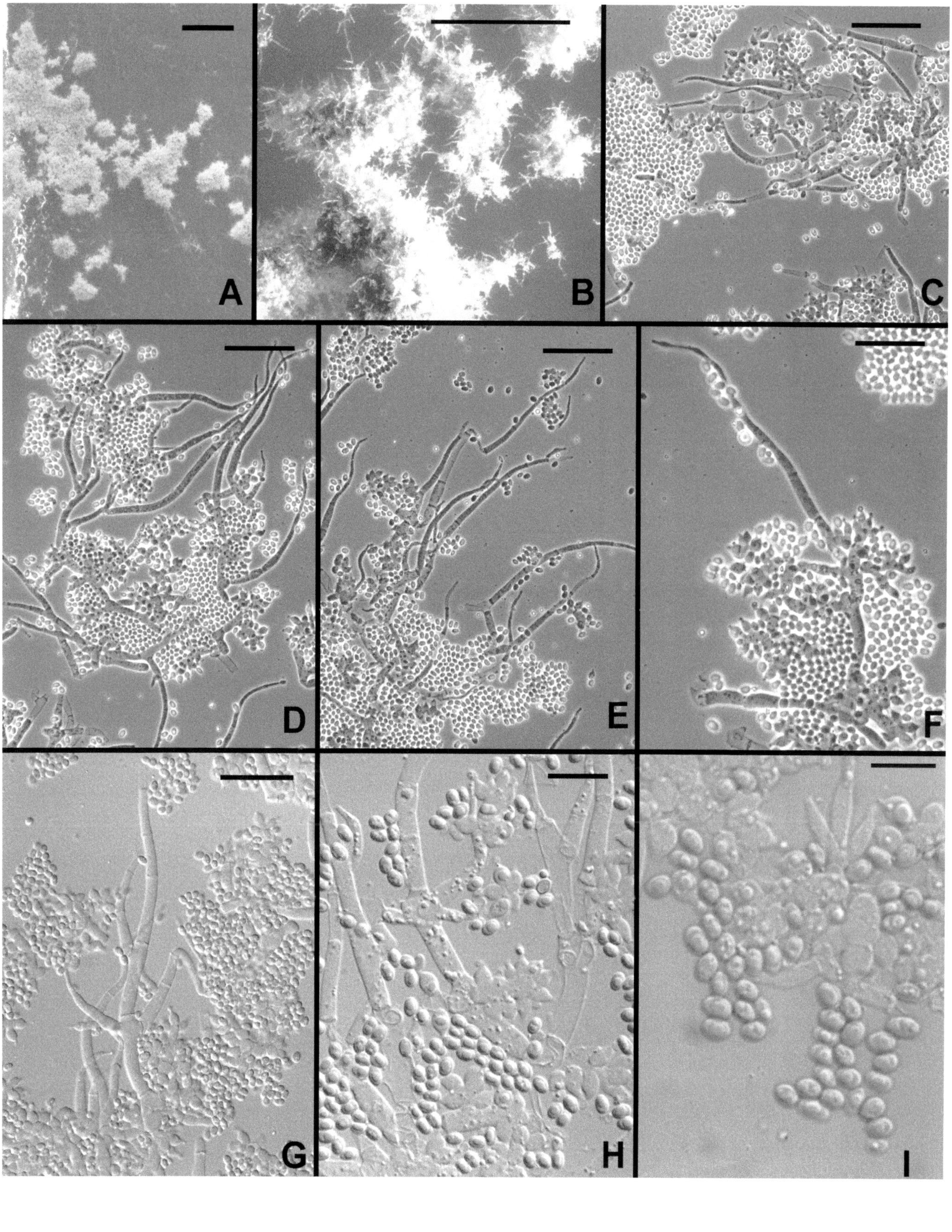
A
B
C
D
E
F
G
H
I

Trichoderma virens

(J. H. Mill., Giddens & A. A. Foster) Arx *in* Arx, Beih. Nova Hedwigia 87:28. 1987. Figs. 6.4G, 6.47.
≡ *Gliocladium virens* J. H. Mill., Giddens & A. A. Foster *in* Miller et al., Mycologia 49:792. 1957.
= *Trichoderma flavofuscum* (J. H. Mill., Giddens & A. A. Foster) Bissett *in* Bissett, Can. J. Bot. 69:2385. 1991 [1992].
≡ *Gliocladium flavofuscum* J. H. Mill., Giddens & A. A. Foster *in* Miller et al., Mycologia 49:793. 1957.
= *Hypocrea virens* P. Chaverri, Samuels & E. L. Stewart *in* Chaverri et al., Mycologia 93:1120. 2001.

Ex-type culture: ATCC 13213 = DAOM 167651 = CBS 249.59
GenBank: *rpb2*: AF545558; *tef1*: GU591800

Clade: Virens

Colony radius at 25°C: PDA (24–)30–65(–70). **SNA** (33–)45–65(–70). On PDA no distinctive odor; often a diffusing yellow pigment, sometimes conidia yellow.

Colony radius at 35°C: PDA (21–)24–38. **SNA** (32–)38–45.

Conidiation: Conidiophores forming abundantly in the aerial mycelium; at the periphery of the colony conidiophores densely disposed, sometimes in fascicles, often forming a continuous lawn. Conidia held in a glistening globose drop of clear, dark green liquid at the tip of each conidiophore.

Microscopic characteristics: Branching pattern Gliocladium type; conidiophores with a distinct stipe and a head of phialides; sometimes conidiophores disposed in a cymose fashion from aerial hyphae; stipe unbranched or producing short lateral branches; lateral branches arising at or near right angles to the stipe, often increasing in length with distance from the tip of the stipe; the longer lateral branches also branching; branches arising near the tip of the stipe closely spaced and arising at a small angle relative to the stipe; the conidiophore and its lateral branches each terminating in a head of phialides; the phialides in a head parallel to each other and tightly appressed. **Phialides** lageniform to ampulliform, often asymmetric with one side flattened, (5.5–)8.7–9.2(–15.0) µm long, (2.5–)3.5–4.5(–5.0) µm at the widest point, (1.5–)2.2–2.5(–3.5) µm at the base, L/W (1.5–)2.2–2.3(–3.4). **Supporting cell** (metula) cylindrical, (7–)9–13(–18) µm long, (2.2–)4.0–4.3(–5.2) µm wide; ratio of length of phialide to width of supporting cell (1.3–)1.8–2.2(–4.5); ratio of width of phialide to width of supporting cell (0.5–)0.8–1.2(–1.7); phialide tip often with green periclinal thickening. **Conidia** variable in shape, subglobose to ellipsoidal or ovoidal, (2.5–)3.5–4.2(–4.7) × (3.6–)4.0–5.2(–6.5) µm, L/W (0.9–)1.1–1.3(–1.6), dark green, smooth (finely warted in scanning electron microscopy [SEM]). **Chlamydospores** abundant, subglobose, terminal and intercalary.

Distinguishing characters: Readily recognizable by penicillately branched, *Gliocladium*-like conidiophores and conidia that are borne in drops of clear, watery liquid and by the abundant chlamydospores, seen even in young cultures. Compare with *T. crassum*, *T. gelatinosum*, and *T. pleuroti*. *Trichoderma virens* is a species complex that will doubtless be subdivided in the future.

Habitat: Soil; teleomorph found once, on wood.

Distribution: Cosmopolitan.

Literature: Bissett (1991b), Chaverri and Samuels (2003), Chaverri et al. (2001), Gams and Bissett (1998). An important biocontrol species (see references in Harman and Kubicek, 1998; Kubicek and Harman, 1998).

Figure 6.47. *Trichoderma virens*. **A–C**, Conidiation. Many conidiophores are densely clustered into pustules in **A**. Note the individual drops of conidia in **C**. **D–J**, Conidiophores. **K**, Conidia. **L**, Chlamydospores. **A–L**, From Spezieller Nährstoffarmer agar (SNA). Scale bars: **A** = 1 mm; **B, C** = 0.5 mm; **D, E, I, L** = 20 µm; **F–H, J, K** = 10 µm.

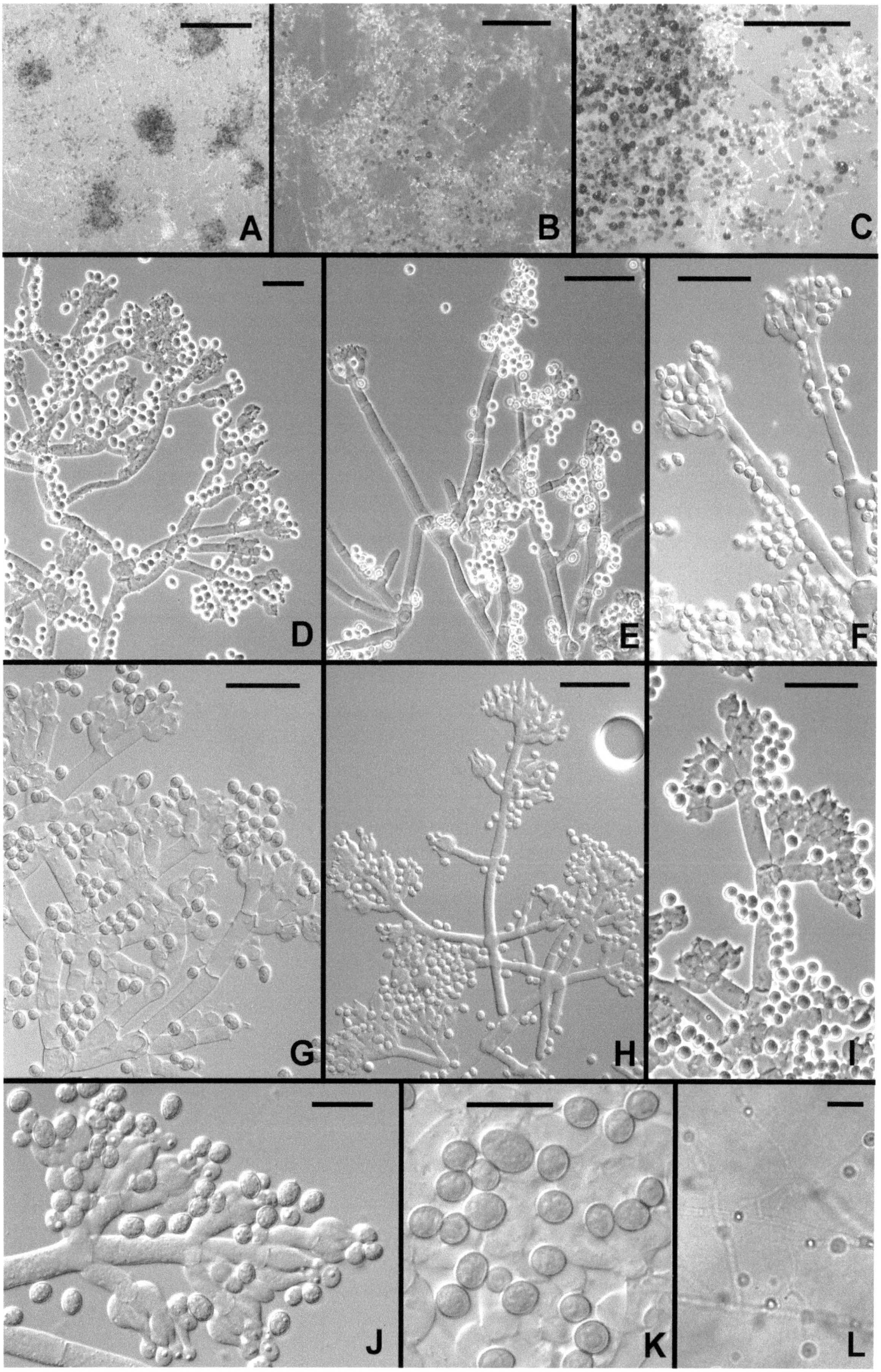
A
B
C
D
E
F
G
H
I
J
K
L

Trichoderma viride

Pers. *in* Persoon, Neues Mag. Bot. 1:92. 1794. Figs. 6.4H, 6.48.

= *Trichoderma lignorum* (Tode) Harz *in* Harz, Linig. Hyph. p. 29. 1871.

= *Trichoderma glaucum* E. V. Abbott, Iowa State Coll. J. Sci. 1:27. 1927.

= *Hypocrea rufa* (Pers.:Fr.) Fr. *in* Fries, Summa Veg. Scand., Section Post. p. 383. 1849.

Ex-epitype culture: CBS 119325 (Jaklitsch et al., 2006)
GenBank: *rpb2*: EU711362; *tef1*: DQ672617

Clade: Viride

Colony radius at 25°C: PDA 30–40. **SNA** 20–35. On PDA diffusing pigment not noted; slight coconut odor rarely noted.

Colony radius at 35°C: PDA and **SNA** 0.

Conidiation: Pustules forming in the center of the colony, radiating outwardly, discrete, pulvinate to hemispherical, 1–2 mm in diameter or confluent in aggregates to approximately 4 mm in diameter, compact to cottony, lacking protruding sterile hairs or conidiophores, becoming green. Few effuse conidiophores forming apart from pustules.

Microscopic characteristics: Branching pattern highly variable, often consisting of a distinct central axis with many Longibrachiatum-like solitary phialides near the tip, branches forming lower down paired or not, increasing in length with distance from the tip and rebranching in similar pattern; often no discernible central axis, fertile branches often arising from swollen nodes, often sinuous or curved, terminating in divergent to verticillate whorls of two or three phialides, solitary phialides common; in all cases branches tending to be widely spaced. **Phialides** lageniform, at most swollen in the middle, straight or often hooked, (5.0–)6.5–11.0(–18) μm long, (1.5–)2.5–3.5(–4.0) μm at the widest point, L/W (1.2–)2.0–4.5(–7.8). **Supporting cell** (1.5–)2.2–3.2(–4.5) μm wide; ratio of length of phialide to width of supporting cell (1.4–)2.4–4.6(–7.5); ratio of width of phialide to width of supporting cell (0.5–)0.9–1.4(–2.0). **Intercalary phialides** infrequent; percurrent proliferation of phialides not seen. **Conidia** subglobose, (2.7–)3.2–4.2(–5.0) × (3.0–)3.5–4.5(–5.5) μm, L/W (0.8–)1.1–1.3(–1.5), green, coarsely warted to tuberculate. **Chlamydospores** infrequent, terminal and intercalary.

Distinguishing characters: The large, subglobose, warted conidia and frequently solitary phialides are distinctive. *Trichoderma martiale* (Hanada et al., 2008), an uncommon tropical species, is closely related and physically very similar. *Trichoderma viridescens* and its close relatives have large, warted subglobose to ovoidal conidia but phialides are more conspicuously whorled and on rich media such as PDA percurrently proliferating phialides are common. *Trichoderma asperelloides* and *T. asperellum* have finely warted conidia and uniformly branched conidiophores.

Habitat: Soil, wood, and bark; sometimes isolated from orchids (Herrera et al., 2010, as "unidentified Hypocreales").

Distribution: North and south temperate region; not very common. Infrequently reported from tropical regions.

Literature: Gams and Bissett (1998), Hanada et al. (2008, 2009), Jaklitsch (2011), Jaklitsch et al. (2006), Lieckfeldt et al. (1999), Samuels et al. (1999).

Figure 6.48. *Trichoderma viride*. **A and B,** Conidial pustules. **C–K,** Conidiophores. An intercalary phialide can be seen in **K** (arrow). **L,** Conidia in surface view, showing low warts. **A–C, E, F, and H–J,** From cornmeal dextrose agar. **D, G, K, and L,** From Spezieller Nährstoffarmer agar (SNA). Scale bars: **A** = 1 mm; **B** = 0.5 mm; **C–K** = 20 μm; **L** = 10 μm. (A, C, G, and I–K reproduced, by permission, from Jaklitsch et al., 2006)

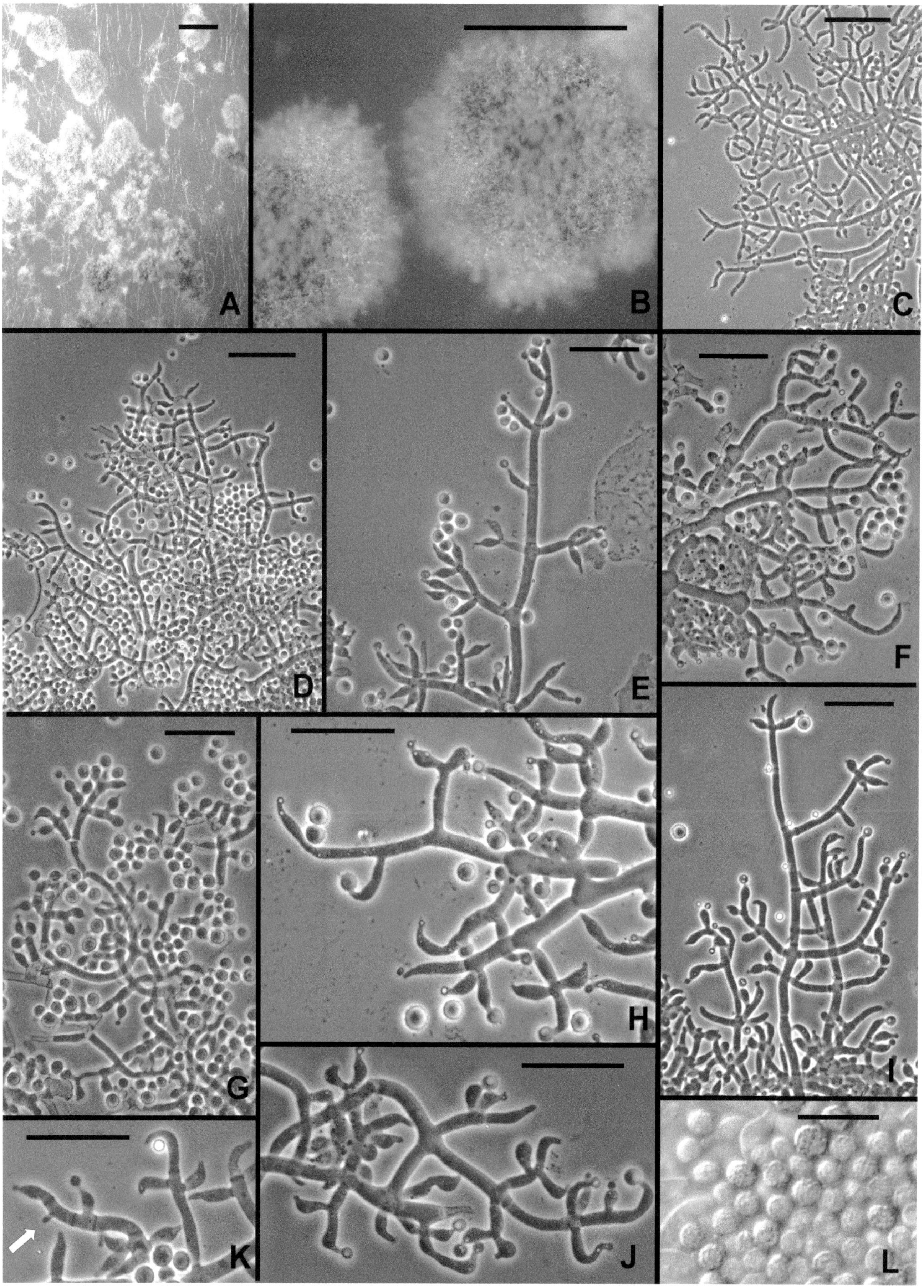
A
B
C
D
E
F
G
H
I
K
J
L

Trichoderma viridescens

(A. S. Horne & H. S. Williamson) Jaklitsch & Samuels *in* Jaklitsch et al., Stud. Mycol. 55:156. 2006. *emend.* Jaklitsch et al., Persoonia 31:125. 2013. Figs. 6.41, 6.49.

≡ *Eidamia viridescens* A. S. Horne & H. S. Will. *in* Horne and Williamson, Ann. Bot. 37:396. 1923.

Ex-epitype culture: CBS 132573 (Jaklitsch et al., 2013)
GenBank: *rpb2*: KC285758; *tef1*: KC285646

Clade: Viride

Colony radius at 25°C: PDA 28–33. **SNA** 24–26. On PDA and cornmeal dextrose agar (CMD) a coconut odor typically present; no diffusing pigment observed.

Colony radius at 35°C: PDA and **SNA** 0.

Conidiation: Conidial production at first effuse on aerial hyphae, later in loose tufts or in approximately 1-mm-diameter pustules; confluent pustules to 8 mm at the margin, becoming green from 4 days, conidia in pustules green, often yellow (pigment not observed on SNA). On PDA and CMD cultures often remaining sterile.

Microscopic characteristics: Conidiophores of two types, 1 and 2, the two types typically not forming in the same culture; conidiophores of both types found on CMD and PDA, type 1 conidiophores alone found on SNA. Type 1: conidiophores at the surface of the pustule protruding from the pustule, typically with one or few terminal phialides, sometimes producing widely spaced phialides along the length; conidiophores forming at the interior of pustules typically with a straight to sinuous, branched main axis, the branches paired, increasing in length with distance from the tip, each branch terminating in one to three phialides in a cruciate whorl. **Phialides from type 1** conidiophores lageniform, slightly swollen in the middle when crowded, straight, less frequently sinuous or hooked, (4.7–)7.2–11.5(–20.0) μm long, (1.7–)2.5–3.2(–4.2) μm at the widest point, L/W (1.5–)2.5–4.5(–9.5), base (1.0–)1.7–2.5(–3.5) μm wide. **Supporting cell** (1.1–)2.0–3.0(–4.5) μm wide; ratio of length of phialide to width of supporting cell (1.2–)2.5–5.0(–9.0); ratio of width of phialide to width of supporting cell (0.6–)0.9–1.3(–2.1). **Phialides from type 2** conidiophores frequently developing in PDA and CMD; a main axis of conidiophores not formed or barely discernible, branching occurring at acute angles with respect to the parent hypha, branches rebranching or not, each branch terminating in one or two phialides; phialides appearing to proliferate percurrently, often resulting in chains of two to six cells, chains terminating in a phialide; phialides and subtending cells distinctive, swollen in the middle and more-or-less conspicuously constricted above and below the middle, the chain of cells and phialides appearing moniliform, phialides sometimes with a long, cylindrical neck. Conidiophores often arising from variously distorted and swollen cells in pustules, often submoniliform chains of proliferated phialides conspicuous in older pustules. **Conidia** subglobose, ovoidal or ellipsoidal, (2.2–)3.0–3.7(–4.7) × (2.7–)3.5–4.5(–8.5) μm, L/W 1.0–1.4, green, finely warted. **Intercalary phialides** present but not common on all media. **Chlamydospores** present or not, terminal or intercalary, subglobose.

Distinguishing characters: *Trichoderma viridescens*, as recognized by Jaklitsch et al. (2006) is a species complex within which several morphologically indistinguishable species are now recognized (Jaklitsch et al., 2013). The description and illustrations presented here are *T. viridescens* in the broad sense. *Trichoderma viridescens* is most readily distinguished from *T. viride* by the coconut odor, proliferating phialides on "distorted" (type 2) conidiophores found on PDA and CMD cultures, and variably shaped conidia in the former. Conidia of *T. viride* are strictly subglobose. *Trichoderma viridescens* and *T. viride* overlap in their distribution. *Trichoderma olivascens*, *T. paraviridescens*, *T. viridarium*, *T. viridescens* s. str., and *T. virilente* are common and widely distributed. *Trichoderma trixiae*, *T. viridialbum*, and *T. composticola* are found in America and/or Europe but are not common. *Trichoderma appalachiense* and *T. neosinense* are only known from their teleomorphs, from southeastern Australia and Taiwan, respectively. *Trichoderma composticola* is distinguished by its fast rate of growth at 30°C (colony radius on PDA 40 mm after 3 days).

Note: *Hypocrea viridescens* in the sense of Jaklitsch et al. (2006) is *T. paraviridescens* (Jaklitsch et al., 2013).

Habitat: Soil, wood, and bark (teleomorph).

Distribution: Members of the *T. viridescens* complex are common in north and south temperate regions; less common in tropical regions.

Literature: Jaklitsch (2011), Jaklitsch et al. (2006, 2013).

Figure 6.49. *Trichoderma viridescens* s. str. **A and B,** Conidial pustules. **C–F,** Conidiophores. Conidiophores in **C** and **D** are type 1; conidiophores in **E** and **F** are type 2, showing percurrent proliferation and swollen phialides on potato dextrose agar and cornmeal dextrose agar, respectively. **G and H,** Conidia (optical section in **G**; surface view in **H**). **I,** Chlamydospores. **A–D, F–I,** From cornmeal dextrose agar. **E,** From potato dextrose agar. Scale bars: **A** = 1 mm; **B** = 0.5 mm; **C, D, F–H** = 10 μm; **E, I** = 20 μm. (A, B, and G–I reproduced, by permission, from Jaklitsch et al., 2006) ➤

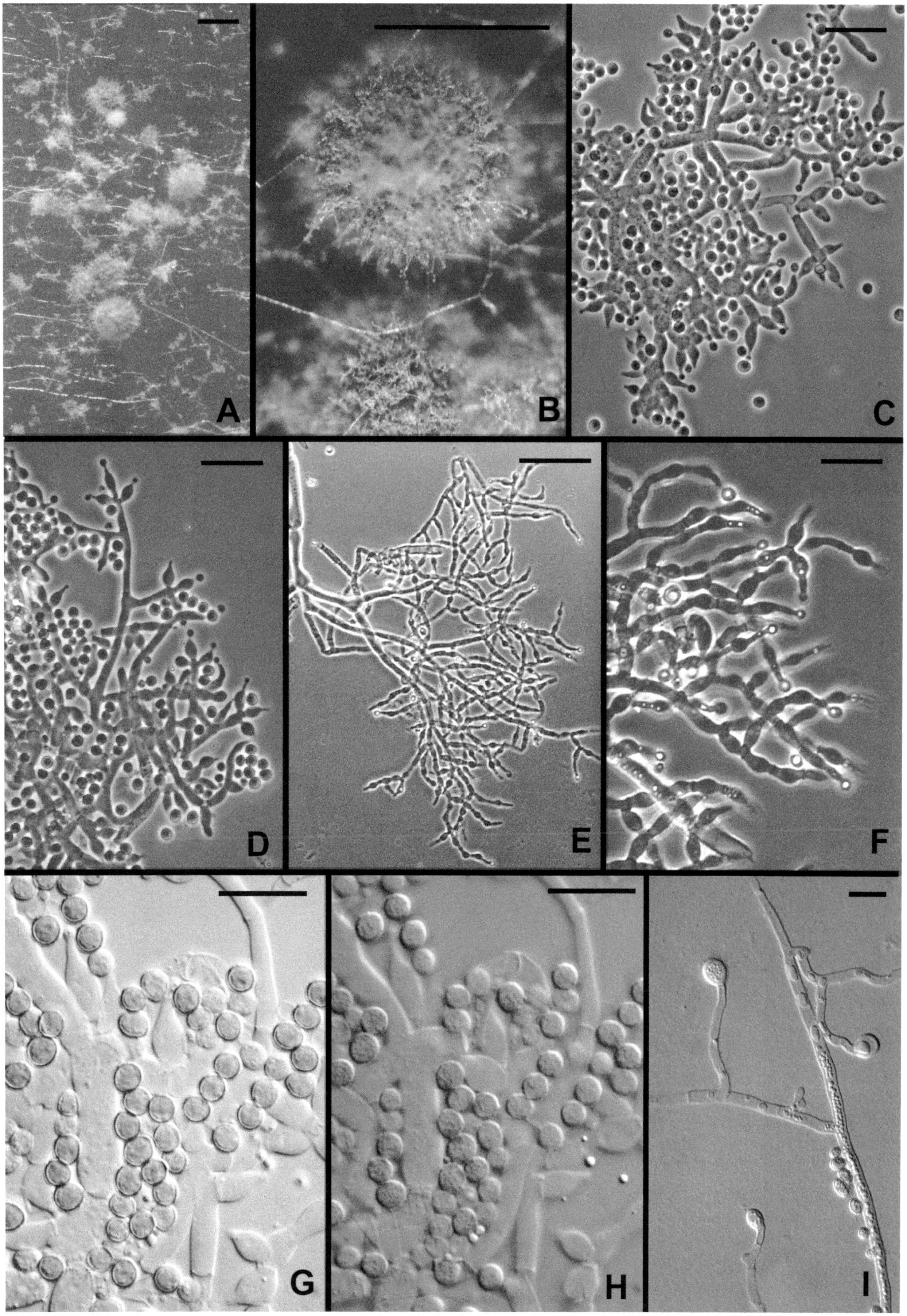
A
B
C
D
E
F
G
H
I

Literature Cited

Abbasi, P. A., Miller, S. A., Meulia, T., Hoitink, H. A. J., and Kim, J.-M. 1999. Precise detection and tracing of *Trichoderma hamatum* 382 in compost-amended potting mixes by using molecular markers. Appl. Environ. Microbiol. 65:5421-5426.

Agosin, E., and Aguillera, J. M. 1998. Industrial production of active propagules of *Trichoderma* for agricultural uses. Pages 205-226 in: *Trichoderma & Gliocladium*, Vol. 2. Enzymes, Biological Control and Commercial Applications. G. E. Harman and C. P. Kubicek, eds. Taylor & Francis, London.

Ahmad, J. S., and Baker, R. 1987. Competitive saprophytic ability and cellulolytic activity of rhizosphere-competent mutants of *Trichoderma harzianum*. Phytopathology 77:358-362.

Alfano, G., Lewis Ivey, M. L., Cakir, C., Bos, J. I. B., Miller, S. A., Madden, L. V., Kamoun, S., and Hoitink, H. A. J. 2007. Systemic modulation of gene expression in tomato by *Trichoderma hamatum* 382. Phytopathology 97: 429-437.

Altomare, C., Norvell, W. A., Björkman, T., and Harman, G. E. 1999. Solubilization of phosphates and micronutrients by the plant-growth-promoting and biocontrol fungus *Trichoderma harzianum* Rifai 1295-22. Appl. Environ. Microbiol. 65:2926-2933.

Andris, H., Michailides, T., and Johnson, S. 2002. Control of *Armillaria* (oak root fungus) with the biological control agents *Trichoderma viride* and *Trichoderma harzianum*. Research Report 2002. California Stone Fruit Research Database. Online. http://ucanr.edu/sites/ctfa/category/Armillaria_Root_-_Crown_Rot/

Anees, M., Tronsmo, A., Edel-Hermann, V., Hjeljord, L. G., Héraud, C., and Steinberg, C. 2010. Characterization of field isolates of *Trichoderma* antagonistic against *Rhizoctonia solani*. Fungal Biol. 114:691-701.

Aneja, M., Gianfagna, T. J., and Hebbar, P. K. 2005. *Trichoderma harzianum* produces nonanoic acid, an inhibitor of spore germination and mycelial growth of two cacao pathogens. Physiol. Mol. Plant Pathol. 67:304-307.

Argumento-Delira, R., Alarcón, A., Ferrera-Cerrato, R., Almaraz, J. J., and Peña-Cabrales, J. J. 2012. Tolerance and growth of 11 *Trichoderma* strains to crude oil, naphthalene, phenanthrene and benzo[α]pyrene. J. Environ. Manage. 95:S291-S299. doi: 10.1016/j.jenvman.2010.08.011

Arnold, A. E., Maynard, Z., and Gilbert, G. S. 2001. Fungal endophytes in dicotyledonous neotropical trees: Patterns of abundance and diversity. Mycol. Res. 105:1502-1507.

Askew, D. J., and Laing, M. D. 1993. An adapted selective medium for the quantitative isolation of *Trichoderma* species. Plant Pathol. 42:686-690.

Atanasova, L., Le Crom, S., Gruber, S., Coulpier, F., Seidl-Seiboth, V., Kubicek, C. P., and Druzhinina, I. S. 2013. Comparative transcriptomics reveals different strategies of *Trichoderma* mycoparasitism. BMC Genomics 14:121.

Babalola, O. L., and Glick, B. R. 2012. The use of microbial inoculants in African agriculture: Current practice and future prospects. J. Food Agric. Environ. 10:540-549.

Bae, Y.-S., and Knudsen, G. R. 2000. Cotransformation of *Trichoderma harzianum* with β-glucuronidase and green fluorescent protein genes provides a useful tool for monitoring fungal growth and activity in natural soils. Appl. Environ. Microbiol. 66:810-815.

Bae, H., Sicher, R. C., Kim, S.-H., Strem, M. D., Melnick, R. L., and Bailey, B. A. 2009. The beneficial endophyte *Trichoderma hamatum* isolate DIS 219b promotes growth and delays the onset of the drought response in *Theobroma cacao*. J. Exp. Bot. 60:3279-3295.

Bae, H., Roberts, D. P., Lim, H.-S., Strem, M. D., Park, S.-C., Ryu, C.-M., Melnick, R. L., and Bailey, B. A. 2011. Endophytic *Trichoderma* isolates from tropical environments delay disease onset and induce resistance against *Phytophthora capsici* in hot pepper using multiple mechanisms. Mol. Plant-Microbe Interact. 24:336-351.

Bahkali, A. H., Adb-Elsalam, K. A., Guo, J.-R., Khiyami, M. A., and Verreet, J.-A. 2012. Microsatellite primers suitable for genotyping various plant pathogenic fungi with special emphasis on fusaria and *Mycosphaerella graminicola*. Int. J. Mol. Sci. 13:2951-2964.

Bailey, B. A., and Lumsden, R. D. 1998. Direct effects of *Trichoderma* and *Gliocladium* on plant growth and resistance to pathogens. Pages 185-204 in: *Trichoderma & Gliocladium*, Vol. 2. Enzymes, Biological Control and Commercial Applications. G. E. Harman and C. P. Kubicek, eds. Taylor & Francis, London.

Bailey, B. A., Bae, H., Strem, M. D., Roberts, D. P., Thomas, S. E., Crozier, J., Samuels, G. J., Choi, I.-Y., and Holmes, K. A. 2006. Fungal and plant gene expression during colonization of cacao seedlings by endophytic isolates of four *Trichoderma* species. Planta 224:1449-1464.

Bailey, B. A., Bae, H., Strem, M. D., Crozier, J., Thomas, S. E., Samuels, G. J., Vinyard, B. T., and Holmes, K. A. 2008. Antibiosis, mycoparasitism, and colonization success for endophytic *Trichoderma* isolates with biological control potential in *Theobroma cacao*. Biol. Control 46:24-35.

Baker, R. 1991. Diversity in biological control. Crop Prot. 10:85-94.

Baker, K. F., and Cook, R. J. 1982. Biological Control of Plant Pathogens. American Phytopathological Society, St. Paul, MN.

Baker, R., Elad, Y., and Chet, I. 1984. The controlled experiment in the scientific method with special emphasis on biological control. Phytopathology 74:1019-1021.

Barrasa, J. M., Martínez, A. T., and Moreno, G. 1985. *Aphysiostroma*, a new nonostiolate hypocrealean fungus. Can. J. Bot. 63:2439-2443.

Bateman, R. 2004. Constraints and enabling technologies for mycopesticide development. Outlooks Pest Manage. 15:64-69. doi: 10.1564/15apl07

Bateman, R. 2014. Application of biopesticides. Pages 411-427 in: Pesticide Application Methods, 4th ed. G. A. Matthews, P. C. H. Miller, and R. P. Bateman, eds. John Wiley & Sons, West Sussex, United Kingdom.

Bateman, R. P., Hidalgo, E., García, J., Arroyo, C., ten Hoopen, G. M., Adonijah, V., and Krauss, U. 2005. Application of chemical and biological agents for the management of frosty pod rot (*Moniliophthora roreri*) in Costa Rican cocoa (*Theobroma cacao*). Ann. Appl. Biol. 147:129-138.

Beezhold, D. H., Green, B. J., Blachere, F. M., Schmechel, D., Weissman, D. N., Veickoff, D., Hogan, M. B., and Wilson, N. W. 2008. Prevalence of allergic sensitization to indoor fungi in West Virginia. Allergy Asthma Proc. 29:29-34.

Benhamou, N., and Chet, I. 1993. Hyphal interactions between *Trichoderma harzianum* and *Rhizoctonia solani*: Ultrastructure and gold cytochemistry of the mycoparasitic process. Phytopathology 83:1062-1071.

Bennett, A. J., Mead, A., and Whipps, J. M. 2009. Performance of carrot and onion seed primed with beneficial microorganisms in glasshouse and field trials. Biol. Control 51:417-426.

Berg, G. 2009. Plant–microbe interactions promoting plant growth and health: Perspectives for controlled use of microorganisms in agriculture. Appl. Microbiol. Biotechnol. 84:11-18. doi: 10.1007/s00253-009-2092-7

Bernal-Vicente, A., Ros, M., and Pascual, J. A. 2012. Inoculation of *Trichoderma harzianum* during maturation of vineyard waste compost to control muskmelon *Fusarium* wilt. Bioresources 7:1948-1960.

Bernard, E., Larkin, R. P., Tavantzis, S., Erich, M. S., Alyokhin, A., Sewell, G., Lannan, A., and Gross, S. D. 2012. Compost, rapeseed rotation, and biocontrol agents significantly impact soil microbial communities in organic and conventional potato production systems. Appl. Soil Ecol. 52:29-41.

Berrocal-Tito, G., Sametz-Baron, L., Eichenberg, K., Horwitz, B. A., and Herrera-Estrella, A. 1999. Rapid blue light regulation of a *Trichoderma harzianum* phytolase gene. J. Biol. Chem. 274:14288-14294.

Besoain, X. A., Pérez, L. M., Araya, A., Lefever, L., Sanguinetti, M., and Montealegre, J. R. 2007. New strains obtained after UV treatment and protoplast fusion of native *Trichoderma harzianum*: Their biocontrol activity on *Pyrenochaeta lycopersici*. Electron. J. Biotechnol. Vol. 10, No. 4, October 15, 2007.

Bezerra, J. L., Costa, J. C., Bastos, C. N., and Faleiro, F. G. 2003. *Hypocrea stromatica* sp. nov. teleomorfo de *Trichoderma stromaticum*. Fitopatol. Bras. 28:408-412.

Bhagat, S., and Pan, S. 2007. Mass multiplication of *Trichoderma harzianum* on agricultural byproducts and their evaluation against seedling blight (*Rhizoctonia solani*) of mungbean and collar rot (*Sclerotium rolfsii*) of groundnut. Indian J. Agric. Sci. 77:583-588.

Bisby, G. R. 1939. *Trichoderma viride* Pers. ex Fries, and notes on *Hypocrea*. Trans. Br. Mycol. Soc. 23:149-168.

Bissett, J. 1984. A revision of the genus *Trichoderma*. I. Section *Longibrachiatum*, sect. nov. Can. J. Bot. 62:924-931.

Bissett, J. 1991a. A revision of the genus *Trichoderma*. II. Infrageneric classification. Can. J. Bot. 69:2357-2372.

Bissett, J. 1991b. A revision of the genus *Trichoderma*. III. Section *Pachybasium*. Can. J. Bot. 69:2373-2416.

Bissett, J. 1991c. A revision of the genus *Trichoderma*. IV. Additional notes on section *Longibrachiatum*. Can. J. Bot. 69:2418-2420.

Bissett, J. 1992. *Trichoderma atroviride*. Can. J. Bot. 70:639-641.

Bissett, J., Szakacs, G., Nolan, C. A., Druzhinina, I., Gradinger, C., and Kubicek, C. P. 2003. New species of *Trichoderma* from Asia. Can. J. Bot. 81:570-586.

Bittel, P., and Robatzek, S. 2007. Microbe-associated molecular patterns (MAMPs) probe plant immunity. Curr. Opin. Plant Biol. 10:335-341.

Blaszczyk, L., Popiel, D., Chelkowski, J., Koczyk, G., Samuels, G. J., Sobieralski, K., and Silwulski, M. 2011. Species diversity of *Trichoderma* in Poland. J. Appl. Genet. 52: 233-243.

Boehm, M. J., and Hoitink, H. A. J. 1992. Sustenance of microbial activity in potting mixes and its impact on severity of Pythium root rot of poinsettia. Phytopathology 82: 259-264.

Bolar, J. P., Norelli, J. L., Wong, K.-W., Hayes, C. K., Harman, G. E., and Aldwinckle, H. S. 2000. Expression of endochitinase from *Trichoderma harzianum* in transgenic apple increases resistance to apple scab and reduces vigor. Phytopathology 90:72-77.

Bowen, J. K., Crowhurst, R. N., Templeton, M. D., and Stewart, A. 1996. Molecular markers for a *Trichoderma harzianum* biological control agent: Introduction of the hygromycin B resistance gene and the β-glucuronidase gene by transformation. N. Z. J. Crop Hortic. Sci. 24:219-228.

Brotman, Y., Kapuganti, J. G., and Viterbo, A. 2010. *Trichoderma*. Curr. Biol. 20(9):R390-R391.

Brožová, J. 2004. Mycoparasitic fungi *Trichoderma* spp. in plant protection. Plant Prot. Sci. 40:63-74.

Buée, M., Reich, M., Murat, C., Morin, E., Nilsson, R. H., Uroz, S., and Martin, F. 2009. 454 Pyrosequencing analyses of forest soils reveal an unexpectedly high fungal diversity. New Phytol. 184:449-456.

Caballero, M. L., Gómez, M., González-Muñoz, M., Reinoso, L., Rodríguez-Pérez, R., Alday, E. A., and Moneo, I. 2007. Occupational sensitization to fungal enzymes used in animal feed industry. Int. Arch. Allergy Immunol. 144:231-239.

Canham, S. C. 1969. Taxonomy and morphology of *Hypocrea citrina*. Mycologia 51:315-331.

Carbone, I., and Kohn, L. M. 1999. A method for designing primer sets for speciation studies in filamentous Ascomycetes. Mycologia 91:553-556.

Carey, S. T., and Rogerson, C. T. 1976. Taxonomy and morphology of a new species of *Hypocrea* on *Marasmius*. Brittonia 28:381-389.

Carroll, G. C., and Carroll, F. E. 1978. Studies on the incidence of coniferous needle endophytes in the Pacific Northwest. Can. J. Bot. 56:3034-3043.

Carsiolo, C., Gutierrez, A., Jiminez, B., Van Mantagu, M., and Herrera-Estrella, A. 1994. Characterization of *ech*-42a, a *Trichoderma harzianum* endochitinase gene expressed during mycoparasitism. Int. Arch. Allergy Immunol. 91: 10903-10907.

Carsiolo, C., Benhamou, N., Haran, S., Cortes, C., Cutierrez, A., Chet, I., and Herrera-Estrella, A. 1999. Role of the *Trichoderma harzianum* endochitinase gene, *ech*-42a, in mycoparasitism. Appl. Environ. Microbiol. 65:929-935.

Chang, Y.-C., Chang, Y.-C., Baker, R., Kleifeld, O., and Chet, I. 1986. Increased growth of plants in the presence of the biological control agent *Trichoderma harzianum*. Plant Dis. 70:145-148.

Chaverri, P., and Gazis, R. 2010. Diversity of fungal endophytes in leaves and stems of wild rubber trees (*Hevea brasiliensis*) in Peru. Fungal Ecol. 3:240-254.

Chaverri, P., and Samuels, G. J. 2003. *Hypocrea/Trichoderma* (*Ascomycota*, *Hypocreales*, *Hypocreaceae*): Species with green ascospores. Stud. Mycol. 48:1-116.

Chaverri, P., and Samuels, G. J. 2013. Evolution of host affiliation and substrate preference in the cosmopolitan fungal genus *Trichoderma* with evidence of interkingdom host jumps. Evolution 67:2823-2837.

Chaverri, P., Samuels, G. J., and Stewart, E. L. 2001. *Hypocrea virens* sp. nov., the teleomorph of *Trichoderma virens*. Mycologia 93:1113-1124.

Chaverri, P., Castlebury, L. A., Overton, B. E., and Samuels, G. J. 2003a. *Hypocrea/Trichoderma*: Species with conidiophore elongations and green conidia. Mycologia 95: 1100-1140.

Chaverri, P., Castlebury, L. A., Samuels, G. J., and Geiser, D. M. 2003b. Multilocus phylogenetic structure of *Trichoderma harzianum/Hypocrea lixii* complex. Mol. Phylogenet. Evol. 27:302-313.

Chaverri, P., Gazis, R., and Samuels, G. J. 2011. *Trichoderma amazonicum*, a new endophytic species on *Hevea brasiliensis* and *H. guianensis* from the Amazon basin. Mycologia 103:139-151.

Chaverri, P., Branco-Rocha, F., Jaklitsch, W., Degenkolb, T., and Samuels, G. J. 2015. Systematics of the *Trichoderma harzianum* species complex including the identification of commercial biocontrol strains. Mycologia 107:558-559. doi: 10.3852/14-147

Chaves, N. P., Pocasangre, L. E., Elango, F., Rosales, F. E., and Sikora, R. 2009. Combining endophytic fungi and bacteria for the biocontrol of *Rhodopholus similis* (Cobb) Thorne and for effects on plant growth. Sci. Hortic. (Amsterdam, Neth.) 122:472-478.

Chet, I. 1987. *Trichoderma*—Application, mode of action, and potential as a biocontrol agent of soil-borne plant pathogenic fungi. Pages 137-160 in: Innovative Approaches to Plant Disease Control. I. Chet. ed. John Wiley & Sons, New York.

Chet, I., and Baker, R. 1980. Induction of suppressiveness to *Rhizoctonia solani* in soil. Phytopathology 70:994-998.

Chet, I., Benhamou, N., and Haran, S. 1998. Mycoparasitism and lytic enzymes. Pages 153-172 in: *Trichoderma & Gliocladium*, Vol. 2. Enzymes, Biological Control and Commercial Applications. G. E. Harman and C. P. Kubicek, eds. Taylor & Francis, London.

Chutrakul, C., Alcocer, M., Bailey, K., and Peberdy, J. E. 2008. The production and characterization of trichotoxin peptaibols, by *Trichoderma asperellum*. Chem. Biodiversity 5:1694-1706.

Contreras-Cornejo, H. A., Macías-Rodríguez, L., Cortés-Penagos, C., and López-Bucio, J. 2009. *Trichoderma virens*, a plant beneficial fungus, enhances root biomass production and promotes lateral root growth through an auxin-dependent mechanism in *Arabidopsis*. Plant Physiol. 149:1579-1592.

Cook, R. J., and Baker, K. F. 1983. The Nature and Practice of Biological Control of Plant Pathogens. American Phytopathological Society, St. Paul, MN.

Cordier, C., Edel-Hermann, V., Martin-Laurent, F., Blal, B., Steinberg, C., and Alabouvette, C. 2007. SCAR-based real time PCR to identify a biocontrol strain (T1) of *Trichoderma atroviride* and study its population dynamics in soils. J. Microbiol. Methods 68:60-68.

Cotxarrera, L., Trillas-Gay, M. I., Steinberg, C., and Alabouvette, C. 2002. Use of sewage sludge compost and *Trichoderma asperellum* isolates to suppress Fusarium wilt of tomato. Soil Biol. Biochem. 34:467-476.

Craft, C. M., and Nelson, E. B. 1996. Microbial properties of composts that suppress damping-off and root rot of creeping bentgrass caused by *Pythium graminicola*. Appl. Environ. Microbiol. 62:1550-1557.

Cumagun, C. J. R. 2012. Managing plant diseases and promoting sustainability and productivity with *Trichoderma*: The Philippine experience. J. Agric. Sci. Technol. (Tehran, Islamic Repub. Iran) 14:699-714.

Cutler, H. G., Crumley, F. G., and Cole, P. D. 1986. 6-pentyl-α-pyrone from *Trichoderma harzianum*: Its plant growth inhibitory and antimicrobial properties. Agric. Biol. Chem. 50:2943-2945.

Davet, P. 1979. A technique for analyzing soil populations of *Trichoderma* spp. and *Gliocladium virens*. Annu. Rev. Phytopathol. 11:529-534.

De Ceuster, T. J. J., and Hoitink, H. A. J. 1999. Prospects for composts and biocontrol agents as substitutes for methyl bromide in biological control of plant diseases. Compost Sci. Util. 7:6-15.

Degenkolb, T., Gräfenhan, T., Berg, A., Nirenberg, H. I., Gams, W., and Brückner, H. 2006a. Peptaibiomics: Screening for polypeptide antibiotics (peptaibiotics) from plant-protective *Trichoderma* species. Chem. Biodiversity 3:593-610.

Degenkolb, T., Gräfenhan, T., Nirenberg, H. I., Gams, W., and Brückner, H. 2006b. *Trichoderma brevicompactum* complex: Rich source of novel and recurrent plant-protective polypeptide antibiotics (peptaibiotics). J. Agric. Food Chem. 54:7047-7061.

Degenkolb, T., Dieckmann, R., Nielsen, K. F., Gräfenhan, T., Theis, C., Zafari, D., Chaverri, P., Ismaiel, A., Brückner, H., von Döhren, H., Thrane, U., Petrini, O., and Samuels, G. J. 2008a. The *Trichoderma brevicompactum* clade: A separate lineage with new species, new peptaibiotics, and mycotoxins. Mycol. Prog. 7:177-209.

Degenkolb, T., von Döhren, H., Nielsen, K. F., Samuels, G. J., and Brückner, H. 2008b. Recent advances and future prospects in peptaibiotics, hydrophobins, and mycotoxin research, and their importance for chemotaxonomy of *Trichoderma* and *Hypocrea*. Chem. Biodiversity 5:671-680.

de Melo, I. S., and Faull, J. L. 2004. Scanning electron microscopy of conidia of *Trichoderma stromaticum*, a biocontrol agent of witches' broom disease of cocoa. Braz. J. Microbiol. 35:330-332.

Dennis, C., and Webster, J. 1970. Antagonistic properties of species-groups of *Trichoderma*: II. Production of volatile antibiotics. Trans. Br. Mycol. Soc. 57:41-48.

De Respinis, S., Vogel, G., Benagli, C., Tonolla, M., Petrini, O., and Samuels, G. J. 2010. MALDI-TOF MS of *Trichoderma*: A model system for the identification of microfungi. Mycol. Prog. 9:79-100.

De Souza, J. T., Bailey, B. A., Pomella, A. W. V., Erbe, E. F., Murphy, C. A., Bae, H., and Hebbar, P. K. 2008. Colonization of cacao seedlings by *Trichoderma stromaticum*, a mycoparasite of the witches' broom pathogen, and its influence on plant growth and resistance. Biol. Control 46:36-45.

Dettman, J., Jacobson, D. J., and Taylor, J. W. 2003. A multilocus genealogical approach to the phylogenetic species recognition in the model eukaryote *Neurospora*. Evolution 57:2703-2720.

Diener, S. E., Chellappan, M. K., Mitchell, T. K., Dunn-Coleman, N., Ward, M., and Dean, R. A. 2004. Insight into *Trichoderma reesei*'s genome content, organization and evolution revealed through BAC library characterization. Fungal Genet. Biol. 41:1077-1087.

Dingley, J. M. 1957. Life history studies in the genus *Hypocrea*. Trans. R. Soc. N. Z. 84:689-693.

Di Pietro, A., Lorito, M., Hayes, C. K., Broadway, R. M., and Harman, G. E. 1993. Endochitinase from *Gliocladium virens*: Isolation, characterization, and synergistic antifungal activity in combination with gliotoxin. Phytopathology 83:308-313.

Dodd, S. L., Lieckfeldt, E., and Samuels, G. J. 2003. *Hypocrea atroviridis* sp nov. the teleomorph of *Trichoderma atroviride*. Mycologia 95:27-40.

Dodd, S. L., Hill, R. A., and Stewart, A. 2004. A duplex-PCR bioassay to detect a *Trichoderma virens* isolate in nonsterile soil. Soil Biol. Biochem. 36:1955-1965.

Doi, Y. 1969. Revision of the Hypocreales with cultural observations IV. The genus *Hypocrea* and its allies in Japan (1) General part. Bull. Natl. Sci. Mus. Tokyo 12:693-724.

Doi, Y. 1972. A revision of the Hypocreales with cultural observations IV. The genus *Hypocrea* and its allies in Japan (2) Enumeration of the species. Bull. Natl. Sci. Mus. Tokyo 15:649-751.

Doi, Y. 1975. Revision of the Hypocreales with cultural observations VII. The genus *Hypocrea* and its allied genera in South America (1). Bull. Natl. Sci. Mus. Ser. B (Bot.) 1:1-33.

Doi, Y. 1976. Revision of the Hypocreales with cultural observations IX. The genus *Hypocrea* and its allied genera in South America (2). Bull. Natl. Sci. Mus. Ser. B (Bot.) 2:119-131.

Domsch, K. H., Gams, W., and Anderson, T.-H. 2007. Compendium of Soil Fungi, 2nd ed. IHW-Verlag, Eching, Germany.

Douglas, J. S. 1985. Advanced Guide to Hydroponics (Soilless Cultivation). Pelham Books, London.

Douthwaite, B., Langewald, J., and Harris, J. 2001. Development and commercialization of the Green Muscle biopesticide. International Institute of Tropical Agriculture, Ibadan, Nigeria. www.fao.org/docs/eims/upload/166444/Douthwaite,Lungeward,Harris.pdf

Drepper, W. J., and Renfro, B. L. 1990. Comparison of methods of inoculation of ears and stalks of maize with *Fusarium moniliforme*. Plant Dis. 74:952-956.

Druzhinina, I. S., Chaverri, P., Fallah, P., Kubicek, C. P., and Samuels, G. J. 2004. *Hypocrea flaviconidia*, a new species from Costa Rica with yellow conidia. Stud. Mycol. 50:401-407.

Druzhininia, I. S., Kopchinskiy, A., and Kubicek, C. P. 2006. The first 100 *Trichoderma* species characterized by molecular data. Mycoscience 47:55-64.

Druzhinina, I. S., Komoń-Zelazowska, M., Kredics, L., Hatvani, L., Antal, Z., Belayneh, T., and Kubicek, C. P. 2008. Alternative reproductive strategies of *Hypocrea orientalis* and genetically close but clonal *Trichoderma longibrachiatum*, both capable of causing invasive mycoses of humans. Microbiology 154:3447-3459.

Druzhinina, I. S., Komoń-Zelazowska, M., Atanasova, L., Seidl, V., and Kubicek, C. P. 2010a. Evolution and ecophysiology of the industrial producer *Hypocrea jecorina* (anamorph *Trichoderma reesei*) and a new sympatric agamospecies related to it. PLoS ONE 5(2):1-15.

Druzhinina, I. S., Kubicek, C. P., Komoń-Zelazowska, M., Belayneh Mulaw, T., and Bissett, J. 2010b. The *Trichoderma harzianum* demon: Complex speciation history resulting in coexistence of hypothetical biological species, recent agamospecies and numerous relict lineages. BMC Evol. Biol. 10:94. doi: 10.1186/1471-2148-10-94

Druzhinina, I. S., Seidl-Seiboth, V., Herrera-Estrella, A., Howritz, B. A., Kenerley, C. M., Monte, E., Mukherjee, P. K., Zeilinger, S., Grigoriev, I. V., and Kubicek, C. P. 2011. *Trichoderma*: The genomics of opportunistic success. Nat. Rev. Microbiol. 9:749-759.

Druzhinina, I. S., Komoń-Zelazowska, M., Ismaiel, A., Jaklitsch, W., Mulaw, T., Samuels, G. J., and Kubicek, C. P. 2012a. Founder effect and asexual speciation are the major mechanisms in the evolution of the Longibrachiatum clade of *Trichoderma*. Fungal Genet. Biol. 49:358-368.

Druzhinina, I. S., Shelest, E., and Kubicek, C. P. 2012b. Novel traits of *Trichoderma* predicted through analysis of its secretome. FEMS Microbiol. Lett. 337:1-9.

Duclohier, H. 2007. Peptaibiotics and peptaibols: An alternative to classical antibiotics? Chem. Biodiversity 4:1023-1026.

Dubey, S. C., and Singh, B. 2013. Integrated management of major diseases of mungbean by seed treatment and foliar application of insecticide, fungicides and bioagent. Crop Prot. 47:55-60.

Dubois, T., Coyne, D., Kahangi, E., Turoop, L., and Nsubuga, E. W. N. 2006. Endophyte-enhanced banana tissue culture: Technology transfer through public-private partnerships in Kenya and Uganda. Afr. Technol. Dev. Forum J. 3:18-24.

Elad, Y. 2000a. *Trichoderma harzianum* T39 preparation for biocontrol of plant diseases—Control of *Botrytis cinerea*, *Sclerotinia sclerotiorum* and *Cladosporium fulvum*. Biocontrol Sci. Technol. 10:499-507.

Elad, Y. 2000b. Biological control of foliar pathogens by means of *Trichoderma harzianum* and potential modes of action. Crop Prot. 19:709-714.

Elad, Y., and Chet, I. 1983. Improved selective medium for isolation of *Trichoderma* spp. or *Fusarium* spp. from soil. Phytoparasitica 11:55-58.

Elad, Y., Kalfon, A., and Chet, I. 1982. Control of *Rhizoctonia solani* in cotton by seed-coating with *Trichoderma* spp. spores. Plant Soil 66:279-281.

Evans, H. C. 2008. The endophyte-enemy release hypothesis: Implications for classical biological control and plant invasions. Pages 20-25 in: Proceedings of the XII International Symposium on Biological Control of Weeds. M. H. Julien, R. Sforza, M. C. Bon, H. C. Evans, P. E. Hatcher, H. L. Hinz, and B. G. Rector, eds. CABI, Egham, United Kingdom.

Evans, H. C., Holmes, K. A., and Thomas, S. E. 2003. Endophytes and mycoparasites associated with an indigenous forest tree, *Theobroma gileri*, in Ecuador and a preliminary assessment of their potential as biocontrol agents of cocoa diseases. Mycol. Prog. 2:149-160.

Ezzi, M. I., and Lynch, J. M. 2005. Plant microcosm studies demonstrating bioremediation of cyanide toxicity by *Trichoderma* and *Fusarium* spp. Biol. Fertil. Soils 42:40-44.

Feng, X. M., Johnsson Holmberg, A.-I., Sundh, I., Ricard, T., and Melin, P. 2011. Specific SCAR markers and multiplex real-time PCR for quantification of two *Trichoderma* biocontrol strains in environmental samples. Biol. Control 56:903-913.

Foley, M. F., and Deacon, J. W. 1985. Isolation of *Pythium oligandrum* and other necrotrophic mycoparasites from soil. Trans. Br. Mycol. Soc. 85:631-639.

Fravel, D. R. 2005. Commercialization and implementation of biocontrol. Annu. Rev. Phytopathol. 43:337-359.

Fravel, D. R., Marois, J. J., Lumsden, R. D., and Connick, W. J., Jr. 1985. Encapsulation of potential biocontrol agents in an alginate-clay matrix. Phytopathology 75:774-777.

Fravel, D. R., Rhodes, D. J., and Larkin, R. P. 1999. Production and commercialization of biocontrol products. Pages 365-376 in: Integrated Pest and Disease Management in Greenhouse Crops. R. Albajes, M. Lodovica Gullino, J. C. van Lenteren, and Y. Elad, eds. Kluwer Academic Publishers, Dordrecht, Netherlands.

Freeman, S., Minz, D., Kolesnik, I., Barbul, O., Zveibil, A., Maymon, M., Nitzani, Y., Kirschner, B., Rav-David, D., Bilu, A., Dag, A., Shafir, S., and Elad, Y. 2004. *Trichoderma* biocontrol of *Colletotrichum acutatum* and *Botrytis cinerea* and survival in strawberry. Eur. J. Plant Pathol. 110:361-370.

Friedl, M. A., and Druzhinina, I. S. 2012. Taxon-specific metagenomics of *Trichoderma* reveals a narrow community of opportunistic species that regulate each other's development. Microbiology 158:69-83.

Frost, C. J., Mescher, M. C., Carlsono, J. E., and De Moraes, C. M. 2008. Plant defense priming against herbivores: Getting ready for a different battle. Plant Physiol. 146:818-824.

Fujii, K., Kuwahara, A., Nakamura, K., and Yamashita, Y. 2011. Development of a simple cultivation method for isolating hitherto uncultured cellulose-producing microbes. Appl. Microbiol. Biotechnol. 91:1183-1192.

Fujita, T., Takaishi, Y., Moritoki, H., Ogawa, T., and Tokimoto, K. 1984. Fungal metabolites. I. Isolation and biological activities of hypelcins A and B (growth inhibitors against *Lentinus edodes*) from *Hypocrea peltata*. Chem. Pharm. Bull. 32:1822-1828.

Funck-Jensen, D., and Lumsden, R. D. 1999. Biological control of soilborne pathogens. Pages 319-332 in: Integrated Pest and Disease Management in Greenhouse Crops. R. Albajes, M. Lodovica Gullino, J. C. van Lenteren, and Y. Elad, eds. Kluwer Academic Publishers, Dordrecht, Netherlands.

Gams, W. 1971. *Cephalosporium*-artige Schimmelpilze (Hyphomycetes). Gustav Fischer, Stuttgart, Germany.

Gams, W., and Bissett, J. 1998. Morphology and identification of *Trichoderma*. Pages 3-34 in: *Trichoderma* & *Gliocladium*, Vol. 1. Basic Biology, Taxonomy and Genetics. C. P. Kubicek and G. E. Harman, eds. Taylor & Francis, London.

Gams, W., and Meyer, W. 1998. What exactly is *Trichoderma harzianum*? Mycologia 90:904-915.

Gazis, R., and Chaverri, P. 2010. Diversity of fungal endophytes in leaves and stems of wild rubber trees (*Hevea brasiliensis*) in Peru. Fungal Ecol. 3:240-254.

Gil, S. V., Pastor, S., and March, G. J. 2009. Quantitative isolation of biocontrol agents *Trichoderma* spp., *Gliocladium* spp. and actinomycetes from soil with culture media. Microbiol. Res. 164:196-205.

Gräfenhan, T., Schroers, H. J., Nirenberg, H. I., and Seifert, K. A. 2011. An overview of the taxonomy, phylogeny, and typification of nectriaceous fungi in *Cosmospora*, *Acremonium*, *Fusarium*, *Stilbella*, and *Volutella*. Stud. Mycol. 68:79-113.

Gravel, V., Martinez, C., Antoun, H., and Tweddell, R. J. 2006. Control of greenhouse tomato root rot [*Pythium ultimum*] in hydroponic systems, using plant-growth promoting microorganisms. Can. J. Plant Pathol. 28:475-483.

Grosch, R., Scherwinski, K., Lottmann, J., and Berg, G. 2006. Fungal antagonists of the plant pathogen *Rhizoctonia solani*: Selection, control efficacy and influence on the indigenous microbial community. Mycol. Res. 110:1464-1474.

Ha, T. N. 2010. Using *Trichoderma* species for biological control of plant pathogens in Viet Nam. J. ISSAAS (Int. Soc. Southeast Asian Agric. Sci.) 16:17-21.

Hadar, Y., Harman, G. E., and Taylor, A. G. 1984. Evaluation of *Trichoderma koningii* and *T. harzianum* from New York soils for biological control of seed rot caused by *Pythium* spp. Phytopathology 74:106-110.

Hadibarata, T., and Tachibana, S. 2009. Microbial degradation of *n*-eicosane by filamentous fungi. Pages 323-329 in: Interdiscipliary Studies on Environmental Chemistry—Environmental Research in Asia. Y. Obayashi, T. Isobe, A. Subramanian, S. Suzuki, and S. Tanabe, eds. Terrapub, Tokyo.

Hanada, R. E., De Souza, J. T., Pomella, A. W. V., Hebbar, K. P., Pereira, J. O., Ismaiel, A., and Samuels, G. J. 2008. *Trichoderma martiale* sp. nov., a new endophyte from sapwood of *Theobroma cacao* and a potential agent of biological control. Mycol. Res. 112:1335-1343. doi: 10.1016/j.mycres.2008.06.022

Hanada, R. E., Pomella, A. W. V., Soberanis, W., Loguercio, L. L., and Pereira, J. O. 2009. Biocontrol potential of *Trichoderma martiale* against the black-pod disease (*Phytophthora palmivora*) of cacao. Biol. Control 50:143-149.

Hanada, R. E., Pomella, A. W. V., Costa, H. S., Bezerra, J. L., Loguercio, L. L., and Pereira, J. O. 2010. Endophytic fungal diversity in *Theobroma cacao* (cacao) and *T. grandiflorum* (cupuaçu) trees and their potential for growth promotion and biocontrol of black-pod disease. Fungal Biol. 114:901-910.

Hanlin, R. T. 1965. Morphology of *Hypocrea schweinitzii*. Am. J. Bot. 52:570-579.

Harman, G. E. 2000. Myths and dogmas of biocontrol: Changes in perceptions derived from research on *Trichoderma harzianum* T-22. Plant Dis. 84:377-393.

Harman, G. E. 2011. Multifunctional fungal plant symbionts: New tools to enhance plant growth and productivity. New Phytol. 189:649-652.

Harman, G. E., and Björkman, T. 1998. Potential and existing uses of *Trichoderma* and *Gliocladium* for plant disease control and plant growth enhancement. Pages 229-265 in: *Trichoderma* & *Gliocladium*, Vol. 2. Enzymes, Biological Control and Commercial Applications. G. E. Harman and C. P. Kubicek, eds. Taylor & Francis, London.

Harman, G. E., and Kubicek, C. P., eds. 1998. *Trichoderma* & *Gliocladium*, Vol. 2. Enzymes, Biological Control and Commercial Applications. Taylor & Francis, London.

Harman, G. E., Hayes, C. K., Lorito, M., Broadway, R. M., Di Pietro, A., Peterbauer, C., and Tronsmo, A. 1993. Chitinolytic enzymes of *Trichoderma harzianum*: Purification of chitobiosidase and endochitinase. Phytopathology 83:313-318.

Harman, G. E., Hayes, C. K., and Ondik, K. L. 1998. Asexual genetics in *Trichoderma* and *Gliocladium*: Mechanisms and implications. Pages 243-270 in: *Trichoderma* & *Gliocladium*, Vol. 1. Basic Biology, Taxonomy and Genetics. C. P. Kubicek and G. E. Harman, eds. Taylor & Francis, London.

Harman, G. E., Howell, C. R., Viterbo, A., Chet, I., and Lorito, M. 2004. *Trichoderma* species—Opportunistic, avirulent plant symbionts. Nat. Rev. Microbiol. 2:43-56.

Harman, G. E., Obregón, M. A., Samuels, G. J., and Lorito, M. 2010. Changing models for commercialization and implementation of biocontrol in the developing and developed world. Plant Dis. 94:928-939.

Hatvani, L., Antal, Z., Manczinger, L., Szekeres, A., Druzhinina, I. S., Kubicek, C. P., Nagy, A., Nagy, E., Vágvölgyi, C., and Kredics, L. 2007. Green mold diseases of *Agaricus* and *Pleurotus* spp. are caused by related but phylogenetically different *Trichoderma* species. Phytopathology 97:532-537.

Hayes, C. K., Klemsdal, S., Lorito, M., Di Pietro, R. M., Peterbauer, C. K., Nakas, J. P., Tronsmo, A., and Harman, G. E. 1994. Isolation and sequence of an endochitinase gene from a cDNA library of *Trichoderma harzianum*. Gene 138:143-148.

Hermosa, M. R., Grondona, I., Díaz-Mínguez, J. M., Iturriaga, A., and Monte, E. 2001. Development of a strain-specific SCAR marker for the detection of *Trichoderma atroviride* 11, a biological control agent against soilborne fungal pathogens. Curr. Genet. 38:343-350.

Hermosa, R., Viterbo, A., Chet, I., and Monte, E. 2012. Plant-beneficial effects of *Trichoderma* and its genes. Microbiology 158:17-25.

Herrera, P., Suárez, J. P., and Kottke, I. 2010. Orchids keep the ascomycetes outside: A highly diverse group of ascomycetes colonizing the velamen of epiphytic orchids from a tropical mountain rainforest in southern Ecuador. Mycology 1:262-268.

Hohmann, P., Jones, E. E., Hill, R. A., and Stewart, A. 2011. Understanding *Trichoderma* in the root system of *Pinus radiata*: Associations between rhizosphere colonization and growth promotion for commercially grown seedlings. Fungal Biol. 115:759-767.

Hoitink, H. A. J. 1980. Composted bark, a lightweight growth medium with fungicidal properties. Plant Dis. 64:142-147.

Hoitink, H. A. J., and Boehm, M. J. 1999. Biocontrol within the context of soil microbial communities: A substrate-dependent phenomenon. Annu. Rev. Phytopathol. 37:427-446.

Hoitink, H. A. J., Inbar, Y., and Boehm, M. J. 1991. Status of compost-amended potting mixes naturally suppressive to soilborne diseases of floricultural crops. Plant Dis. 75:869-873.

Hoitink, H. A. J., Stone, A. G., and Han, D. Y. 1997a. Suppression of plant diseases by composts. Hortscience 32:184-187.

Hoitink, H. A. J., Zhang, W., Han, D., and Dick, W. 1997b. Making a compost to suppress plant disease. Biocycle 38:40-42.

Hoitink, H. A. J., Madden, L. V., and Dorrance, A. E. 2006. Systemic resistance induced by *Trichoderma* spp.: Interactions between the host, the pathogen, the biocontrol agent, and soil organic matter quality. Phytopathology 96:186-189.

Holmes, K., Schroers, H.-J., Thomas, S. E., Evans, H. C., and Samuels, G. J. 2004. Taxonomy and biocontrol potential of a new species of *Trichoderma* from the Amazon basin of South America. Mycol. Prog. 3:199-210.

Horst, L. E., Locke, J., Krause, C. R., McMahon, R. W., Madden, L. V., and Hoitink, H. A. J. 2005. Suppression of Botrytis blight of begonia by *Trichoderma hamatum* 382 in peat and compost-amended potting mixes. Plant Dis. 89:1195-1200.

Howell, C. R. 1998. The role of antibiosis in biocontrol. Pages 173-184 in: *Trichoderma* & *Gliocladium*, Vol. 2. Enzymes, Biological Control and Commercial Applications. G. E. Harman and C. P. Kubicek, eds. Taylor & Francis, London.

Howell, C. R. 1999. Selective isolation from soil and separation in vitro of P and Q strains of *Trichoderma virens* with differential media. Mycologia 91:930-934.

Howell, C. R. 2003. Mechanisms employed by *Trichoderma* species in the biological control of plant diseases: The history and evolution of current concepts. Plant Dis. 87:4-10.

Howell, C. R. 2006. Understanding the mechanisms employed by *Trichoderma virens* to effect biological control of cotton diseases. Phytopathology 96:178-180.

Hoyos-Carvajal, L., Orduz, S., and Bissett, J. 2009a. Genetic and metabolic biodiversity of *Trichoderma* from Colombia and adjacent neotropic regions. Fungal Genet. Biol. 46:615-631.

Hoyos-Carvajal, L., Orduz, S., and Bissett, J. 2009b. Growth stimulation in bean (*Phaseolus vulgaris* L.) by *Trichoderma*. Biol. Control 51:409-416.

Huelsenbeck, J. P., and Ronquist, F. 2001. Mr Bayes: Bayesian inference of phylogeny. Bioinformatics 17:754-755.

Hujslová, M., Kubátová, A., Chudíčková, M., and Kolařík, M. 2010. Diversity of fungal communities in saline and acidic soils in the Soos National Reserve, Czech Republic. Mycol. Prog. 9:1-15.

Huong, T. T., Padgham, J. L., and Sikora, R. A. 2009. Biological control of the rice root-knot nematode *Meloidogyne graminicola* on rice, using endophytic rhizosphere fungi. Int. J. Pest Manage. 55:31-36.

Inbar, J., Abramsky, M., Cohen, D., and Chet, I. 1994. Plant growth enhancement and disease control in vegetable seedlings grown under commercial conditions. Eur. J. Plant Pathol. 100:337-346.

Jaklitsch, W. M. 2009. European species of *Hypocrea*. Part I: The green-spored species. Stud. Mycol. 63:1-91.

Jaklitsch, W. M. 2011. European species of *Hypocrea*. Part II: Species with hyaline ascospores. Fungal Diversity 48:1-250.

Jaklitsch, W. M., and Voglmayr, H. 2013. New combinations in *Trichoderma* (*Hypocreaceae*, *Hypocreales*). Mycotaxon 126:143-156.

Jaklitsch, W. M., and Voglmayr, H. 2015. Biodiversity of *Trichoderma* (*Hypocreaceae*) in southern Europe and Macaronesia. Stud. Mycol. 80:1-87. doi: 10.1016/j.simyco.2014.11.001

Jaklitsch, W. M., Samuels, G. J., Dodd, S. L., Lu, B.-S., and Druzhinina, I. S. 2006. *Hypocrea rufa*/*Trichoderma viride*: A reassessment, and description of five closely related species with and without warted conidia. Stud. Mycol. 56: 137-177.

Jaklitsch, W. M., Põldmaa, K., and Samuels, G. J. 2008. Reconsideration of *Protocrea* (Hypocreales, Hypocreaceae). Mycologia 100:962-984.

Jaklitsch, W. M., Samuels, G. J., Ismaiel, A., and Voglmayr, H. 2013. Disentanglement of the *Trichoderma viridescens* complex. Persoonia 31:112-146.

Jaklitsch, W. M., Lechat, C., and Voglmayr, H. 2014. The rise and fall of *Sarawakus* (Ascomycetes, Hypocreales). Mycologia 106:133-144.

Jeger, M. J., Jeffries, P., Elad, Y., and Xu, X.-M. 2009. A generic theoretical model for biological control of foliar plant parasites. J. Theor. Biol. 256:201-214.

Jensen, M. H. 2002. Deep flow hydroponics—Past, present, and future. Proc. Natl. Agric. Plastics Congr. 30:40-53.

Jin, X., and Custis, D. 2011. Microencapsulating aerial conidia of *Trichoderma harzianum* through spray drying at elevated temperatures. Biol. Control 56:202-209.

Jin, X., Harman, G. E., and Taylor, A. G. 1991. Conidial biomass and desiccation tolerance of *Trichoderma harzianum* produced at different medium water potentials. Biol. Control 1:237-243.

Jin, X., Taylor, A. G., and Harman, G. E. 1996. Development of media and automated liquid fermentation methods to produce desiccation-tolerant propagules of *Trichoderma harzianum*. Biol. Control 7:267-274.

John, R. P., Tyagi, R. D., Prévost, D., Brar, S. K., Pouleur, S., and Surampalli, R. Y. 2010. Mycoparasitic *Trichoderma viride* as a biocontrol agent against *Fusarium oxysporum* f. sp. *adzuki* and *Pythium arrhenomanes* and as a growth promoter of soybean. Crop. Prot. 29:1452-1459.

Jumpponen, A., and Jones, K. L. 2009. Massively parallel 454 sequencing indicates hyperdiverse fungal communities in temperate *Quercus macrocarpa* phyllosphere. New Phytol. 184:438-448.

Kaewchai, S., Soytong, K., and Hyde, K. D. 2009. Mycofungicides and fungal biofertilizers. Fungal Diversity 38:25-50.

Kantarcioglu, A. S., Celkan, T., Yucel, A., Mikami, Y., Kurugoglu, S., Mitani, H., and Altas, K. 2009. Fatal *Trichoderma harzianum* infection in a leukemic pediatric patient. Med. Mycol. 47:207-215.

Kavoo-Mwangi, A. M., Kahangi, E. M., Ateka, E., Onuso, J., Mukhongo, R. W., Mwangi, E. K., and Jefwa, J. M. 2013. Growth effects of microorganisms based commercial products inoculated to tissue cultured banana cultivated in three different soils in Kenya. Appl. Soil Ecol. 64:152-162.

Khan, J., Ooka, J. J., Miller, S. A., Madden, L. V., and Hoitink, H. A. J. 2004. Systemic resistance induced by *Trichoderma hamatum* 382 in cucumber against Phytophthora crown rot and leaf blight. Plant Dis. 88:280-286.

Khan, M. R., Majid, S., Mohidin, F. A., and Khan, N. 2011. A new bioprocess to produce low cost powder formulations of biocontrol bacteria and fungi to control fusarial wilt and root-knot nematode of pulses. Biol. Control 59:130-140.

Kleifeld, O., and Chet, I. 1992. *Trichoderma*-plant interaction and its effect on increased growth response. Plant Soil 144:267-272.

Klein, D., and Eveleigh, D. E. 1998. Ecology of *Trichoderma*. Pages 57-74 in: *Trichoderma & Gliocladium*, Vol. 1. Basic Biology, Taxonomy and Genetics. C. P. Kubicek and G. E. Harman, eds. Taylor & Francis, London.

Knudsen, G. R., and Bin, L. 1990. Effects of temperature, soil moisture, and wheat bran on growth of *Trichoderma harzianum* from alginate pellets. Phytopathology 80:724-727.

Komatsu, M. 1976. Studies on *Hypocrea*, *Trichoderma* and allied fungi antagonistic to shiitake, *Lentinus edodes* (Berk.) Sing. Rep. Tottori Mycol. Inst. 13:1-113.

Komoń-Zelazowska, M., Bissett, J., Zafari, D., Hatvani, L., Manczinger, L., Woo, S., Lorito, M., Kredics, L., Kubicek, C. P., and Druzhinina, I. S. 2007. Genetically closely related but phenotypically divergent *Trichoderma* species cause green mold disease in oyster mushroom farms worldwide. Appl. Environ. Microbiol. 73:7415-7426.

Kraus, G. F., Druzhinina, I., Gams, W., Bissett, J., Zafari, D., Szakacs, G., Koptchinski, A., Prillinger, H., Zare, R., and Kubicek, C. P. 2004. *Trichoderma brevicompactum* sp. nov. Mycologia 96:1059-1073.

Kraus, C., Kirschbaum, J., and Brückner, H. 2006. Peptaibiomics: An advanced, rapid and selective analysis of peptaibiotics/peptaibiols by SPE/L-ES-MS. Amino Acids (Vienna) 30:435-443.

Krauss, U., and Soberanis, W. 2001. Biocontrol of cocoa pod diseases with mycoparasite mixtures. Biol. Control 22:149-158.

Krauss, U., Bidwell, R., and Ince, J. 1998. Isolation and preliminary evaluation of mycoparasites as biocontrol agents against crown rot of banana. Biol. Control 13:111-119.

Krauss, U., Matthews, P., Bidwell, R., Hocart, M., and Anthony, F. 2001. Strain discrimination by fungal antagonists of *Colletotrichum musae*: Implications for biocontrol of crown rot of banana. Mycol. Res. 105:67-76.

Krauss, U., Hidalgo, E., Bateman, R., Adonijah, V., Arroyo, C., García, J., Crozier, J., Brown, N. A., ten Hoopen, G. M., and Holmes, K. A. 2010. Improving the formulation and timing of application of endophytic biocontrol and chemical agents against frosty pod rot (*Moniliophthora roreri*) in cocoa (*Theobroma cacao*). Biol. Control 54:230-240.

Kredics, L., Anntal, Z., Dóczi, I., Manczinger, L., Kevei, F., and Nagy, E. 2003. Clinical importance of the genus *Trichoderma*. Acta Microbiol. Immunol. Hung. 50:105-117.

Kubicek, C. P., and Harman, G. E., eds. 1998. *Trichoderma & Gliocladium*, Vol. 1. Basic Biology, Taxonomy and Genetics. Taylor and Francis, London.

Kubicek, C. P., Bissett, J., Druzhinina, I. S., Kullnig-Gradinger, C., and Szakacs, G. 2003. Genetic and metabolic diversity of *Trichoderma*: A case study on South-East Asian isolates. Fungal Genet. Biol. 38:310-319.

Kubicek, C. P., Mikus, M., Schuster, A., Schmoll, M., and Seiboth, B. 2009. Metabolic engineering strategies for the improvement of cellulase production by *Hypocrea jecorina*. Biotechnol. Biofuels 2:19. doi: 10.1186/1754-6834-2-19

Kubicek, C. P., Herrera-Estrella, A., Seidl-Seiboth, V., Martinez, D. A., Druzhinina, I., Thon, M., Zeilinger, S., Casas-Flores, S., Horwitz, B. A., Mukherjee, P. K., Mukherjee, M., Kredics, L., Alcaraz, L. D., Aerts, A., Antal, Z., Atanasova, L., Cervantes-Badillo, M. G., Challacombe, J., Chertkov, O., McCluskey, K., Coulpier, F., Deshpande, N., von Döhren, H., Ebbole, D. J., Esquivel-Naranjo, E. U., Fekete, E., Flipphi, M., Glaser, F., Gómez-Rodríguez, E. Y., Gruber, S., Han, C., Henrissat, B., Hermosa, R., Hernández-Oñate, M., Karaffa, L., Kosti, I., Le Crom, S., Lindquist, E., Lucas, S., Lübeck, M., Lübeck, P. S., Margeot, A., Metz, B., Misra, M., Nevalainen, H., Omann, M., Packer, N., Perrone, G., Uresti-Rivera, E. E., Salamov, A., Schmoll, M., Seiboth, B., Shapiro, H., Sukno, S., Tamayo-Ramos, J. A., Tisch, D., Wiest, A., Wilkinson, H. H., Zhang, M., Coutinho, P. M., Kenerley, C. M., Monte, E., Baker, S. E., and Grigoriev, I. V. 2011. Comparative genome sequence analysis underscores mycoparasitism as the ancestral life style of *Trichoderma*. Genome Biol. 12:R40. doi: 10.1186/gb-2011-12-4-r40

Kuhls, K., Lieckfeldt, E., Samuels, G. J., Börner, T., Meyer, W., and Kubicek, C. P. 1997. Revision of *Trichoderma* sect. *Longibrachiatum* including related teleomorphs based on analysis of ribosomal DNA internal transcribed spacer sequences. Mycologia 89:442-460.

Kuhls, K., Lieckfeldt, E., Börner, T., and Guého, E. 1999. Molecular re-identification of human pathogenic *Trichoderma* isolates as *Trichoderma longibrachiatum* and *Trichoderma citrinoviride*. Med. Mycol. 37:25-33.

Kullnig, C. M., Szakacs, G., and Kubicek, C. P. 2000. Molecular identification of *Trichoderma* species from Russia, Siberia and the Himalaya. Mycol. Res. 104:1117-1125.

Kullnig, C. M., Krupica, T., Woo, S. L., Mach, R. L., Rey, M., Lorito, M., and Kubicek, C. P. 2001. Confusion abounds over identities of *Trichoderma* biocontrol isolates. Mycol. Res. 105:770-771.

Kullnig-Gradinger, C. M., Szakacs, G., and Kubicek, C. P. 2002. Phylogeny and evolution of the genus *Trichoderma*: A multigene approach. Mycol. Res. 106:757-767.

Kuter, G. A., Nelson, E. B., Hoitink, H. A. J., and Madden, L. V. 1983. Fungal populations in container media amended with composted hardwood bark suppressive and conducive to Rhizoctonia damping-off. Phytopathology 73:1450-1456.

Lantz, S. E., Goedegebuur, F., Hommes, R., Kaper, T., Kelemen, B. R., Mitchinson, C., Wallace, L., Stahlberg, J., and Larenas, E. A. 2010. *Hypocrea jecorina* CEL6A protein engineering. Biotechnol. Biofuels 3:20. doi: 10.1186/1754-6834-3-20

Le, H. T. T., Padgham, J. L., and Sikora, R. A. 2009. Biological control of the rice root-knot nematode *Meloidogyne graminicola* on rice, using endophytic and rhizosphere fungi. Int. J. Pest. Manage. 55:31-36.

Leandro, L. F. S., Guzman, T., Ferguson, L. M., Fernandez, G. E., and Louws, F. J. 2007. Population dynamics of *Trichoderma* in fumigated and compost-amended soil and on strawberry roots. Appl. Soil Microbiol. 35:237-246.

Lee, S. K., Soh, H. B., Kim, G. G., and Chung, Y. R. 2006. Enhancement of biological control of *Botrytis cinerea* on cucumber by foliar sprays and bed potting mixes of *Trichoderma harzianum* YC459 and its application on tomato in the greenhouse. Plant Pathol. J. 22:283-288.

Lewis, J. A., and Papavizas, G. C. 1983. Production of chlamydospores and conidia by *Trichoderma* spp. liquid and solid growth media. Soil Biol. Biochem. 15:351-357.

Lewis, J. A., and Papavizas, G. C. 1984. A new approach to stimulate population proliferation of *Trichoderma* species and other potential biocontrol fungi introduced into natural soils. Phytopathology 74:1240-1244.

Lieckfeldt, E., Samuels, G. J., Börner, T., and Gams, W. 1998. *Trichoderma koningii*: Neotypification and *Hypocrea* teleomorph. Can. J. Bot. 76:1507-1522.

Lieckfeldt, E., Samuels, G. J., and Nirenberg. H. I. 1999. A morphological and molecular perspective of *Trichoderma viride*: Is it one or two species? Appl. Environ. Microbiol. 65:2418-2428.

Lieckfeldt, E., Cavignac, Y., Fekete, C., and Börner, T. 2000. Endochitinase gene-based phylogenetic analysis of *Trichoderma*. Microbiol. Res. 155:1-9.

Lieckfeldt, E., Kullnig, M., Kubicek, C. P., Samuels, G. J., and Börner, T. 2001. *Trichoderma aureoviride*: Phylogenetic position and characterization. Mycol. Res. 105:313-322.

Lifshitz, R., Windham, M. T., and Baker, R. 1986. Mechanism of biological control of preemergence damping-off of pea by seed treatment with *Trichoderma* spp. Phytopathology 76:720-725.

Limón, M. C., Chacón, M. R., Mejías, R., Delgado-Jarana, J., Rincón, A. M., Codón, A. C., and Benítez, T. 2004. Increased antifungal and chitinase specific activities of *Trichoderma harzianum* CECT 2413 by addition of a cellulose binding domain. Appl. Microbiol. Biotechnol. 64:675-685.

Liu, Y. J., Whelen, S., and Hall, B. D. 1999. Phylogenetic relationships among Ascomycetes: Evidence from an RNA polymerase II subunit. Mol. Biol. Evol. 16:1799-1808.

Lo, C.-T., Nelson, E. B., and Harman, G. E. 1996. Biological control of turfgrass diseases with a rhizosphere competent strain of *Trichoderma harzianum*. Plant Dis. 80:736-741.

Lo, C.-T., Nelson, E. B., and Harman, G. E. 1997. Improved biocontrol efficacy of *Trichoderma harzianum* 1295-22 for foliar phass of turf diseases by use of spray applications. Plant Dis. 81:1132-1138.

Lo, C.-T., Nelson, E. B., Hayes, C. K., and Harman, G. E. 1998. Ecological studies of transformed *Trichoderma harzianum* strain 1295-22 in the rhizosphere and on the phylloplane of creeping bentgrass. Phytopathology 88:129-136.

Lopes, F. A. C., Steindorff, A. S., Geraldine, A. M., Brandao, R. S., Monteiro, V. N., Lobo, M., Jr., Coelho, A. S. G., Ulhoa, C. J., and Silva, R. N. 2012. Biochemical and metabolic profiles of *Trichoderma* strains isolated from common bean crops in the Brazilian Cerrado, and potential antagonism against *Sclerotinia sclerotiorum*. Fungal Biol. 116:815-824.

López-Quintero, C., Atanasova, L., Franco-Molano, A. E., Gams, W., Komon-Zelazowska, M., Theelen, B., Müller, W. H., Boekhout, T., and Druzhinina, I. 2013. DNA barcoding survey of *Trichoderma* diversity in soil and litter of the Colombian lowland Amazonian rainforest reveals *Trichoderma strigosellum* sp. nov. and other species. Antonie van Leeuwenhoek 104:657-674. doi: 10.1007/s10482-013-9975-4

Lorito, M. 1998. Chitinolytic enzymes and their genes. Pages 73-99 in: *Trichoderma* & *Gliocladium*, Vol. 2. Enzymes, Biological Control and Commercial Applications. G. E. Harman and C. P. Kubicek, eds. Taylor & Francis, London.

Lorito, M., Harman, G. E., Hayes, C. K., Broadway, R. M., Tronsmo, A., Woo, S. L., and Di Pietro, A. 1993. Chitinolytic enzymes produced by *Trichoderma harzianum*: Antifungal activity of purified endochitinase and chitobiosidase. Phytopathology 83:302-307.

Lorito, M., Mach, R. L., Sposato, P., Strauss, J., Peterbauer, C. K., and Kubicek, C. P. 1996a. Mycoparasitic interaction relieves binding of the Cre1 carbon catabolite repressor protein to promoter sequences of the *ech42* (endochitinase-encoding) gene in *Trichoderma harzianum*. Proc. Natl. Acad. Sci. U.S.A. 93:14868-14872.

Lorito, M., Farkas, V., Rebuffat, S., Bodo, B., and Kubicek, C. P. 1996b. Cell wall synthesis is a major target of mycoparasitic antagonism by *Trichoderma harzianum*. J. Bacteriol. 178:6382-6385.

Lorito, M., Woo, S. L., Fernandez, I. G., Colucci, G., Harman, G. E., Pintor-Toro, J. A., Filippone, E., Mucifora, S., Lawrence, C. B., Zoina, A., Tuzun, S., and Scala, F. 1998. Genes from mycoparasitic fungi as a source for improving plant resistance to fungal pathogens. Proc. Natl. Acad. Sci. U.S.A. 95:7860-7865.

Lorito, M., Woo, S. L., Harman, G. E., and Monte, E. 2010. Translational research on *Trichoderma*: From 'omics' to the field. Annu. Rev. Phytopathol. 48:1-19.

Lu, B. S., and Samuels, G. J. 2003. *Hypocrea stilbohypoxyli* and its *Trichoderma koningii*-like anamorph: A new species from Puerto Rico on *Stilbohypoxylon muelleri*. Sydowia 55:255-266.

Lu, B. S., Druzhinina, I. S., Fallah, P., Chaverri, P., Gradinger, C., Kubicek, C. P., and Samuels, G. J. 2004a. *Hypocrea*/*Trichoderma* species with pachybasium-like conidiophores: Teleomorphs for *T. minutisporum* and *T. polysporum*, and their newly discovered relatives. Mycologia 96:310-342.

Lu, Z., Tombolini, R., Woo, S., Zeilinger, S., Lorito, M., and Jansson, J. K. 2004b. In vivo study of *Trichoderma*-pathogen-plant interactions, using constitutive and inducible green fluorescent protein reporter systems. Appl. Environ. Microbiol. 70:3073-3081.

Lubeck, M., Poulsen, S. K., Lubeck, P. S., Jensen, D. F., and Thrane, U. 2000. Identification of *Trichoderma* strains from building materials by ITS1 ribotyping and UP-PCR fingerprinting and UP-PCR cross hybridization. FEMS Microbiol. Lett. 185:129-134.

Lumsden, R. D., and Locke, J. C. 1989. Biological control of damping-off caused by *Pythium ultimum* and *Rhizoctonia solani* with *Gliocladium virens* in soilless mix. Phytopathology 79:361-366.

Lumsden, R. D., Locke, J. C., Adkins, S. T., Walter, J. F., and Ridout, C. J. 1992. Isolation and localization of the antibiotic gliotoxin produced by *Gliocladium virens* from alginate prill in soil and soilless media. Phytopathology 82:230-235.

Lumsden, R. D., Lewis, J. A., and Locke, J. C. 1998. Development of SoilGard: From discovery to implementation. Proc. Int. Congr. Plant Pathol. 7:5.2.6S.

Luttrell, E. S. 1951. Taxonomy of the Pyrenomycetes. Univ. Mo. Stud. 24:1-120.

MacKenzie, A. J., Starman, T. W., and Windham, M. T. 1995. Enhanced root and shoot growth of chrysanthemum cuttings propagated with the fungus *Trichoderma harzianum*. HortScience 30:496-498.

Malmierca, M. G., Cardoza, R. E., Alexander, N. J., McCormic, S. P., Hermosa, R., Monte, E., and Gutiérrez, S. 2012. Involvement of *Trichoderma* trichothecenes in biocontrol activity and induction of plant defense-related genes. Appl. Environ. Microbiol. 78:4856-4868.

Mao, W., Lewis, J. A., Lumsden, R. D., and Hebbar, K. P. 1998. Biocontrol of selected soilborne diseases of tomato and pepper plants. Crop Prot. 17:532-542.

Martínez-Medina, A., Roldán, A., and Pascual, J. A. 2009. Performance of a *Trichoderma harzianum* bentonite-vermiculite formulation against Fusarium wilt in seedling nursery melon plants. HortScience 44:2025-2027.

Mastouri, F., Björkman, T., and Harman, G. E. 2010. Seed treatment with *Trichoderma harzianum* alleviates biotic, abiotic, and physiological stresses in germinating seeds and seedlings. Phytopathology 100:1213-1221.

Mastouri, F., Björkman, T., and Harman, G. E. 2012. *Trichoderma harzianum* enhances antioxidant defense of tomato seedlings and resistance to water defecit. Mol. Plant-Microbe Interact. 9:1264-1271.

Mbarga, J. B., Ten Hoopen, G. M., Kuaté, J., Adiobo, A., Ngonkeu, M. E. L., Ambang, Z., Akoa, A., Tondje, P. R., and Begoude, B. A. D. 2012. *Trichoderma asperellum*: A potential biocontrol agent for *Pythium myriotylum*, causal agent of cocoyam (*Xanthosoma sagittifolium*) root rot disease in Cameroon. Crop Prot. 36:18-22.

Mbarga, J. B., Begoude, B. A. D., Ambang, Z., Meboma, M., Kuate, J., Schiffers, B., Ewbank, W., Dedieu, L., and ten Hoopen, G. M. 2014. A new oil-based formulation of *Trichoderma asperellum* for the biological control of cacao black pod disease caused by *Phytophthora megakarya*. Biol. Control 77:15-22.

Medeiros, F. H. V., Pomella, A. W. V., De Souza, J. T., Niella, G. R., Valle, R., Bateman, R. P., Fravel, D., Vinyard, B., and Hebbar, P. K. 2010. A novel, integrated method for management of witches' broom disease in Cacao in Bahia, Brazil. Crop Prot. 29:704-711.

Mejía, L. C., Rojas, E. I., Maynard, Z., Arnold, A. E., Van Bae, S., Samuels, G. J., Robbins, N., and Herre, E. A. 2008. Endophytic fungi as biocontrol agents of *Theobroma cacao* pathogens. Biol. Control 46:4-14.

Miller, T. L., and Churchill, B. W. 1986. Substrates for large scale fermentation. Pages 122-136 in: Manual of Industrial Microbiology and Biotechnology. A. L. Demain and N. A. Solomon, eds. American Society of Microbiology, Washington, DC.

Ming, Q., Han, T., Li, W., Zhang, Q., Zhang, H., Zheng, C., Huang, F., Rahman, K., and Qin, L. 2012. Tanshinone IIA and tanshinone I production by *Trichoderma atroviride* D16, an endophytic fungus in *Salvia miltiorrhiza*. Phytomedicine 19:330-333.

Miyazaki, K., Tsuchiya, Y., and Okuda, T. 2009. Specific PCR assays for the detection of *Trichoderma harzianum* causing green mold diseases during mushroom cultivation. Mycoscience 50:94-99.

Montealegre, J. R., Herrera, R., Velásquez, J. C., Silva, P., Besoaín, X., and Pérez, L. M. 2005. Biocontrol of root and crown rot in tomatoes under greenhouse conditions using *Trichoderma harzianum* and *Paenibacillus lentimorbus*. Additional effect of solarization. Electron. J Biotechnol. Vol. 8 No. 3. December 15, 2005.

Montealegre, J., Valderrama, L, Sánchez, S., Herrera, R., Besoain, X., and Pérez, L. M. 2010. Biological control of *Rhizoctonia solani* in tomatoes with *Trichoderma harzianum* mutants. Electron. J. Biotechnol. Vol. 13 No. 2. March 15, 2010.

Mulaw, T. B., Kubicek, C. P., and Druzhinina, I. S. 2010. The rhizosphere of *Coffea arabica* in its native highland forests of Ethiopia provides a niche for a distinguished diversity of *Trichoderma*. Diversity 2:527-549.

Muniappan, R., Vaughan, L., Rich, M., and Steed, A. 2009. Integrated Pest Management Collaborative Research Support Program. FY 2009. Annual Report. 2008–2009. USAID Cooperative Agreement No. EPP-A-00-0400016-00.

Naef, A., Senatore, M., and Défato, G. 2005. A microsatellite based method for quantification of fungi in decomposing plant material elucidates the role of *Fusarium graminearum* DON production in the saprophytic competition with *Trichoderma atroviride* in maize tissue microcosms. FEMS Microbiol. Ecol. 55:211-220.

Naeimi, S., Okhovvat, S. M., Javan-Nikkah, M., Vágvölgyi, C., Khohsravi, V., and Kredics, L. 2010. Biological control of *Rhizoctonia solani* AG1-1A, the causal agent of rice sheath blight with *Trichoderma* strains. Phytopathol. Mediterr. 49:287-300.

Naeimi, S., Kocsubé, S., Antal, Z., Okhovvat, S. M., Javan-Nikkhah, M., Vágvölgyi, C., and Kredics, L. 2011. Strain-specific SCAR markers for the detection of *Trichoderma harzianum* AS12-2, a biological control agent against *Rhizoctonia solani*, the causal agent of rice sheath blight. Acta Biol. Hung. 62:73-84.

Nagayma, K., Watanbe, S., Kumakura, K., Ichikawa, T., and Makino, T. 2007. Development and commercialization of *Trichoderma asperellum* SKT-1 (Ecochope®), a microbial pesticide. J. Pestic. Sci. 32:141-142.

Nalimova, M. S. 2007. Introducción y eficacia técnica del biocontrol de fitopatógenos con *Trichoderma* spp. en Cuba. Fitosanidad 11:75-79.

Nelson, E. E., Goldfarb, B., and Thies, W. G. 1987. *Trichoderma* species from fumigated Douglas-fir roots decayed by *Phellinus weirii*. Mycologia 9:370-374.

Nelson, E. B., Harman, G. E., and Nash, G. T. 1988. Enhancement of *Trichoderma*-induced biological control of *Pythium* seed rot and pre-emergence damping-off of peas. Soil Biol. Biochem. 20:145-150.

Neuhof, T., Dieckman, R., Druzhinina, I. S., Kubicek, C. P., and von Döhren, H. 2007. Intact-cell MALDI-TOF mass spectrometry analysis of peptaibol formation by the genus *Trichoderma*/*Hypocrea*: Can molecular phylogeny of species predict peptaibol structures? Microbiology 153:3417-3437.

Nielsen, K. F., Gräfenhan, T., Zafari, D., and Thrane, U. 2005. Trichothecene production by *Trichoderma brevicompactum*. J. Agric. Food. Chem. 53:8190-8196.

Nirenberg, H. I. 1976. Untersuchungen über die morphologische und biologische differenzierung in der *Fusarium*-Sektion *Liseola*. Mitt. Biol. Bundesanst. Land- Forstw., Berlin-Dahlem 169:i-v, 1-117.

Norton, G. W., Heinrichs, E. A., Luther, G. C., and Irwin, M. E., eds. 2005. Globalizing Integrated Pest Management. A Participatory Research Process. Blackwell Publishing, Ames, IA.

Okoth, P., Okoth, S., and Jefwd, J. M. 2013. The conservation and use of micro-organisms and invertebrates in root crop-based systems: State of knowledge, trends and future prospects. Commission on Genetic Resources for Food and Agriculture. Background Study Paper No. 63.

Oudemans, C. A. J. A., and Koning, C. J. 1902. Prodrome d'une flore mycologique obetne par la culture sur gelatin prepare de la terre humeuse du Spanderswoud, près de Bussum. Arch. Neerl. Sci. Exactes Nat. Ser. 2 7:266-298.

Overton, B. E., Stewart, E., Geiser, D. M., and Jaklitsch, W. M. 2006a. Systematics of *Hypocrea citrina* and related taxa. Stud. Mycol. 56:1-38.

Overton, B. E., Stewart, E., and Geiser, D. M. 2006b. Taxonomy and phylogenetic relationships of nine species of *Hypocrea* with anamorphs assignable to *Trichoderma* section *Hypocreanum*. Stud. Mycol. 56:39-65.

Ozbay, N., Newman, S. E., and Brown, W. M. 2004. Evaluation of *Trichoderma harzianum* strains to control crown and root rot of greenhouse fresh market tomatoes. Acta Hortic. 235:79-84.

Pal, K. K., and Gardener, B. M. 2006. Biological control of plant pathogens. The Plant Health Instructor. Online. doi: 10.1094/PHI-A-2006-1117-02

Papavizas, G. C. 1970. Colonization and growth of *Rhizoctonia solani* in soil. Pages 108-122 in: *Rhizoctonia solani*, Biology and Pathology. J. R. Parmeter, Jr., ed. University of California Press, Oakland, CA.

Papavizas, G. C., ed. 1981. Biological control in crop production. BARC Symposium, May 1980. Allenheld, Osmun, Totowa, NJ.

Papavizas, G. C. 1985. *Trichoderma* and *Gliocladium*: Biology, ecology and potential for biocontrol. Annu. Rev. Phytopathol. 23:23-54.

Papavizas, G. C., and Lewis, J. A. 1983. Physiological and biocontrol characteristics of stable mutants of *Trichoderma viride* resistant to MBC fungicides. Phytopathology 73:407-411.

Papavizas, G. C., and Lumsden, R. D. 1982. Improved medium for isolation of *Trichoderma* spp. from soil. Plant Dis. 66:1019-1020.

Papavizas, G. C., Dunn, M. T., Lewis, J. A., and Beagle-Ristaino, J. 1984. Liquid fermentation technology for experimental production of biocontrol fungi. Phytopathology 74:1171-1175.

Park, M. S., Bae, K. S., and Yu, S. H. 2006. Two new species of *Trichoderma* associated with green mold of oyster mushroom cultivation in Korea. Mycobiology 34:111-113.

Paulitz, T. C. 1997. Biological contol of root pathogens in soilless and hydroponic systems. HortScience 32:193-196.

Paulitz, T. C., and Bélanger, R. R. 2001. Biological control in greenhouse systems. Annu. Rev. Phytopathol. 39:103-133.

Paulitz, T. C., Ahmad, J. S., and Baker, R. 1990. Integration of *Pythium nunn* and *Trichoderma harzianum* isolate T-95 for the biological control of Pythium damping-off of cucumber. Plant Soil 121:243-250.

Paz, Z., Komoń-Zelazowska, M., Druzhinina, I. S., Aveskamp, M. M., Schnaiderman, A., Aluma, Y., Carmeli, S., Ilan, M., and Yarden, O. 2010. Diversity and potential antifungal properties of fungi associated with a Mediterranean sponge. Fungal Diversity 42:17-26.

Persoon, C. H. 1794. Neuer Versuch einer systematischen Eintheilung der Schwamme. Neues Mag. Bot. 1:63-80.

Peterbauer, C., Lorito, M., Hayes, C. K., Harman, G. E., and Kubicek, C. P. 1996. Molecular cloning and expression of *nag1* (N-acetyl-β-D-glucosaminadase encoding gene from *Trichoderma harzianum* P1. Curr. Genet. 29:812-820.

Phuwiwat, W., and Soytong, K. 1999. Growth and yield response of Chinese radish to application of *Trichoderma harzianum*. Thammasit Int. J. Sci. Technol. 4:68-71.

Phuwiwat, W., Kaewkong, P., and Soytong, K. 2001. Effect of *Trichoderma harzianum* strain PC01 and planting media on growth and yield of Chinese radish. Thammasaat Int. J. Sci. Technol. 6:1-5.

Pieterse, C. M. J., Leon-Reyes, A., Van der Ent, S., and Van Wees, S. C. M. 2009. Networking by small-molecule hormones in plant immunity. Nat. Chem. Biol. 5:308-316.

Polizzi, V., Adams, A., Picco, A. M., Adriaens, E., Lenoir, J., Van Peteghem, C., De Saeger, S., and De Kimpe, N. 2011. Influence of environmental conditions on production of volatiles by *Trichoderma atroviride* in relation with the sick-building syndrome. Build. Environ. 46:945-954.

Pryor, B. M., Davis, R. M., and Gilbertson, R. L. 2000. A toothpick inoculation method for evaluating carrot cultivars for resistance to *Alternaria radicina*. HortScience 35:1099-1102.

Puglese, M., Liu, B. P., Ludovica Gullino, M., and Garibaldi, A. 2011. Microbial enrichment of compost with biological control agents to enhance suppressiveness to four soilborne diseases in greenhouse. J. Plant Dis. Prot. 118:45-50.

Punja, Z. K., and Utkhede, R. S. 2003. Using fungi and yeasts to manage vegetable crop diseases. Trends Biotechnol. 21:400-407.

Quilliam, R. S., and Jones, D. L. 2012. Evidence for specificity of culturable fungal root endophytes from the carnivorous plant *Pinguicula vulgaris* (common butterwort). Mycol. Prog. 11:583-585.

Raghavendra, A. K. H., and Newcombe, G. 2013. The contribution of foliar endophytes to quantitative resistance to *Melampsora* rust. New Phytol. 197:909-918.

Ramão-Dumaresque, A. S., de Araújo, W. L., Talbot, N. J., and Thornton, C. R. 2012. RNA interference of endochitinases in sugarcane endophyte *Trichoderma virens* 223 reduces its fitness as a biocontrol agent of pineapple disease. PLoS ONE 7(10):e47888, art.

Rawat, L., Singh, Y., Shukla, N., and Kumar, J. 2012. Seed biopriming with salinity tolerant isolates of *Trichoderma harzianum* alleviates salt stress in rice: Growth, physiological and biochemical characteristics. J. Plant Pathol. 94: 353-365.

Regliński, T., Rodenburg, N., Taylor, J. T., Northcott, G. L., Ah Chee, A., Spiers, T. M., and Hill, R. A. 2012. *Trichoderma atroviride* promotes growth and enhances systemic resistance to *Diplodia pinea* in radiata pine (*Pinus radiata*) seedlings. For. Pathol. 42:75-78.

Rehner, S. A., and Samuels, G. J. 1994. Taxonomy and phylogeny of *Gliocladium* analyzed by large subunit rDNA sequences. Mycol. Res. 98:625-634.

Rehner, S. A., and Samuels, G. J. 1995. Molecular systematics of the Hypocreales: A teleomorph gene phylogeny and the status of their anamorphs. Can. J. Bot. 73(Suppl. 1): S816-S823.

Reithner, B., Ibarra-Laclette, E., Mach, R. L., and Hererra-Estrella, A. 2011. Identification of mycoparasitism-related genes in *Trichoderma atroviride*. Appl. Environ. Microbiol. 77:4361-4370.

Rifai, M. A. 1969. A revision of the genus *Trichoderma*. Mycol. Pap. 116:1-56.

Rifai, M. A., and Webster, J. 1966a. Culture studies on *Hypocrea* and *Trichoderma* II. *H. aureoviridis* and *H. rufa* f. *sterilis* f. nov. Trans. Br. Mycol. Soc. 49:289-296.

Rifai, M. A., and Webster, J. 1966b. Culture studies on *Hypocrea* and *Trichoderma* III. *H. H. lactea* (=*H. citrina*) and *H. pulvinata*. Trans. Br. Mycol. Soc. 49:297-310.

Röhrich, C. R., Iversen, A., Jaklitsch, W. M., Voglmayr, H., Berg, A., Dörfelt, H., Thrane, U., Vilcinskas, A., Nielsen, K. F., von Döhren, H., Brückner, H., and Degenkolb, T. 2012. Hypopulvins, novel peptaibiotics from the polyporicoloous fungus *Hypocrea pulvinata*, are produced during infection of its natural hosts. Fungal Biol. 30:1-13.

Rossert, P. M. 1997. Cuba: Ethics, biological control, and crisis. Agric. Human Values 14:291-302.

Rossman, A. Y., Seifert, K. A., Samuels, G. J., Minnis, A. M., Schroers, H.-J., Lombard, L., Crous, P. W., Põldmaa, K., Cannon, P. F., Summerbell, R. C., Geiser, D. M., Zhuang, W.-y., Hirooka, Y., Herrera, C., Salgado-Salazar, C., and Chaverri, P. 2013. Genera in the *Bionectriaceae*, *Hypocreaceae*, and *Nectriaceae* (*Hypocreales*) proposed for acceptance or rejection. IMA Fungus 4:41-51. doi: 10.5598/imafungus.2013.04.01.05

Ruiz, N., Wielgosz-Collin, G., Poirier, L., Grovel, O., Petit, K. E., Mohamed-Benkada, M., du Pont, T. R., Bissett, J., Vérité, P., Barnathan, G., and Pouchus, Y. F. 2007. New Trichobrachins, 11-residue peptaibols from a marine strain of *Trichoderma longibrachiatum*. Peptides 28:1351-1358.

Ryder, L. S., Harris, B. D., Soanes, D. M., Kershaw, M. J., Talbot, N. J., and Thornton, C. R. 2012. Saprotrophic competitiveness and biocontrol fitness of a genetically modified strain of the plant-growth-promoting fungus *Trichoderma hamatum* GD12. Microbiology 158:84-97.

Samuels, G. J. 1996. *Trichoderma*: A review of biology and systematics of the genus. Mycol. Res. 100:923-935.

Samuels, G. J. 2006. *Trichoderma*: Systematics, the sexual state, and ecology. Phytopathology 96:195-206.

Samuels, G, J. 2014. (2305–2309) Proposals to conserve the names *Trichoderma catoptron* against *Hypocrea catoptron*, *H. sulfurella*, and *H. flavovirens*; *T. citrinoviride* against *Sphaeria schweinitzii* (*H. schweinitzii*), *S. contorta*, *H. repanda*, and *H. minima*; *H. lutea* against *Gliocladium deliquescens* (*T. deliquescens*) with a recommendation to reject the proposal (cf. Art. 57.2); *H. pezizoides* (*T. pezizoides*) against *T. pezizoideum*; and *T. reesei* against *H. jecorina* (*Ascomycota*: *Pezizomycotina*: *Sordariomycetes*: *Hypocreales*: *Hypocreaceae*). Taxon 63:936-938.

Samuels, G. J., and Ismaiel, A. 2009. *Trichoderma evansii* and *T. lieckfeldtiae*: Two new *T. hamatum*-like species. Mycologia 101:142-156.

Samuels, G. J., and Ismaiel, A. 2011. *Hypocrea peltata*: A mycological Dr. Jeckyll and Mr. Hyde? Mycologia 103:616-630.

Samuels, G. J., Petrini, O., Kuhls, K., Lieckfeldt, E., and Kubicek, C. P. 1998. The *Hypocrea schweinitzii* complex and *Trichoderma* sect. *Longibrachiatum*. Stud. Mycol. 41:1-54.

Samuels, G. J., Lieckfeldt, E., and Nirenberg, H. I. 1999. *Trichoderma asperellum*, a new species with warted conidia, and redescription of *T. viride*. Sydowia 51:71-88.

Samuels, G. J., Pardo-Schultheiss, R., Hebbar, K. P., Lumsden, R. D., Bastos, C. N., Costa, J. C., and Bezerra, J. L. 2000. *Trichoderma stromaticum* sp. nov., a parasite of the cacao witches broom pathogen. Mycol. Res. 104:760-764.

Samuels, G. J., Dodd, S. L., Gams, W., Castlebury, L. A., and Petrini, O. 2002. *Trichoderma* species associated with the green mold epidemic of commercially grown *Agaricus bisporus*. Mycologia 94:146-170.

Samuels, G. J., Dodd, S. L., Lu, B.-S., Petrini, O., Schroers, H.-J., and Druzhinina, I. S. 2006a. The *Trichoderma koningii* aggregate species. Stud. Mycol. 56:67-133.

Samuels, G. J., Suarez, C., Solis, K., Holmes, K. A., Thomas, S. E., Ismaiel, A., and Evans, H. C. 2006b. *Trichoderma theobromicola* and *T. paucisporum*: Two new species from South America. Mycol. Res. 110:381-392.

Samuels, G. J., Ismaiel, A., Bon, M.-C., De Respinis, S., and Petrini, O. 2010. *Trichoderma asperellum* sensu lato consists of two cryptic species. Mycologia 102:944-966.

Samuels, G. J., Ismaiel, A., de Souza, J., and Chaverri, P. (2011 Online.) 2012. *Trichoderma stromaticum* and its overseas relatives. Mycol. Prog. 11:215-254. doi: 10.1007/s11557-011-0743-4

Samuels, G. J., Ismaiel, A., Mulaw, T. B., Szakacs, G., Druzhinina, I. S., Kubicek, C. P., and Jaklitsch, W. M. 2012. The Longibrachiatum clade of *Trichoderma*: A revision with new species. Fungal Diversity 55:77-108.

Sandoval-Denis, M., Sutton, D. A., Cano-Lira, J. F., Gené, J., Fothergill, A. W., Wiederhold, N., and Guarro, J. 2014. Phylogeny of the clinically relevant species of the emerging fungus *Trichoderma* and their antifungal susceptibilities. J. Clin. Microbiol. 52:2112-2125. doi: 10.1128/JCM.00429-14

Santos, C., Paterson, R. R. M., Venâncio, A., and Lima, N. 2010. Filamentous fungal characterizations by matrix-assisted laser desorption/ionization time-of-flight mass spectrometry. J. Appl. Microbiol. 108:375-385.

Santos, A., Garcia, M., Cotes, A. M., and Villamizar, L. 2012. Efecto de la formulación sobre la vida útil de bioplaguicidas a base de dos aislamentos Colombianos de *Trichoderma koningiopsis* Th003 y *Trichoderma asperelum* Th034. Rev. Iberoam. Micol. 29:150-156.

Saravanakumar, K., Arasu, V. S., and Kathiresan, K. 2012. Effect of *Trichoderma* on soil phosphate solubilization and growth improvement of *Avicennia marina*. Aquat. Bot. 104:101-105.

Savazzini, F., Oliveira Longa, C. M., Pertot, I., and Gessler, C. 2008. Real-time PCR for detection and quantification of the biocontrol agent *Trichoderma atroviride* strain SC1 in soil. J. Microbiol. Methods 73:185-194.

Schirmböck, M., Lorito, M., Wang, Y., Hayes, C. K., Arisan-Atac, I., Scala, F., Harman, G. E., and Kubicek, C. P. 1994. Parallel formation and synergism of hydrolytic enzymes and peptaibol antibiotics: Molecular mechanisms involved in the antagonistic action of *Trichoderma harzianum* against phytopathogenic fungi. Appl. Environ. Microbiol. 60:4364-4370.

Schroers, H.-J., and Samuels, G. J. 1998. *Bionectria*: A genus for species of the *Nectria ochroleuca* group. Z. Mykol. 63:149-154.

Seaby, D. 1998. *Trichoderma* as a weed mould or pathogen in mushroom cultivation. Pages 267-287 in: *Trichoderma & Gliocladium*, Vol. 2. Enzymes, Biological Control and Commercial Applications. G. E. Harman and C. P. Kubicek, eds. Taylor & Francis, London.

Segarra, G., Casanova, E., Bellido, D., Odena, M. A., Oliveira, E., and Trillas, I. 2007. Proteome, salicyic acid, and jasmonic acid changes in cucumber plants inoculated with *Trichoderma asperellum* T34. Proteomics 7:3943-3952.

Seidl, V., Song, L., Lindquist, E., Gruber, S., Koptchinskiy, A., Zeilinger, S., Schmoll, M., Martínez, P., Sun, J., Grigoriev, I., Herrera-Estrella, A., Baker, S. E., and Kubicek, C. P. 2009. Transcriptome response of the mycoparasitic fungus *Trichoderma atroviride* to the presence of a fungal prey. BMC Genomics 10:567.

Seifert, K. A., Morgan-Jones, G., Gams, W., and Kendrick, B. 2011. The genera of hyphomycetes. CBS Biodiversity Ser. 9:1-997.

Sharon, E., Bar-Eyal, M., Chet, I., Herrera-Estrella, A., Kleifeld, O., and Spiegel, Y. 2001. Biological control of the root-knot nematode *Meloidogyne javanica* by *Trichoderma harzianum*. Phytopathology 91:687-693.

Sharon, E., Chet, I., Viterbo, A., Bar-Eyal, M., Nagan, H., Samuels, G. J., and Spiegel, Y. 2007. Parasitism of *Trichoderma* on *Meloidogyne javanica* and role of the gelatinous matrix. Eur. J. Plant Pathol. 118:247-258.

Shoresh, M., and Harman, G. E. 2008. The relationship between increased growth and resistance induced in plants by root colonizing microbes. Plant Signaling Behav. 3:737-739.

Shoresh, M., Yedidia, I., and Chet, I. 2005. Involvement of jasmonic acid/ethylene signaling pathway in the systemic resistance induced in cucumber by *Trichoderma asperellum* T203. Phytopathology 95:76-84.

Shoresh, M., Harman, G. E., and Mastouri, F. 2010. Induced systemic resistance and plant responses to fungal biocontrol agents. Annu. Rev. Phytopathol. 48:21-43.

Shrestha, P., Szaro, T. M., Bruns, T. D., and Taylor, J. D. 2011. Systematic search for cultivatable fungi that best deconstruct cell walls of *Miscanthus* and sugarcane in the field. Appl. Environ. Microbiol. 77:5490-5504.

Shukla, N., Awasthi, R. P., Rawat, L., and Kumar, J. 2012. Biochemical and physiological responses of rice (*Oryza sativa* L.) as influenced by *Trichoderma harzianum* under drought stress. Plant Physiol. Biochem. 54:78-88.

Sikora, R. A., Pocasangre, L., zum Felde, A., Niere, B., Vu Tam, T., and Dababat, A. A. 2008. Mutualistic endophytic fungi and *in-planta* suppressiveness of plant parasitic nematodes. Biol. Control 46:15-23.

Sivan, A., Elad, Y., and Chet, I. 1984. Biological control effects of a new isolate of *Trichoderma harzianum* on *Pythium aphanidermatum*. Phytopathology 74:498-501.

Sivan, A., Harman, G. E., and Stasz, T. E. 1990. Transfer of isolated nuclei into protoplasts of *Trichoderma harzianum*. Appl. Environ. Microbiol. 56:2402-2409.

Sivan, A., Stasz, T. E., Hemmat, M., Hayes, C. K., and Harman, G. E. 1992. Transformation of *Trichoderma* spp. with plasmids conferring hygromycin B resistance. Mycologia 84:687-694.

Sivasithamparam, K., and Ghisalberti, E. L. 1998. Secondary metabolism in *Trichoderma* and *Gliocladium*. Pages 139-191 in: *Trichoderma & Gliocladium*, Vol. 1. Basic Biology, Taxonomy and Genetics. C. P. Kubicek and G. E. Harman, eds. Taylor & Francis, London.

Smith, A., Beltrán, C. A., Kusanoki, M., Cotes, A. M., Motohashi, K., Kondo, T., and Deguchi, M. 2013. Diversity of soil-dwelling *Trichoderma* in Colombia and their potential as biocontrol agents against the phytopathogenic fungus *Sclerotinia sclerotiorum* (Lib.) de Bary. J. Gen. Plant Pathol. 79:74-85.

Sobowale, A. A., Cardwell, K. F., Odebode A. C., Bandyopadhyay, R., and Jonathan, S. G. 2007. Persistence of *Trichoderma* species within maize stem against *Fusarium verticillioides*. Arch. Phytopathol. Plant Prot. 40:215-231.

Sobowale, A. A., Babalola, O. O., Ayansina, A. D. V., and Obisesan, A. O. 2011. Abilities of *Trichoderma* species to persist within maize (*Zea mays*) stem long after inoculation. Br. Microbiol. Res. J. 1:95-103.

Sperry, S., Samuels, G. J., and Crews, P. 1998. Vertinoid polyketides from the saltwater culture of the fungus *Trichoderma longibrachiatum* separated from a *Haliclona* marine sponge. J. Org. Chem. 63:10011-10014.

Stasz, T. E., Harman, G. E., and Weeden, N. F. 1988. Protoplast preparation and fusion in two biocontrol strains of *Trichoderma harzianum*. Mycologia 80:141-150.

Stergiopolus, I., and de Wit, P. J. G. M. 2009. Fungal effector proteins. Annu. Rev. Phytopathol. 47:233-263.

Stone, A. G., Scheuerell, S. J., and Darby, H. M. 2004. Diseases in field agricultural systems: Organic matter management, cover cropping, and other cultural practices. Pages 131-177 in: Soil Organic Matter in Sustainable Agriculture. F. Magdoff, and R. R. Weil, eds. CRC Press, Boca Raton, FL.

Tachapattaworakul, T., Tachon, F., Guiraud, J. P., and Ratomahenina, R. 2008. Application of benomyl resistance and RAPD-PCR for detecting a *Trichoderma atroviride* strain used as biocontrol agent. Biol. Agric. Hortic. 25:369-377.

Tang, P., Mohan, S., Sigler, L., Witterick, I., Summerbell, R., Campbell, I., and Mazzulli, T. 2003. Allergic fungal sinusitis associated with *Trichoderma longibrachiatum*. J. Clin. Microbiol. 41:5333-5336.

Taylor, J. W., Jacobson, D. J., Kroken, S., Kasuga, T., and Geiser, D. M. 2000. Phylogenetic species recognition and species concepts in fungi. Fungal Genet. Biol. 31:21-32.

Thrane, C., Lübeck, M., Green, H., Degefu, Y., Allerup, S., Thrane, U., and Jensen, D. F. 1995. A tool for monitoring *Trichoderma harzianum*: I. Transformation with the GUS gene by protoplast technology. Phytopathology 85:1428-1435.

Thrane, U., Poulsen, S. B., Nirenberg, H. I., and Lieckfeldt, E. 2001. Identification of *Trichoderma* strains by image analysis of HPLC chromatograms. FEMS Microbiol. Lett. 203:249-255.

Tondje, P. R., Roberts, D. P., Bon, M.-C., Widmer, T., Samuels, G. J., Ismaiel, A., Begoude, A. D., Tchana, T., Nyemb-Tschomb, E., Ndoumbe-Nkeng, M., Bateman, R., Fontem, D., and Hebbar, P. K. 2007. Isolation and identification of mycoparasitic isolates of *Trichoderma asperellum* with potential for suppression of black pod disease of cacao in Cameroon. Biol. Control 43:202-212.

Tu, J. C. 1980. *Gliocladium virens*, a destructive mycoparasite of *Sclerotinia sclerotiorum*. Phytopathology 70:670-674.

U'ren, J. M., Dalling, J. W., Gallery, R. E., Maddison, D. R., Davis, E. C., Gibson, C. M., and Arnold, A. E. 2009. Diversity and evolutionary origins of fungi associated with seeds of a neotropical pioneer tree: A case study for analysing fungal environmental samples. Mycol. Res. 113:432-449.

van Hulten, M., Pelser, M., van Loon, L. C., Pieterse, C. M. J., and Ton, J. 2006. Costs and benefits of priming for defense in *Arabidopsis*. Proc. Natl. Acad. Sci. U.S.A. 103:5602-5607.

van Os, E. A., Postma, J., Pettitt, T., and Wohanka, W. 2004. Microbial optimisation in soilless cultivation, a replacement for methyl bromide. Acta Hortic. 635:48-58.

Van Wees, S. C. M. 2012. Unravelling intimacies between plants and their enemies. Plant Biol. 14(Suppl. 1):iii, iv.

Van Wees, S. C. M., Van der Ent, S., and Pieterse, C. M. J. 2008. Plant immune responses triggered by beneficial microbes. Curr. Opin. Plant Biol. 11:443-448.

Vargas, W. A., Mandawe, J. C., and Kenerley, C. M. 2009. Plant-derived sucrose is a key element in the symbiotic association between *Trichoderma virens* and maize plants. Plant Physiol. 151:792-808.

Verma, V. C., Gond, S. K., Kumar, A., Kharwar, R. N., and Strobel, G. A. 2007. The endophytic mycoflora of bark, leaf, and stem tissues of *Azadirachta indica* A. Juss (Neem) from Vanasi (India). Microb. Ecol. 54:119-125.

Vinale, F., Sivasithamparam, K., Ghisalberti, E. L., Marra, R., Woo, S. L., and Lorito, M. 2008. *Trichoderma*–plant–pathogen interactions. Soil Biol. Biochem. 40:1-10.

Vinale, F., Sivasithamparam, K., Ghissalberti, E. L., Woo, S. L., Nigro, M., Marra, R., Lombardi, N., Pascale, A., Ruocco, M., Lanzuise, S., Manganiello, G., and Lorito, M. 2014. *Trichoderma* secondary metabolites active on plants and fungal pathogens. Open Mycol. J. 8:127-139. doi: 10.2174/1874437001408010127

Viterbo, A., and Chet, I. 2006. *TasHyd1*, a new hydrophobin gene from biocontrol agent *Trichoderma asperellum*, is involved in plant root colonization. Mol. Plant Pathol. 7:249-258.

Waksman, S. A. 1916. Soil fungi and their activities. Soil Sci. 2:103-106.

Watanabe, S., Kumakura, K., Kato, H., Iyozumi, H., Togawa, M., and Nagayama, K. 2005. Identification of *Trichoderma* SKT-1, a biological control agent against seed-borne pathogens of rice. J. Gen. Plant Pathol. 71:351-356.

Watanabe, S., Kato, H., Kumakura, K., Ishibashi, E., and Nagayama, K. 2006. Properties and biological control activities of aerial and submerged spores of *Trichoderma asperellum* SKT-1. J. Pestic. Sci. 31:375-379.

Waweru, B. W., Losenge, T., Kahangi, E. M., Dubois, T., and Coyne, D. 2013. Potential biological control of lesion nematodes on banana using Kenyan strains of *Fusarium oxysporum*. Nematology 15:101-107.

Webster, J. 1964. Culture studies of *Hypocrea* and *Trichoderma* I. Comparison of perfect states of *H. gelatinosa* and *H. rufa*, and *Hypocrea* sp. 1. Trans. Br. Mycol. Soc. 47:75-96.

Webster, J., and Rifai, M. 1968. Culture studies of *Hypocrea* and *Trichoderma* IV. *Hypocrea pilulifera* sp. nov. Trans. Br. Mycol. Soc. 51:511-514.

Weindling, R. 1932. *Trichoderma lignorum* as a parasite of other soil fungi. Phytopathology 22:837-845.

Weindling, R. 1934. Studies on a lethal principle effective in the parasitic action of *Trichoderma lignorum* on *Rhizoctonia solani* and other soil fungi. Phytopathology 24:1153-1179.

Weindling, R. 1941. Experimental consideration of the mold toxins of *Gliocladium* and *Trichoderma*. Phytopathology 31:991-1003.

Weindling, R., and Fawcett, H. S. 1936. Experiments in the control of Rhizoctonia damping-off of citrus seedlings. J. Agric. Sci. Calif. Agric. Exp. Stn. 10:1-16.

Wharton, P., and Kirk, W. 2007. Potato Seed Piece Health Management. Michigan Potato Diseases Series. Michigan State Univ. Ext. Bull. E-2995. www.potatodiseases.org/pdf/seed-piece-health-bulletin.pdf

Wharton, P. S., and Kirk, W. W. 2014. Evaluation of biological seed treatments in combination with management practices for the control of Fusarium dry rot of potato. Biol. Control 73:23-30.

Wharton, P. S., Kirk, W. W., Schafer, R. L., and Tumbalam, P. 2012. Evaluation of biological seed treatments in combination with management practices for the control of seed-borne late blight in potato. Biol. Control 63:326-332.

Whipps, J. M. 2001. Microbial interactions and biocontrol in the rhizosphere. J. Exp. Bot. 52:487-511.

White, T. J., Bruns, T., Lee, S., and Taylor, J. W. 1990. Amplification and direct sequencing of fungal ribosomal RNA genes for phylogenetics. Pages 315-322 in: PCR Protocols: A Guide to Methods and Applications. M. A. Innis, D. H. Gelfand, J. J. Sninsky, and T. J. White, eds. Academic Press, San Diego, CA.

Widmer, T. L., and Samuels, G. J. 2010. Screening *Trichoderma asperellum* as a mycoparasite on *Phytophthora ramorum*. Pages 375-384 in: Proceedings of the Sudden Oak Death Fourth Science Symposium. S. J. Frankel, J. T. Kliejunas, and K. M. Palmieri, eds. Gen. Tech. Rep. PSW-GTR-229. USDA, Forest Service, Pacific Southwest Research Station, Albany, CA.

Williams, J., Clarkson, J. M., Mills, P. R., and Cooper, R. M. 2003. A selective medium for quantitative reisolation of *Trichoderma harzianumm* from *Agaricus bisporous* compost. Appl. Environ. Microbiol. 69:4190-4191.

Wilson, P. S., Ahvenniemi, P. M., Lehtonen, M. J., Kukkonen, M., Rita, H., and Valkonen, J. P. T. 2008. Biological and chemical control and their combined use to control different stages of the *Rhizoctonia* disease complex on potato through the growing season. Ann. Appl. Biol. 53:307-320.

Woo, S. L., Ruocco, M., Vinale, F., Nigro, M., Marra, R., Lombardi, N., Pascale, A., Lanzuise, S., Manganiello, G., and Lorito, M. 2014. *Trichoderma*-based products and their widespread use in agriculture. Open Mycol. J. 8:71-126. doi: 10.2174/1874437001408010071

Wuczkowski, M., Druzhinina, I., Gherbawy, Y., Klug, B., Prillinger, H.-J., and Kubicek, C. P. 2003. Species pattern and genetic diversity of *Trichoderma* in a mid-European, primeval floodplain-forest. Microbiol. Res. 158:125-133.

Yedidia, I., Srivastva, A. K., Kapulnik, Y., and Chet, I. 2001. Effect of *Trichoderma harzianum* on microelement concentrations and increased growth of cucumber plants. Plant Soil 235:235-242.

Yedidia, I., Shoresh, M., Kerem, Z., Benhamou, N., Kapulnik, Y., and Chet, I. 2003. Concomitant induction of systemic resistance to *Pseudomonas syringae* pv. *lachrymans* in cucumber by *Trichoderma asperellum* (T-203) and accumulation of phytoalexins. Appl. Environ. Microbiol. 69:7343-7353.

Yobo, K. S., Laing, M. D., and Hunter, C. H. 2011. Effects of single and combined inoculations of selected *Trichoderma* and *Bacillus* isolates on growth of dry bean and biological control of *Rhizoctonia solani*. Afr. J. Biotechnol. 44:8746-8756.

Yogev, A., Raviv, M., Hadar, Y., Cohen, R., Wolf, S., Gil, L., and Katan, J. 2010. Induced resistance as a putative component of compost suppressiveness. Biol. Control 54:46-51.

Yun, H. S., and Kwon, C. 2012. Trafficking at the host cell surface during plant immune responses. J. Plant Biol. 55: 185-190.

Zachow, C., Berg, C., Müller, H., Meincke, R., Komoń-Zelazowska, M., Druzhinina, I. S., Kubicek, C. P., and Berg, G. 2009. Fungal diversity in the rhizosphere of endemic plant species of Tenerife (Canary Islands): Relationship to vegetation zones and environmental factors. ISME J. 3:79-92.

Zhu, Z. X., and Zhuang, W. Y. 2015. *Trichoderma* (*Hypocrea*) species with green ascospores from China. Persoonia 34:113-129.

Zinati, G. M. 2005. Compost in the 20th century: A tool to control plant diseases in nursery and vegetable crops. HortTechnology 15:61-66.

zum Felde, A., Pocasangre, L., and Sikora, R. A. 2005. The potential use of microbial communities inside suppressive banana plants for banana root protection. Pages 169-177 in: Banana Root System: Towards a Better Understanding for Its Productive Management. Proceedings of an International Symposium. D. W. Turner and F. E. Rosales, eds. International Network for the Improvement of Banana and Plantain, Montpellier, France.

zum Felde, A., Pocasangre, L. E., Carñizares Monteros, C. A., Sikora, R. A., Rosales, F. E., and Riveros, A. S. 2006. Effect of combined inoculations of endophytic fungi on the biocontrol of *Radopholus similis*. InfoMusa 15(1-2):12-18.

Index